E. Balázs et al. (Eds.)

Biological Resource Management Connecting Science and Policy

Springer
Berlin
Heidelberg
New York
Barcelona
Hong Kong
London
Milan
Paris
Singapore
Tokyo

Ervin Balázs · Ennio Galante · James M. Lynch
James S. Schepers · Jean-Pierre Toutant
Dietrich Werner · P. A. Th. J. Werry (Eds.)

Biological Resource Management Connecting Science and Policy

With 43 Figures and 47 Tables

Springer

Professor Dr. Ervin Balázs
Agricultural Biotechnology Center
P.O. Box 411
2101 Gödöllő
Hungary

Professor Dr. Ennio Galante
Institute for Plant Biosynthesis
National Research Council
Via Bassini 15
20133 Milan
Italy

Professor Dr. James M. Lynch
School of Biological Sciences
University of Surrey
Guildford, Surrey GU2 5XH
UK

Professor Dr. James S. Schepers
Soil and Water Conservation Research
USDA, ARS, NPA
University of Nebraska
113 Keim Hall
Lincoln, NB 68583-0915
USA

Professor Dr. Jean-Pierre Toutant
Laboratoire de Différenciation Cellulaire
Et Croissance
INRA
2 place Viala
34060 Montpellier Cedex 1
France

Professor Dr. Dietrich Werner
Plant Physiology and General Botany
Phillips-University
Lahnberge
35032 Marburg
Germany

Professor Dr. P. A. Th. J. Werry
Director of International Cooperation
DLO-NL
P. O. Box 59
6700 AB Wageningen
The Netherlands

ISBN 3-540-67117-X Springer-Verlag Berlin Heidelberg New York

Cataloging-in-Publication Data applied for
Die Deutsche Bibliothek – CIP-Einheitsaufnahme
Biological resource management: connecting science and policy; with 47 tables / INRA Ed.; OECD.
Ervin Balázs ... (ed.). – Berlin; Heidelberg; New York; Barcelona; Hong Kong; London; Milan; Paris;
Singapore; Tokyo: Springer, 2000
ISBN 3-540-67117-X

Springer-Verlag is a company in the BertelsmannSpringer publishing group

Camera ready by INRA, France
Cover design: D&P, Heidelberg
Cover Photograph from Kawashima et al., this volume
SPIN 10743189 31/3136 5 4 3 2 1 0 – Printed on acid-free paper

Acknowledgements

The editors wish to thank Mrs Fay Evans, the Executive Secretary of the Programme, for the excellent organisation of the meeting and for her expert assistance through the years.

Preface

Ministers for Agriculture, meeting at the OECD on 5-6 March 1998, stated that most of the issues related to new technologies, food safety and environmental issues had transboundary and trans-sectoral dimensions and that there was a need for further research, a better understanding of current scientific knowledge, and better information to consumers. Indeed, the appearance of new food products, the use of new production methods including genetic engineering, and the recent crises over food-borne diseases, have added to consumer concerns, leading to a growing questioning of science as a basis for developing sound policies. Moreover, in the context of globalisation, developments in one country can have an impact in other countries, through international trade, as can domestic regulations and policies. The public is also increasingly sensitive to the interface between agriculture and the environment, the need to ensure sustainable agriculture, and for food to be produced under environmentally friendly conditions.

The Conference on Biological Resource Management: Connecting Science and Policy provided an opportunity to respond to the call by OECD Agriculture Ministers. In particular, its last session explored the relationships between science, agricultural production, agri-business, policy-making and society's concerns. In so doing, it contributes to the programme of work carried out under the auspices of the OECD Committee for Agriculture. The overall objective of this work is to assist policy-makers in furthering the process of agricultural policy reform and trade liberalisation, taking into account emerging issues such as food safety and quality, resource sustainability, environmental protection, rural development, and animal welfare.

The Conference was the culmination of the current 5-year phase of the OECD Co-operative Research Programme. Launched in 1995, the Programme aims at stimulating international cooperation and networking in basic and applied research, through an annual fellowship programme and the organisation of workshops attended by the world's best experts to address issues within its scope. The Programme has focused on four themes: safe exploitation of microorganisms in plant/soil systems; quality of animal production; utilisation and ecology of new organisms; and surface- and groundwater quality and agricultural practices. The possibility of extending the Research Programme for another 5-year period will be under discussion within OECD over the next few months, and the results of the Conference will contribute to this discussion, in particular by identifying new themes which could be addressed through the cooperative research mechanisms developed by this Programme, which have the potential to contribute a sound scientific foundation for the overall programme of work on agriculture.

Gérard Viatte
Director of Food, Agriculture and Fisheries
OECD, Paris

List of participants

Parts I and II

ADAMS Marshall
Oak Ridge National Laboratory
Environmental Sciences Division,
Bldg. 1505
Oak Ridge, TN 37831-6036, USA

BAREA José-Miguel
Dpt. of Microbiology for Soil and
Symbiotic Systems
Experimental Laboratory of Zaidin
C/ Profesor Albareda, 1
18008 Granada, Spain

BOESCH Donald
University of Maryland
Center for Environmental Sciences,
PO. Box 775
Cambridge, MD 21613, USA

GALANTE Ennio
Institute for Plant Biosynthesis
National Research Council
Via Bassini 15
20133 Milano, Italy

GOSS Michael
Agrifood System Resource & Env.
Program
University of Guelph
Guelph, Ontario N1G 2W1, Canada

MOORE Philip
USDA/ARS, Agronomy Department
University of Arkansas
Fayetteville, AR 72701, USA

O'GARA Fergal
Biomerit Research Laboratory
Microbiology Department
National University of Ireland
UCC-Cork, Ireland

RYDER Maarten
CSIRO, Land & Water
PMB 2, Glen Osmond
SA 5064, Australia

SCHEPERS James
Soil & Water Conservation Research,
USDA/ARS
University of Nebraska
East Campus, 113 Keim Hall
Lincoln, Nebraska 68583-0915, USA

VAN VEEN Hans
Netherlands Institute of Ecology-
Center for Terrestrial Ecology
NIOO-CTO, PO. Box 40
6666 ZG Heteren, The Netherlands

WERNER Dietrich
Plant Physiology & General Botany
Botanisches Institut
Fachbereich Biologie der Philipps-
Universität
Lahnberge, 35032 Marburg, Germany

Panel 1

BLUM Wilfried
International Union of Soil Sciences
Gregor-Mendel-Strasse 33
1180 Vienna, Austria

JOHNSSON Holger
Swedish University of Agr. Sciences
Division of Water Quality Management, Box 7072
75007 Uppsala, Sweden

NUTI Marco
Res. Center for Soil Microbiology
Department of Chemistry
University of Pisa, Via del Borghetto, 80
56100 Pisa, Italy

SUKAN Suha
Ege University
Dept. of Food Engineering
35100, Bornova
Izmir, Turkey

WERRY P.A.Th.J.
International Co-operation, DLO-NL
PO. Box 59, 6700 AB Wageningen
The Netherlands

Parts III and IV

AGUZZI Adriano
Institute of Neuropathology
University Hospital of Zurich
Schmelzbergerstrasse 12, 8091 Zurich
Switzerland

BALÁZS Ervin
Agricultural Biotechnology Center
PO Box 411, 2101 Gödöllö
Hungary

DEMEYER Daniel
University of Ghent
Department of Animal Production
Laboratory of Animal Nutrition & Meat Science
Proefhoevestraat 10, Ghent, Belgium

FISHER Andrew
AgResearch, Ruakura Agricultural Research Centre
Private Bag 3123
Hamilton, New Zealand

HOUDEBINE Louis-Marie
BMC/INRA
Domaine de Vilvert
78352 Jouy-en-Josas cedex, France

KUBICEK Christian
Dept of Technology and Microbiology
Institute of Vienna
1060 Wien, Austria

NAGASHIMA Hiroshi
Meiji University, Faculty of Agriculture
1-1-1 Higashimita, Tama,
Kawasaki 214-8571
Japan

POWELL Richard
Dept. of Microbiology
National University of Ireland
Galway, Ireland

TSAFTARIS Athanasios
Dept. of Genetics & Plant Breeding
Aristoteleian University of Thessaloniki
54006 Thessaloniki, Greece

TOUTANT Jean-Pierre
Laboratoire de Différenciation cellulaire
INRA - 2, place Viala
34060 Montpellier cedex, France

WILMUT Ian
Roslin Institute of Edinburgh
Roslin, Midlothian, EH25 9PS
United Kingdom

Panel 2

FULKA Josef
Institute of animal production, POB 1
CZ-104 01 Prague10, Czech Republic

KING Allan
Dept. of Biomedical Sciences
University of Guelph
Guelph, Ontario Canada N1G 2W1

Senator KERREY Robert
United States Senator
141 Hart Senate Office Building
Washington, D.C. 20510
United States

SOLTI Laszló
University of Veterinary Science
Department of Animal Reproduction
P.O. Box 2, 1400 Budapest
Hungary

TEPFER Mark
Laboratoire de Biologie cellulaire
INRA - Centre de Versailles
78026 Versailles Cedex, France

VAZQUEZ-BOLAND José-Antonio
Microbiology & Immunology Unit
Veterinary Faculty
Complutense University of Madrid
28040 Madrid, Spain

Parts V and VI

ALABOUVETTE Claude
Flore Pathogène du sol
INRA - 17, rue Sully
21034 Dijon Cedex, France

DAVID Andow
Insect Ecology, Dept. of Entomology
University of Minnesota
219 Hodson Hall,
St.Paul, MN 55108, USA

BERINGER John
University of Bristol
School of Biological Sciences, Woodland Road
Bristol BS8 1UG, UK

KAWASHIMA Hiroyuki
International Environmental Economics
Dept. of Global Agricultural Sciences
Graduate School of Agricultural & Life Sciences
The University of Tokyo , 1-1-1 Yayoi Bunkyo
Tokyo 113-8657, Japan

LYNCH James
School of Biological Sciences
Univ. of Surrey, Guildford
Surrey GU2 5XH, UK

MALTBY Edward
Royal Holloway Institute for Environmental Research
Huntersdale, Callow Hill Virginia Water,
Surrey GU 25 4 NL, UK

NANNIPIERI Paolo
Department of Soil Science
University of Florence
P. delle Cascine 28
50144 Firenze, Italy

WALL Diana
Natural Resource Ecology Laboratory
College of Natural Resources
Colorado State University
Fort Collins, CO 80523, USA

Panel 3

LIPA Jerzy J.
Dept. of Biological Control & Quarantine
Institute of Plant Protection
Miczurina 20
60-318 Poznan, Poland

OLSEN Rolf Arnt
Department of Biotechnology
Agricultural University of Norway
P.O. Box 5040
1432 As, Norway

TOMLIN Alan
Agriculture and Agri-Food Canada
Southern Crop Protection & Food Research
Research Farm - Vineland
49002 Victoria Ave. N., PO.Box 6000
Vineland, Ontario - L0R 2E0, Canada

Final panel

DIXON Bernard
130 Cornwall Road
HA4 6AW Ruislip Manor
Middlesex, UK

DOORNBOS Gerald
President of International Federation of Agricultural Producers
60, rue St Lazare
75009 Paris, France

MOE Thorvald
Deputy Secretary-General
OECD, 2, rue André Pascal
75775 Paris Cedex 16, France

POUTIAINEN Esko
Agricultural Research Centre
31600 Jokioinen, Finland

REED Debbie
Legislative Assistant
141 Hart Senate Office Building
Washington, D.C. 20510, USA

SCHOFIELD Geraldine
Foods Unilever Research Colworth
Colworth House, Sharnbrook
Bedford MK44 1LQ, United Kingdom

SHANNON David
Ministry of Agriculture, Fisheries & Food
Whitehall Place (West)
London SW 1A 2HH, United Kingdom

SILVERGLADE Bruce
President IACFO
1875 Connecticut Avenue N.W. ,Suite 300
Washington, DC 20009, USA

TOET Dirk
Nestlé Limited
Dept of Biotechnology
Vevey, Switzerland

VIALLE Paul
Directeur Général/INRA
147 rue de l'Université
75338 Paris Cedex 07, France

VIATTE Gérard
Director for Food, Agriculture and Fisheries
OECD, 2, rue André Pascal
75016 Paris, France

VON WARTBURG Walter
Novartis Communication
Lichtstrasse 35, P.O. Box 4002
Basel, Switzerland

Contents

Part III Managing Quality of Production (JP. Toutant, France)

Part IV Biotechnology (E. Balázs, Hungary)

Panel on Production Systems

Part V Beneficial Organisms (J. Lynch, England)

Part I
Ecotoxicology

Dietrich WERNER

M.H. RYDER et al.
S.M. ADAMS
M. GOSS et al.
K. BECKER and J. TARADELLA

Soil Microbiota: a Gold Mine and a Minefield for Biotechnology

MH. Ryder[1], Herdina[1], AL. Juhasz[1], PR. Harvey[1], IL. Ross[1], KM. Ophel-Keller[2], DK. Roget[1]

Biotechnology has been successfully applied to the management of soils and soil organisms with the aim of improving plant productivity and remediation of contaminated soils. The ability to isolate nucleic acids from soil enables the detection of particular soil organisms. Plant pathogens are being quantified in soil using specific DNA probes and farmers are able to use the information in crop management. Biological control of soil-borne diseases by soil bacteria and fungi is being practised. Genetic manipulation has led to the improvement of some commercial biological control agents such as Agrobacterium for the control of crown gall disease. The activities of both naturally-occurring microbial communities and inoculants can be harnessed for the remediation of soil contaminated by organic pollutants. The future development of biotechnology applied to soil organisms will continue to require contributions from a combination of scientific disciplines.

Introduction

Soil biotechnology has been defined as "the study of the soil biota and their metabolic processes to optimize and sustain crop productivity and industrial processes, and to combat pollution in the environment" (Lynch 1997). This science has enabled the soil and its biological resources to be managed for more efficient, sustainable agriculture while at the same time helping to reduce negative impacts on the environment (Pankhurst et al. 1994). It is also leading to the alleviation of negative impacts on the environment through bioremediation.

Soil biotechnology has provided the means to utilize soil organisms and their metabolites in medicine, agriculture and industry. Historic examples are the discovery of medically important antibiotics from soil microorganisms and the use of inoc-

1. CSIRO Land and Water. Glen Osmond, SA 5064, Australia
2. South Australian Research and Development Institute. Glen Osmond, SA 5064, Australia

ulants in agriculture, such as *Rhizobium* for symbiotic nitrogen fixation and *Agrobacterium* for biological control.

Large-scale screening of environmental samples, taken from a large range of environments around the globe from polar to arid regions is an ongoing process which is still leading to the discovery of new microbes capable of catalyzing novel processes and producing new metabolites. *Rhizobium* inoculants are being improved (Maier and Triplett 1996) and new microbial inoculants are being developed to enhance the efficiency of agriculture by controlling plant disease and stimulating yield (Ogoshi et al. 1997). It is possible to track inoculants using DNA-based techniques to assess their survival and spread (Ryder 1995), while microbial activity can be assessed using promoterless bioluminescence (*lux*) genes (Prosser 1994) and green fluorescent protein (*gfp*) genes (Egener et al. 1998; Reinhold-Hurek and Hurek 1998).

New microbial indicators of ecotoxicity have been developed, based on the ability of contaminants to inhibit metabolic rates and gene expression (Paton et al. 1995). DNA-based methods to study soil microbial communities are being developed, and these are being applied to the study of microbially mediated soil processes such as nutrient cycling (Stephen et al. 1998; Bruns et al. 1998). In addition, large groups of non-culturable organisms present in soil are being revealed and characterized by molecular phylogeny, but the function(s) of many of these organisms remain generally unknown at this stage (Liesack and Stackebrandt 1992).

Soil biotechnology has been used to solve complex problems and has led to advances in agricultural practice and to environmental applications. Three successful examples are described in this short review. These are (1) the commercial application of DNA probe-based diagnostics for the prediction of root disease severity in cereal crops in southern Australia, (2) the commercial use of a genetically manipulated bacterium (*Agrobacterium*) for biological control of a plant disease (crown gall), and (3) research on the manipulation of soil microbial communities for biodegradation of organic contaminants in soil. There have been considerable difficulties to overcome in the achievement of success, (1) in developing new methodology, (2) at the scientific-commercial interface and (3) in dealing with government regulatory bodies. The major difficulties (mines in the goldfield) are discussed for each example.

Applications of Soil Biotechnology

Detection and Prediction of Take-All Disease of Cereals

The Development and Use of a Specific DNA Probe for the Pathogen Causing take-all Disease of Wheat

The soil-borne fungus *Gaeumannomyces graminis* var. *tritici* (*Ggt*) is the causal agent of take-all, an economically important root disease of cereal crops in many countries. Harvey (1993) isolated and identified a genomic clone of *Ggt* (pG158) that specifically hybridizes to DNA of the pathogenic *Ggt* and the related pathogen *G. g.* var. *avenae*, but not to the DNA of non-pathogenic *G. g.* var. *graminis*. This clone, which is based on highly repeated DNA sequences, was later shown not to hybridize to other soil-borne fungi (Herdina et al. 1996), and is now being used routinely to specifically detect the DNA of *Ggt* and *Gga* in soils (Ophel-Keller et al. 1999). Previously, a seedling bioassay was used to estimate *Ggt* inoculum in soils as a commercial service to cereal growers in southern Australia. The data from the slot-blot DNA hybridization method using pG158 agreed well with the seedling assay (Herdina et al. 1997a) and the slot-blot assay can be completed in 4 to 6 days. The seedling assay requires considerably more time (6 weeks) and substantial amounts of space in controlled environment conditions. The DNA probe assay uses DNA extracted from soil organic matter (Herdina et al. 1996). With current protocols, this approach yields more DNA and is therefore a more sensitive method than extracting DNA from whole soil. The DNA probe assay for *Ggt/Gga* using pG158 requires extraction of DNA of sufficient purity and, importantly, has been developed to the stage of being quantitative (Herdina et al. 1997a). Based on the DNA test result, soil samples from farmers'fields are divided into four categories: non-detectable, low, medium and high levels of *Ggt/Gga* inoculum. Herdina and Roget (1999) investigated the requirements for field sampling in detail and showed that composite soil samples taken at the rate of one sample per ha (in 10-ha fields) provide adequate precision for disease estimation.

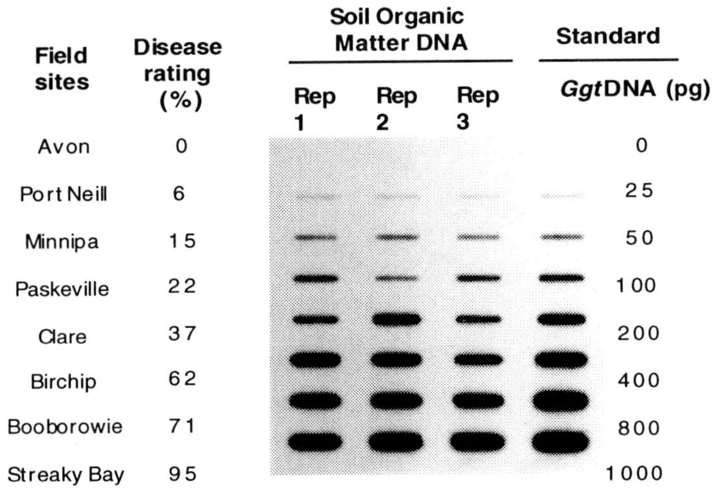

Fig. 1: Use of a specific DNA probe for quantitative detection of *Gaeumannomyces graminis* var. *tritici* (*Ggt*) in soil samples and the correlation with a wheat seedling bioassay to assess pathogen inoculum density in soil. Soil samples were collected from wheat-growing areas of South Australia in the summer of 1995-1996. Disease rating = proportion (%) of primary roots of wheat seedlings affected by the characteristic black vascular lesions caused by the take-all fungus, *Ggt*. DNA was extracted from soil organic matter (0.35-1.4 -mm fraction) and probed with radio-labeled pG158. (Data Herdina et al. 1997b)

The Prediction of Disease Severity

Take-all severity has long been known to vary from season to season and has been thought to be unpredictable (Hornby 1978). Roget and Rovira (1991) developed a model that allows prediction of take-all severity in southern Australia and this model has been refined further (D. K. Roget, unpubl). The model is based on the initial level of pathogen inoculum in the previous season and the spring rainfall from that year (note that in southern Australia, wheat is usually sown in late autumn and harvested in early summer). It provides predicted yield losses (expressed as a range) for different possible rainfall patterns for the current season. A current season high spring rainfall would give the highest risk of serious yield loss.

This ability to predict likely yield loss associated with low, medium and high levels of pathogen inoculum density in soil has now been combined very effectively with the DNA assay which is performed on a commercial basis by SARDI. The results from the DNA probe test for *Ggt*/*Gga* inoculum level give information that is useful to farmers in managing their cropping systems.

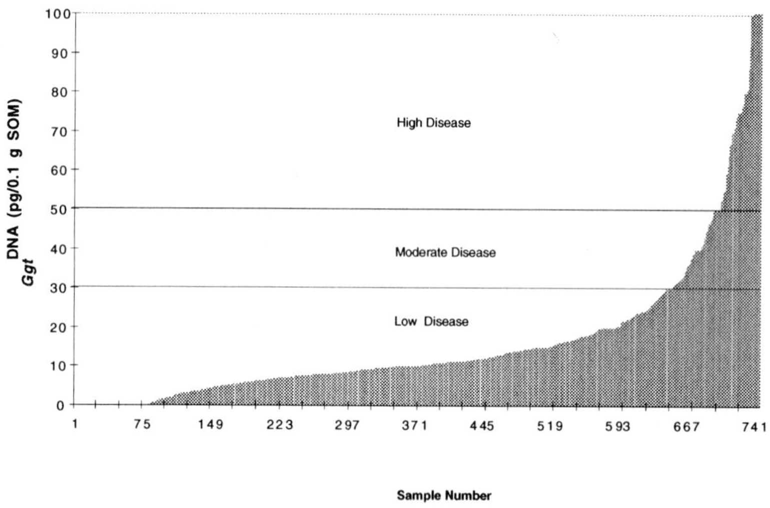

Fig. 2: Results of DNA testing for level of *Ggt* inoculum in 741 soil samples from farmers' fields. Soil samples were collected from across southeastern Australia in the summer of 1996-1997. DNA was extracted from soil organic matter (SOM, 0.35-1.4 mm fraction) and probed with radio-labeled pG158. The results from soils with detectable levels of *Ggt* DNA are divided into three categories, i.e. high, medium and low risk of serious take-all occurring in the current season wheat crop. (Data Herdina and Roget 1997)

Integration of the System and Delivery of Information to Farmers

The DNA probe test has been commercialized by the SARDI (Ophel-Keller et al. 1999). This has involved establishment of new business partnerships. The service is closely linked to fertilizer companies with their network of advisers and agronomists. The DNA assay for plant pathogen levels is performed on the same soil samples that are normally taken for nutrient analysis, so that no extra sampling is necessary. The recommended sampling density and pattern is suitable for both nutrient and DNA analysis, i.e. a minimum of one soil core per hectare for fields of 10 ha.

The information is delivered to farmers via accredited agronomists and extension officers. Accreditation requires successful completion of a 1 day training course run by SARDI. The course content includes sampling strategy, general root pathology, disease identification, interpretation of results of the DNA test and decision trees for the management strategies for fields with high or low risk of serious disease (take-all and several other root pathogens). Hundreds of agronomists working across southern Australia have now gained accreditation (Ophel-Keller et al. 1999).

In addition to the test for the take-all pathogen, the commercial SARDI Root Disease Testing Service also offers routine tests for several other root pathogens. These are *Rhizoctonia solani* AG-8, which causes bare-patch of cereals, cereal cyst nematode (*Heterodera avenae*), and the root lesion nematodes *Pratylenchus neglectus* and *P. thornei*. The tests for nematode populations are based on PCR.

Mines in the Goldfield

There have been several major challenges in the establishment of the Root Disease Testing Service. Firstly, on a technical level, the development of a protocol for the extraction of good-quality DNA from an enormous variety of field soils has been a formidable challenge. DNA of sufficient purity for PCR amplification is required, and the range of soil types encountered in the commercial service is very large. Research to improve DNA extraction and purification protocols in the first 2 years of the service has decreased the failure rate from 10% of soils to less than 1%. Technical aspects of the work are constantly being evaluated for improvements in precision and efficiency.

Secondly, at the organizational level, the development of strong working links between government scientists and business and an organizational structure were required.

Thirdly, a model for organization of the service was developed, focusing particularly on the delivery of results and information at the interface with the client farmers. The organizational framework was established with the utilization of trained, accredited agronomists who work in association with fertilizer companies. This has enabled effective sample collection and delivery of test results, as well as guidance by trained staff on how to use this information for the management of the farming system.

The Registration and Use of *Agrobacterium* Strain K1026 for Biological Control of Crown Gall Disease of Plants

The Method, the Problem and the Genetic Engineering

Crown gall is a plant disease that affects many dicotyledonous plants, including the economically important plants such as stone fruits, almonds, roses and grapevines. The disease is caused by pathogenic strains of the soil-borne bacterium *Agrobacterium* and results from the transfer of bacterial T-DNA from the Ti plasmid of the pathogen to the host plant. Biological control of the disease caused by most species of *Agrobacterium* has been achieved by the treatment of planting material with a non-pathogenic strain of *Agrobacterium* (Kerr 1980), which has been registered as a pesticide in Australia since 1976. It has been used commercially in many countries round the world since that time. One of the mechanisms by which strain K84 controls crown gall is the production of antibiotics. Two nucleoside analogues are produced (Tate et al. 1979; Donner et al. 1993) and both play a major role in the prevention of crown gall disease (McClure et al. 1998). The two molecules, agrocin 84 and agrocin 434, are synthesized via metabolic pathways encoded on plasmids carried by strain K84.

There are several possible ways in which biological control of crown gall could break down (Ryder and Jones 1991). One possibility is that the 48-kb plasmid which encodes agrocin 84 production could move into a pathogenic strain of *Agrobacterium* (Panagopoulos et al. 1979). Since the plasmid codes not only for production of agrocin 84 but also for resistance to the antibiotic (Ryder et al. 1987), the new strain would be a resistant pathogen. This led to the development of a genetically manipulated derivative of strain K84 which could no longer transfer its agrocin 84 plasmid to other agrobacteria. A total of 5.9 kb of DNA, including part of the region encoding plasmid transfer, was deleted (Jones et al. 1988). The resultant strain, K1026, was identical to the parent strain K84 in all other respects and controlled crown gall as effectively as the parent strain (Jones and Kerr 1989). Strain K1026 was registered for use as a pesticide in Australia in June 1989. Application for registration in the USA has been applied for and may be granted in 1999.

As a footnote to this short review of the genetic manipulation of *Agrobacterium* strain K84, an alternative way of spontaneously generating a new pathogenic strain of *Agrobacterium* that is resistant to biological control by K84 and K1026 was suggested by Ryder and Jones (1991). Lopez-Lopez et al. (1999) have now reported the occurrence of this event, i.e. transfer of a Ti plasmid from a pathogenic *Agrobacterium* into strain K84, which then became resistant to control. Therefore, some further genetic manipulation of strain K1026 may be required to prevent or reduce the frequency of acquisition of Ti plasmids for the maintenance of effective biological control.

Mines in the Goldfield

In registering the genetically manipulated strains K1026 in Australia, BioCare technology Pty Ltd. proceeded with state-by-state registration, beginning with the state of New South Wales in 1988. Major delays were experienced in obtaining federal clearance in Australia. This was not given until 1991, 3 years after initial application. The process of registration of K1026 was critically reviewed as part of an Australian Government inquiry into genetic manipulation (Commonwealth of Australia 1992). Several recommendations were made to improve the process and the lengthy time delay for federal clearance was strongly criticized.

The registration of strain K1026 in the USA is imminent (G. Bullard, BioCare Pty Ltd., pers. comm.). The process has so far taken over 5 years. In applying for registration of strain *Agrobacterium* K1026 in the USA, BioCare was required to conclusively demonstrate identity of the engineered strain with the parent strain K84, which had been sold in that country for many years previously. The emphasis of the authorities has been on this process, rather than on any perceived risk of using the engineered organism. This is presumably due to the nature of the genetic modification (deletion of DNA) and the long history of successful, safe use of the parent strain. The extremely long time taken to pass through registration processes is not assisting the development and use of new biotechnologies.

One of the difficulties in registering genetically manipulated organisms for a worldwide market is that the regulations vary from country to country. The Australian Government report on genetic manipulation (Commonwealth of Australia 1992) was published in the same year as an OECD report on safety in biotechnology (OECD 1992). These publications followed several other government reports and papers which were published in the late 1980s after the initial releases of genetically modified organisms, which at that time were predominantly bacteria. The OECD report (OECD 1992) outlines the criteria for good industrial large-scale practice for microorganisms and cell cultures, as well as good developmental principles for designing field experiments with genetically modified plants and microorganisms. Since the early 1990s, releases of genetically engineered organisms into the environment have predominantly involved plants. There does not appear to have been any review within the plast 5 years that specifically focuses on the release of genetically modified microorganisms in the environment.

Inoculants and Biostimulation for the Remediation of Organically Polluted Soil

Definitions and the Use of Bioremediation

Environmental contamination by potentially harmful chemicals and other products of human industrial origin is a worldwide problem. Many of the common organic contaminants can be degraded by microorganisms in the environment (Allard and Neilson 1997). Microorganisms play an important role in the environment, as they serve as biogeochemical agents for the conversion of organic compounds to simple inorganic compounds or their constituent elements. The conversion of organic compounds to carbon dioxide with the concomitant reduction in molecular oxygen is facilitated by a wide variety of bacteria, fungi and algae (Cerniglia 1992). The ability of microorganisms to degrade environmental pollutants, such as petroleum hydrocarbons (Atlas 1981), polycyclic aromatic hydrocarbons (PAHs) (Cerniglia 1992), polychlorinated biphenyls (PCBs) (Abramowicz 1990), pentachlorophenol (PCP) (McAllister et al. 1996) and pesticides (Aislabie et al. 1997), has generated considerable interest in the use of microorganisms for waste minimization and bioremediation of contaminated soils and waste streams. In addition, microorganisms have been shown to transform some inorganic contaminants to different oxidation states and have been used to accumulate metal ions for the remediation of heavy metal-contaminated waste streams (Naidu et al. 1999). However, some pollutants are recalcitrant to microbial attack (e.g. high molecular weigh PAHs, highly chlorinated PCBs, 1,1,1-trichloro-2,2-*bis*[*p*-chlorophenyl]ethane {DDT}) and may persist in the environment indefinitely.

Bioremediation, the biological degradation of organic hazardous wastes, has emerged over the past decade as a viable alternative to physical and chemical remediation processes. Bioremediation exploits the metabolic diversity of microorganisms, in particular those organisms displaying an ability to catabolize xenobiotic substances. The advantage of microbial degradation of pollutants over other processes is that it is a "green technology" whereby contaminants may be degraded to innocuous by-products. Furthermore, the process can be economical, reducing operating costs for treatment. Table 1 shows the cost effectiveness for the removal of arsenic and DDT from contaminated soil in Australia (Van Zwieten and Grieve 1995).

There are several different ways to approach bioremediation of organically polluted soils. Biostimulation is a form of bioremediation in which additives (e.g. nutrients) are used to enhance the activity of the indigenous microflora for the degradation of the contaminants. Bioaugmentation, on the other hand, involves the addition (inoculation) of non-indigenous microorganisms with specific degradative abilities towards the target compounds (Alexander 1991). In phytoremediation, which is usually used for the removal of heavy metals from soils, plants (hyperaccumulators) that can absorb high levels of the contaminant with their roots and concentrate them either in their shoots or leaves are utilized (Watanabe 1997). Rhizoremediation is a variant of phytoremediation in which microorganisms at the root surface, as well as the plant itself, play a role in the remediation process.

Table 1. Cost-effectiveness of bioremediation. Removal of arsenic and DDT from contaminated agricultural soils. (After Van Zwieten and Grieve 1995)

Technique	Cost[(a)]	Suitability[(b)]
Bioremediation	< A$100 t^{-1}	23
Thermal desorption	A$100-300 t^{-1}	13
Solvent extraction	A$100-300 t^{-1}	18
Membrane partitioning	A$100-300 t^{-1}	18

(a) cost per tonne of contaminated soil (1 ECU = A$1.75)
(b) Suitability rating out of 29, based on a range of factors

Biostimulation and Bioaugmentation for the Reduction of Polycyclic Aromatic Hydrocarbons in Soil

An example of the application of biostimulation and bioaugmentation for the clean up of PAH-contaminated soil from an industrially polluted site is shown in Fig. 3. In this laboratory study, the rate of PAH degradation was measured after biostimulation of the indigenous microflora with nutrients (nitrogen and phosphorus or yeast extract) and bioaugmentation with a PAH-degrading microbial community (Juhasz 1998). In all trials, the rate of PAH degradation decreased when the molecular weight and complexity of the compounds increased from two to five rings. When the activity of the indigenous microflora was stimulated by the addition of N and P or yeast extract, the rate of degradation was greater for every class size of PAH compared to non-stimulated soils. However, bioaugmentation of contaminated soils with the PAH-degrading microbial community was much more effective in removing PAHs than biostimulation under these conditions. Nevertheless, addition of yeast extract increased the effectiveness of the inoculum, especially in the degradation of phenanthrene (three-ring PAH). The results show that both biostimulation and bioaugmentation increased the rate of PAH degradation and the combined approach of the addition of bacteria and yeast extract was better than either method alone.

Mines in the Goldfield

One of the major limitations of bioremediation is the existence of contaminants that are highly resistant to catabolism by microorganisms, such as high molecular weight PAHs, highly chlorinated PCBs and some pesticides. In addition, bioremediation may not be effective at some sites due to physical and chemical characteristics of the soil (Pollard et al. 1994).

A second problem is the potential of biodegradative processes to generate toxic intermediates or by-products. The biodegradation of a number of compounds may be incomplete, depending on environmental conditions and the metabolic capabilities of the microorganisms. Degradation products, which may be more mobile and toxic than the parent compound (Dasappa and Loer 1991), may be generated, and, as such, pose a greater threat to human health. This is of concern as relatively little is known about whether biotransformations of environmental pollutants reduce the toxicity of the parent compound. For example, under aerobic conditions, DDT may be transformed to 2,2-*bis*(*p*-chlorophenyl)-1,1-dichlorethylene (DDE) while under anaerobic conditions the major transformation product of DDT is 1,1-dichloro-2,2-*bis*(*p*-chlorophenyl)ethane (DDD). Both these transformation products are highly recalcitrant to microbial attack and DDD is more toxic than DDT (Aislabie et al. 1997). Any program to remediate DDT-contaminated soil needs to take into consideration the problems associated with DDD and DDE accumulation.

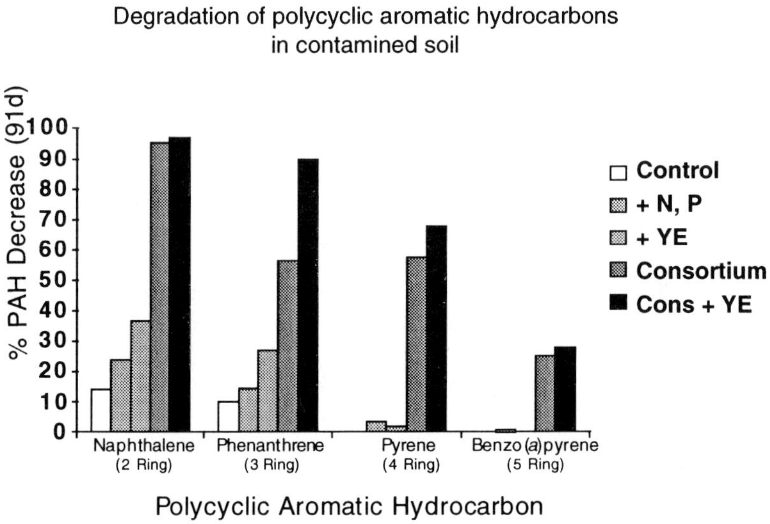

Fig. 3: Degradation of polycyclic aromatic hydrocarbons (PAHs) of different levels of complexity in contaminated soil. Results are expressed as % decrease in concentration of the particular PAH after 91 days of incubation of soil in the laboratory. Starting PAH concentration was approx. 820 mg kg^{-1} soil, with naphthalene, phenanthrene, fluoranthrene and pyrene as major components. Treatments (additions to soil): *Control* no addition; *+N, P* nitrogen and phosphorus (800 mg kg^{-1} soil); *+YE* yeast extract (1 g kg^{-1} soil); *Consortium* PAH-degrading microbial community (7.5 x 10^7 cells g^{-1} soil with N and P); *Cons + YE* combination of bacterial consortium and yeast extract (1 g kg^{-1} soil). (Data Juhasz 1998)

The bioavailability of environmental pollutants plays a critical role in the microbial degradation of these compounds. Many contaminants are readily degraded in situ; however, their persistence in soil can be attributed to the compound being inaccessible to the microorganism. Hydrophobic compounds tend to adsorb onto soil constituents and over long contact time may diffuse into the inorganic and organic matrix to form bound residues (Bouwer et al. 1994). As most evidence indicates that biotransformations occur in the liquid phase (Wodzinski and Bertolini 1972), adsorption of compounds to humic material or the clay fraction of the soil reduces the concentration of the compound available to the liquid phase and ultimately reduces the degradation rate. Bioavailability is particularly important when inoculants are to be added to the contaminated soil.

Lastly, the delivery of specific inocula to contaminated soil, as practised in bioaugmentation, can be limited by the lack of movement and survival of the microorganisms. The rate of contaminant degradation can also be limited if the introduced organisms are in a state or situation where they have low metabolic activity. This may be avoided by combining biostimulation and bioaugmentation to ensure adequate nutrient supply to the introduced microorganisms.

Conclusions

Soil biotechnology is being applied successfully to solve complex environmental problems. In the first example discussed here, specific DNA probes are being used commercially for the detection and, importantly, the quantification of plant pathogenic nematodes and soil borne fungal plant pathogens in soil. The development of techniques for routinely extracting clean DNA from soil has been achieved and a commercial diagnostic service, based on five specific DNA probes, now operates commercially in Australia at the South Australian Research and Development Institute. This technology can clearly be used on a routine basis with a very large variety of soil types. The DNA test result for the fungus that causes the root disease take-all of wheat is used together with a disease prediction model to enable cereal farmers to minimize the risk of serious yield losses caused by the disease. The future development of specific DNA probes for other soil organisms is in progress. The successful application of these probes will require continued cooperative and multidisciplinary research, to enable the quantification of the organism to be related to its effect in soil or on plants.

The regulatory and registration frameworks that apply to the release and commercial use of both native and genetically manipulated microbial pesticides are complex, time-consuming and can be costly. The testing and commercial registration of the genetically modified *Agrobacterium* strain K1026 for biocontrol of crown gall disease is one example. This situation is discouraging the development and commercial application of new biological treatments for agricultural use. In many countries, biological control agents are regulated in the same way as chemical pesticides. The development of separate, simpler guidelines for biological products would again encourage their development, because the markets for biological products are often limited and, as a consequence, there are limited financial resources available for the registration process.

The successful commercial use of a number of microorganisms as inoculants in agriculture and bioremediation is clear evidence that they can indeed work. However, inoculant microbes often do not establish adequate numbers in soil to effectively carry out their specific functions. Research on the fate of inoculants and the ecology of the environments into which they are introduced will be beneficial. The management of soil microbial communities to produce beneficial effects is an alternative, perhaps more ecologically based and therefore robust, way to proceed. One example is biostimulation to achieve the degradation of organic contaminants in soil. Another example, from agriculture, is the development of plant disease-suppressive soils (Alabouvette 1990; Roget 1995).

From the examples and references cited above, it is evident that there have been major advances in both techniques and applications in soil biotechnology. The advances have been enabled by interdisciplinary research. The future for soil biotechnology appears bright. For continued success, it will be important for soil biotechnology to remain a multidisciplinary science and to remain closely linked to other disciplines.

Acknowledgments. Research on the DNA probe for the take-all fungus was supported by the Grains Research and Development Corporation (PH, H, DKR) and the Co-operative Research Centre for Soil and Land Management (KMOK, H, DKR). The work on bioremediation (ALJ) was funded by the Australian Research Council, in conjunction with Australian Defence Industries, Victoria Education Foundation and Centre for Bioprocessing and Food Technology and Victoria University of Technology.

References

Abramowicz DA (1990) Aerobic and anaerobic biodegradation of PCBs: a review. Crit Rev Biotechnol 10:241-251

Aislabie JM, Richards NK, Boul HL (1997) Microbial degradation of DDT and its residues – a review. N Z J Agric Res 40:269-282

Alabouvette C (1990) Biological control of fusarium wilt pathogens in suppressive soils. In: Hornby D (ed) Biological control of soilborne plant pathogens. CAB International Wallingford, Oxon, UK, pp 27-44

Alexander M (1991) Research needs in bioremediation. Environ Sci Technol 25:1972-1973

Allard AS, Neilson AH (1997) Bioremediation of organic waste sites: A critical review of microbiological aspects. Int Biodeterior Biodegrad 39:253-285

Atlas RM (1981) Microbial degradation of petroleum hydrocarbons: an environmental perspective. Microbiol Rev 45:180-209

Bouwer E, Durant N, Wilson L, Zhang W, Cunningham A (1994) Degradation of xenobiotic compounds in situ: Capabilities and limits. FEMS Microbiol Rev 15:307-317

Bruns MA, Fries MR, Tiedje JM, Paul EA (1998) Functional gene hybridization patterns of terrestrial ammonia-oxidizing bacteria. Microb Ecol 36:293-302

Cerniglia CE (1992) Biodegradation of polycyclic aromatic hydrocarbons. Biodegradation 3:351-368

Commonwealth of Australia (1992) Genetic manipulation: the threat or the glory? Report of the Standing Committee on Industry, Science and Technology. Australian Government Publishing Service, Canberra

Dasappa SM, Loehr RC (1991) Toxicity reduction in contaminated soil bioremediation processes. Water Res 9:1121-1130

Donner SC, Jones DA, McClure NC, Rosewarne GM, Tate ME, Kerr A, Fajardo NN, Clare BG (1993) Agrocin 434, a new plasmid encoded agrocin from the biocontrol *Agrobacterium* strains K84 and K1026, which inhibits biovar 2 agrobacteria. Physiol Mol Plant Pathol 42:185-194

Egener T, Hurek T, Reinhold-Hurek B (1998) Use of green fluorescent protein to detect expression of *nif* genes of *Azoarcus* sp. BH72, a grass-associated diazotroph, on rice roots. Mol Plant-Microbe Interact 11:71-75

Harvey PR (1993) Genetic diversity among populations of the take-all fungus, *Gaeumannomyces graminis*. PhD Thesis, University of Adelaide, Adelaide, Australia

Herdina, Roget DK (1997) Detection and prediction of take-all. In: Proc 11th biennial conf of the Australasian Society of Plant Pathology, Perth

Herdina, Roget DK (1999) Suitability of a DNA-based assay for assessment of take-all risk in commercial fields. In: Magarey RC (ed) Proc Australasian Conf on Soilborne Root Diseases. Bureau of Sugar Experiment Stations, Brisbane Australia, pp. 61-62

Herdina, Harvey P, Ophel-Keller K (1996) Quantification of *Gaeumannomyces graminis* var. *tritici* in infected roots and soil using slot-blot hybridization. Mycol Res 100:962-970

Herdina, Yang HA, Ophel-Keller K (1997a) Correlation of take-all disease severity and inoculum level of *Gaeumannomyces graminis* var. *tritici* using a slot-blot hybridization assay. Mycol Res 101:1311-1317

Herdina, Ophel-Keller K, Roget D, Harvey P (1997b) Comparison between a DNA-based assay and a soil bioassay in quantifying the amount of *Gaeumannomyces graminis* var. *tritici* in soil. In: Dehne HW, Adam G, Diekmann M, Frahm J, Mauler-Machnik A, van Halteren P (eds) Diagnosis and Identification of Plant Pathogens. Kluwer, Dordrecht, pp 503-505

Hornby D (1978) The problems of trying to forecast take-all. In: Scott PR, Bainbridge A (eds) Plant disease epidemiology. Blackwell Scientific, Oxford, pp 151-158

Jones DA, Kerr A (1989) *Agrobacterium radiobacter* K1026, a genetically engineered derivative of strains K84, for biological control of crown gall. Plant Disease 73:15-18

Jones DA, Ryder MH, Clare BG, Farrand SK, Kerr A (1988) Construction of a Tra-deletion mutant of pAgK84 to safeguard the biological control of crown gall. Mol Gen Genet 212:207-214

Juhasz AL (1998) Microbial degradation of high molecular weight polycyclic aromatic hydrocarbons. Ph.D. Thesis, Victoria University of Technology, Melbourne, Australia

Kerr A (1980) Biological control of crown gall through production of agrocin 84. Plant Disease 64:25-30

Liesack W, Stackebrandt E (1992) Occurrence of novel groups of the domain Bacteria as revealed by analysis of genetic material isolated from an Australian terrestrial environment. J Bacteriol 174:5072-5078

Lopez-Lopez MJ, Vicedo B, Orellana N, Piquer J, Lopez MM (1999) Behavior of a virulent strain derived from *Agrobacterium radiobacter* strain K84 after spontaneous Ti plasmid acquisition. Phytopathology 89:286-292

Lynch JM (1997) Has biotechnology a role in soil science? Aust J Soil Res 35:1049-1060

Maier RJ, Triplett EW (1996) Toward more productive, efficient, and competitive nitrogen fixing symbiotic bacteria. Crit Rev Plant Sci 15:191-234

McAllister KA, Lee H, Trevors JT (1996) Microbial degradation of pentachlorophenol. Biodegradation 7:1-40

McClure NC, Ahmadi AR, Clare BG (1998) Construction of a range of derivatives of the biological control strain *Agrobacterium rhizogenes* K84 – a study of factors involved in biological control of crown gall disease. Appl Environ Microbiol 64:3977-3982

Naidu R, Smith E, Mallavarapu M, Smith LH, Sreedaran R, Churchman GJ, Kookana RS, Juhasz AL, Gates W, Oliver D, Ragusa S, Taylor G, Peter P (1999) Options for remediating metal-contaminated soils: an overview. Contaminated Site Remediation Conference, Freemantle, Western Australia, March 1999

OECD (1992) Safety considerations for biotechnology. Organisation for Economic Co-operation and Development, Paris

Ogoshi A, Kobayashi K, Homma Y, Kodama F, Kondo N, Akino S (Eds) (1997) Plant growth-promoting rhizobacteria: present status and future prospects. Hokkaido University, Sapporo, Japan

Ophel Keller K, McKay A, Driver F, Curran J (1999) The cereal root disease testing service. In: Magarey RC (ed) Proc st Australasian Conf on Soilborne Root Diseases. Bureau of Sugar Experiment Stations, Brisbane, Australia, pp 63-64

Panagopoulos CG, Psallidas PG, Alivizatos AS (1979) Evidence for a breakdown in the effectiveness of biological control of crown gall. In: Schippers B, Gams W (eds) Soil-borne plant pathogens. Academic Press, London, pp 569-578

Pankhurst CE, Doube BM, Gupta VVSR, Grace PR (Eds) (1994) Soil biota: management in sustainable farming systems. CSIRO Australia, Melbourne

Paton GI, Campbell CD, Glover LA, Killham K (1995) Soil ecotoxicity assessment using *lux*-modified constructs of *Pseudomonas fluorescens*. Soil Use and Manage 11:153-154

Pollard SJT, Hurdey SE, Fedorak PM (1994) Bioremediation of petroleum- and creosote-contaminated soils: a review of constraints. Waste Manage Res 12:173-194

Prosser JI (1994) Molecular marker systems for detection of genetically engineered micro-organisms in the environment. Microbiology (Reading) 140:5-17

Reinhold-Hurek B, Hurek T (1998) Life in grasses: diazotrophic endophytes. Trends Microbiol 6:139-144

Roget DK (1995) Decline in root rot (*Rhizoctonia solani* AG-8) in wheat in a tillage and rotation experiment at Avon, South Australia. Aust J Expe Agric 35:1009-1013

Roget DK, Rovira AD (1991) The relationship between incidence of infection by the take-all fungus (*Gaeumannomyces graminis* var. *tritici*), rainfall and yield of wheat in South Australia. Aust J Exp Agric 21:509-513

Ryder MH (1995) Monitoring of biocontrol agents and genetically engineered microorganisms in the environment: biotechnological approaches. In: Singh US, Singh RP (eds) Molecular methods in plant pathology. Lewis Publishers/CRC Press, Boca Raton, Florida, pp 475-492

Ryder MH, Jones DA (1991) Biological control of crown gall using *Agrobacterium* strains K84 and K1026. Aust J Plant Physiol 18:571-579

Ryder MH, Slota JE, Scarim A, Farrand SK (1987) Genetic analysis of agrocin 84 production and immunity in *Agrobacterium* spp. J Bacteriol 169:4184-4189

Stephen JR, Kowalchuk GA, Bruns MAV, McCaig AE, Phillips CJ, Embley TM, Prosser JI (1998) Analysis of beta-subgroup proteobacterial ammonia oxidizer populations in soil by denaturing gradient gel electrophoresis analysis and hierarchical phylogenetic probing. Appl Environ Microbiol 64:2958-2965

Tate ME, Murphy PJ, Roberts WP, Kerr A (1979) Adenine N^6-substituent of agrocin 84 determines its bacteriocin-like specificity. Nature (Lond) 280:697-699

Van Zwieten L, Grieve AM (1995) Arsenic and DDT contaminated cattle tick dipsites: a review of remediation technologies. Report for the dipsite management committee, NSW Agriculture, Wollongbar, New South Wales

Watanabe ME (1997) Phytoremediation on the brink of commercialisation. Environ Sci Technol 31:182-186

Wodzinski RS, Bertolini D (1972) Physical state in which naphthalene and biphenyl are utilised by bacteria. Appl Microbiol 23:1077-1081

Assessing Sources of Stress to Aquatic Ecosystems Using Integrated Biomarkers

SM. Adams[1]

Establishing causal relationships between sources of environmental stressors and aquatic ecosystem health is difficult because of the many biotic and abiotic factors which can influence or modify responses of biological systems to stress, the orders of magnitude involved in extrapolation over both spatial and temporal scales, and compensatory mechanisms such as density-dependent responses that operate in populations. To address the problem of establishing causality between stressors and effects on aquatic systems, a diagnostic approach, based on exposure-response profiles for various anthropogenic activities, was developed to help identify sources of stress responsible for effects on aquatic systems at ecological significant levels of biological organization (individual, population, community). To generate these exposure-effects profiles, biomarkers of exposure were plotted against bioindicators of corresponding effects for several major anthropogenic activities including petrochemical, pulp and paper, domestic sewage, mining operations, land-development activities, and agricultural activities. Biomarkers of exposure to environmental stressors varied depending on the type of anthropogenic activity involved. Bioindicator effects, however, including histopathological lesions, bioenergetic status, individual growth, reproductive impairment, and community-level responses, were similar among many of the major anthropogenic activities. This approach is valuable to help identify and diagnose sources of stressors in environments impacted by multiple stressors. By identifying the types and sources of environmental stressors, aquatic ecosystems can be more effectively protected and managed to maintain acceptable levels of environmental quality and ecosystem fitness.

Introduction

Environmental scientists are continually confronted with issues related to assessing and evaluating the effects of environmental stressors on the health of aquatic eco-

1. Environmental Sciences Division, Oak Ridge National Laboratory*, Oak Ridge, Tennessee 37830, USA

* Managed by Lockheed Martin Energy Research Corp. Under contract DE-AC05-960R22464 with the US Department of Energy

systems. Some of the more challenging issues which have proven to be the most problematic are (1) evaluating the relevance of laboratory studies, particularly toxicity testing, for application to field situations, (2) assessing the ecological significance of the sensitive early-warning indicators of ecosystem health, (3) importance of temporal and spatial variability in assessing aquatic ecosystem health, and (4) establishing cause-and-effect relationships between specific stressors and environmental damage.

Laboratory studies, employing standard toxicological testing, have been one of the primary approaches by which the effects of stressors have been assessed on the health of aquatic organisms. Laboratory studies typically involve short-term exposures of one or more contaminants on standard test organisms, and the effects of these contaminants are evaluated using lethal or other simple endpoints such as survival, growth, or reproductive potential. However, it is evident that an organism ceases to function normally long before these critical endpoints are reached (Larsson et al. 1985). Furthermore, the test conditions during lab studies seldom reflect the environment of natural populations (Cairns 1981) lacking ecological realism (NRCC 1985; Lagadic et al. 1994). Many of these laboratory studies have focused on short-term exposures using lethal and other nonspecific endpoints. Thus, much of the progress in aquatic toxicology had been due to an increase in data rather than an increase in knowledge (Wester and Vos 1994).

Over the past two decades there have been many sensitive "early-warning" indicators (biomarkers) measured on sentinel organisms collected at field sites which have been subjected to pollutants and other environmental stressors. Some of the more well-known and sensitive biomarkers of pollutant exposure are the mixed function oxidase (MFO) P450 detoxification enzymes (Stegeman et al. 1992; Jimenez and Stegeman 1990), heat shock (HSP70 stress proteins) (Pyza et al. 1997; Bradley 1993), antioxidant enzymes (DiGiulio 1992; Doyotte et al. 1997), and various measures of DNA damage (Shugart et al. 1992; Theodorakis et al. 1992). In many cases, these biomolecular and biochemical measures have proven to be excellent biomarkers of exposure to a variety of environmental stressors. These exposure biomarkers cannot serve, however, as ecologically relevant indicators of stress effects at higher levels of biological organization. Studies which include biomarkers of exposure must also be conducted, however, to help understand the mechanisms or relationships between stressors and ecologically relevant endpoints.

Notwithstanding these first two issues related to the relevance of laboratory studies to field applications and the ecological significance of early warning biomarkers of stress, perhaps the most important challenge facing environmental scientists is establishing causal relationships between environmental stressors and ecological relevant effects (Adams et al. 1999). Relating stressors to biomolecular and biochemical markers of exposure and ultimately to some relevant ecological effect is particularly problematic in natural field situations, primarily because of the many biotic and abiotic factors which can influence or modify responses of biological systems to environmental stressors (McCarthy and Munkittrick 1996; Wolfe 1996), the orders of magnitude involved in extrapolation over both spatial and temporal scales, and compensatory mechanisms such as density-dependent responses that operate in natural populations (Power1997).

An observable biomolecular or biochemical response such as induction of the P450 system from PAH exposure or inhibition of acethycholinerase from pesticide exposure will not necessarily be manifested as measurable biological effects at higher levels of organization. For effects to be realized at increasing higher levels, the stressor(s) must be of sufficient magnitude and duration to overwhelm the normal homeostatic capacity of specific biological systems (Schlenk et al. 1996). For example, when the capacity of protein systems such as the HSP70 stress proteins is exceeded, pathological damage can occur to important tissues or organs such as liver, gill, or kidney. Structural damage to liver tissue can, in turn, compromise the ability of this organ to produce vitellogenin, a critical component of egg development, and ultimately compromise reproductive success of individuals. Starting with normal individuals, exposure to contaminants and other stressors results in a progressive deterioration in organism health which may ultimately compromise the success of populations and communities (Fig. 1). Departures from the healthy state in organisms are associated with the initiation of compensatory responses with little change in disability (zone 1, Fig. 1). Additional impairment beyond the compensation limit may become associated with increased disability and overt disease (zone 2, Fig. 1). With additional environmental challenges, the survival potential of organisms declines because of their decreased ability to respond to increased challenges. Beyond the limit of compensation, it is unlikely that organisms could successfully mount any response at all to additional environmental challenges (zone 3, Fig. 1). Provided, however, that environmental conditions improve sufficiently and rapidly enough, an organism may be able to recover somewhat and repair damaged systems and restore compensatory responses. Monitoring impairment of biochemical, physiological, and behavioral responses, however, should provide early warning of the onset of disabilities (Depledge 1989). On the disability scale (Fig. 1), measures of stress effects are usually not detected until after the loss of compensation, while on the impairment scale, effects of stressors are detected much earlier and can be reversible and curable.

Several approaches have been applied to the problem of establishing causal relationships between stressors and biological effects. Except for field studies that have attempted to relate spatial patterns in contaminant loading with spatial patterns in biological responses over several levels of organization, usually from point sources of contamination (Adams et al. 1996; Soimasuo et al. 1995; Ericson et al. 1998), there are no reliable and proven techniques that have been applied in field studies to address the issue of establishing causal relationships between stressors and ecologically significant effects. Because it is difficult to establish causality between environmental stressors and ecologically relevant effects in field situations, approaches are needed which can help identify sources and causes of environmental damage at higher levels of biological organization. The primary objective of this study, therefore, is to develop and demonstrate a diagnostic approach that can help identify sources of anthropogenic stress in aquatic systems which are impacted by multiple stressors.

Approach

The health of aquatic ecosystems can be compromised by a variety of stressors which are related to anthropogenic activities including domestic sewage, atmospheric deposition, agricultural activities, mining operations, land-use activities

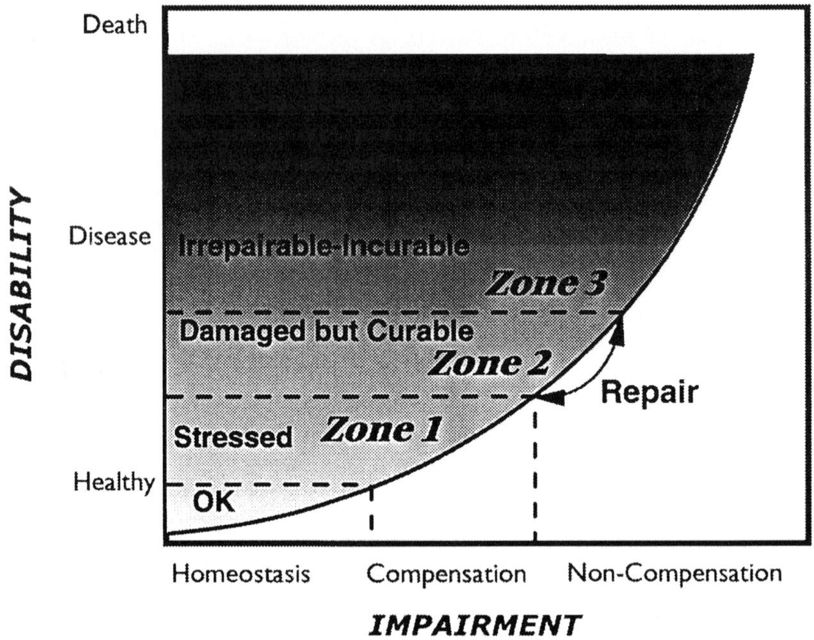

Fig. 1: The responses of aquatic organisms to environmental stressors are characterized by three major stress-response zones which correspond to the level of environment damage incurred relative to the disability-impairment curve

including urban development, logging, and clear-cutting, heavy industry including pulp and paper mill discharges, and petrochemical operations including drilling, refining, and exploration. Each of these activities produces a specific suite of environmental stressors, such as contaminants which characterize that activity. These characteristic sets of stressors can be used to separate or identify these activities from each other. For example, point-source discharges from paper mill operations are typically characterized by chlorophenolic and resin acid compounds, dioxin-type contaminants, and high nutrient loading. In contrast, non point source agricultural activities can contribute pesticides, nutrients, and sediment to receiving aquatic ecosystems. At the lower levels of biological organization, these various environmental stressors, which are related to specific anthropogenic activities, can also produce a characteristic set or profile of biomolecular, biochemical, or physiological responses in organisms, which are generally referred to as biomarkers of exposure to environmental stressors (Huggett et al. 1992). The primary advantage of biomarkers is that they can generally respond rapidly and directly to stressors and can serve, therefore, as early-warning indicators of effects at higher levels of biological organization. The principle biomarkers generally used as sensitive early-warning indicators and which characterize zone 1 of the disability-impairment curve (Fig. 1) are shown in Table 1.

Because certain stressors can cause specific types of biological responses, it is relatively straightforward to relate cause (the particular stressor) to lower level responses or to these biomarkers of exposure (i.e biochemical, biomolecular markers). For example, the principle types of exposure responses elicited by PAH-type compounds originating from petrochemical activities are generally high inductions of the P450 system and fluorescent aromatic metabolites in the bile. Agricultural activities, however, may result in inhibition of acetylcholinesterase activity from pesticide exposure. Relating biological responses at higher levels of organization such as at the individual, population, or community level to specific environmental stressors from specific anthropogenic activities, however, is much more difficult because of the modifying effects of biotic and abiotic factors in the environment, the high temporal and spatial variability of natural systems, and compensatory mechanisms that operate in populations and communities. Responses to environmental stressors at these more ecologically relevant endpoints are typically referred to as bioindicators of effects (Adams 1990).

Table 1. Biomarkers of exposure (zone 1) which respond relatively rapidly to environment stressors and bioindicators of higher-level biological effects which correspond to Zones 2 and 3 in Fig. 1

Zone 1 (rapid responses)	Zone 2 (intermediate responses)	Zone 3 (population-community responses)
Detoxification enzymes	Selected histopathologies	Size-frequency distributions
DNA damage	Immune system dysfunction	Altered sex ratios
Biliary metabolites	Bioenergetic impairment	Food-web alterations
Antioxidant enzymes	Reproductive integrity	Trophic-level relationships
Acetylcholinesterase	Growth	Community diversity, richness
Stress proteins		

To construct diagnostic exposure-response profiles for the various stressor exposure-biological effects relationships corresponding to each major type of anthropogenic activity, the principle types of stressors produced or characterized by each activity were first determined and then matched with the corresponding biomarker exposure response (Table 2). Since certain stressors, and in particular various types of contaminants, are associated with specific responses at the biomolecular, biochemical, or physiological levels, this analysis matched each major type of stressor to responses at these lower levels of biological organization. Once the major exposure responses (biomarkers) associated with each anthropogenic activity were identified, a stress exposure-biological response profile was generated for each of these activities by plotting these exposure biomarkers on one axis and bioindicator responses at higher levels of organization on the other. A literature review was conducted in order to identify which major types of biological responses at the higher levels of organization (bioindicators) were typically associated with each major type of anthropogenic activity. Those bioindicators which reflect effects at the individual, population, and community levels are generally used to characterize zones 2 and 3 of the disability-impairment curve (Fig. 1) and are listed in Table 1.

Table 2. Major types of anthropogenic activities and stressors that can compromise the integrity of aquatic systems and their corresponding biomarkers of exposure and bioindicators of significant ecological effects

Anthropogenic activity	Major types of stressors	Representative exposure responses	Representative ecological relevant responses	Principal references
Agriculture	Pesticides, herbicides, nutrients, sediments	Inhibition of acelylcholinesterase	Reproductive impairment, increased growth, disease	Bass et al. (1977) Kaur and Dhawan (1996) Coulliard et al. (1997) Sanchez et al. (1997)
Pulp and Paper	Nutrients, dioxins, resin acids, color, chlorophenolics	Mild MFO induction, bile metabolites,	Reproductive impairment, increased growth, gill histopathology	Owens (1991) Sandstrom (1996) Axelsson and Norrgren (1991) Andersson et al. (1988)
Petrochemical	PAHs, heavy metals	High MFO induction, aromatic bile metabolites	Impaired reproduction, decreased growth, liver tumors	Spies et al. (1996) Baumann et al. (1991) Moles and Norcross (1998) Vetemaa et al. (1997)
Mining	Heavy metals, sediments	Metallothiones, antioxidant enzymes, DNA damage	Reduced growth, gill histopathology, metabolic impairment, genetic diversity	Larsson et al. (1985) Farag et al. (1995) Mallatt (1985) Woodward et al. (1995)
Domestic Sewage	Chorine, nutrients, detergents	Mild MFO induction, antioxidant enzymes	Behavior changes, gill and liver histopath, bioenergetic impair	Mitz and Giesy (1985) Bass et al. (1977) Osborne et al. (1981)
Land Development, Clearcutting	Sediment and temperature increases, altered hydraulics	Stress proteins	Decreased growth, gill abnormalities, changes in feeding guilds	Newcombe and McDonald (1991) Alabaster and Lloyd (1982) Bergstedt and Bergersen (1997) Bradley (1993)
Power Plants	Temperature increases, chlorine	Stress proteins, antioxidant enzymes	Bioenergetic impairment, behavioral changes	Coutant (1997)

Results and Discussion

In Fig. 2, biomarkers of exposure are plotted against bioindicators of corresponding effects to generate biomarker-bioindicator response profiles which are characteristic of each major type of anthropogenic activity. Cross marks within the exposure-response profile for each activity indicate those specific biomarkers of exposure that are associated with various bioindicators of effects based on both field and laboratory studies. The principle biomarkers of exposure for petrochemical and pulp and paper activities are induction of the P450 mixed function oxidase (MFO) system and production of aromatic and chlorophenolic biliary metabolites, respectively. Effluent discharges from both petrochemical and paper mills have been reported to cause various gill lesions in fish and impair reproductive function in aquatic organisms. Growth, however, has been found to decrease under petrochemical exposure and actually increase in systems receiving paper mill effluents, due primarily to nutrient enrichment and increased productivity of receiving waters. In addition, organisms inhabiting systems impacted by PAH compounds typically have relatively high incidences of liver tumors, a situation not normally observed in aquatic systems receiving paper mill effluents. The exposure biomarkers related to mining activities and resulting heavy metal contamination of the environment, are the metallothione-type protein compounds, antioxidant enzymes, and DNA damage (genotoxicants). With this activity, decreased growth of fish has been observed along with severe gill damage and bioenergetic impairment. The indicators of exposure for domestic sewage are primarily the antioxidant enzymes which are produced from exposure to chlorine and detergents and some genotoxic indicators of DNA integrity. Organisms living downstream of municipal sewage outfalls have been reported to have increased growth (due to nutrient enrichment), both gill and liver damage, and various levels of bioenergetic impairment. The major environmental stressors associated with agricultural activities are pesticide and herbicides, nutrient additions, and sediment inputs into aquatic systems. The principle indicator of exposure in this situation is usually acetylcholinesterase inhibition. The major bioindicators of effects reported for agricultural activities are reproductive dysfunction, decreased growth, gill damage, and bioenergetic impairment. Land use and development activities resulting from urban development, logging, and clearcutting can result in a variety of stressors to aquatic systems including increased loading of sediment and nutrients, destruction of spawning and feeding habitat for aquatic organisms, and increased temperature regimes of aquatic systems. Even though there might not be contaminants involved in this type of activity, increased temperatures can trigger induction of the heat shock (stress proteins), which is a biomarker of increased temperature exposure. Bioindicators of environmental effects characteristic of this type of activity are gill damage (from suspended sediment), decreased growth, and changes in community richness and diversity which result from the ecological effects of sedimentation and increased water temperature. Because of elevated water temperatures, stress proteins may also be a characteristic biomarker of exposure in systems subjected to power plant thermal discharges. Even though many organisms can avoid thermally enriched areas, both bioenergetic impairment and changes in community structure and diversity have been reported for this activity.

Biomarkers of exposure to environmental stressors vary depending on the type of anthropogenic activity (stressor) involved. The MFO enzymes and bile metabo-

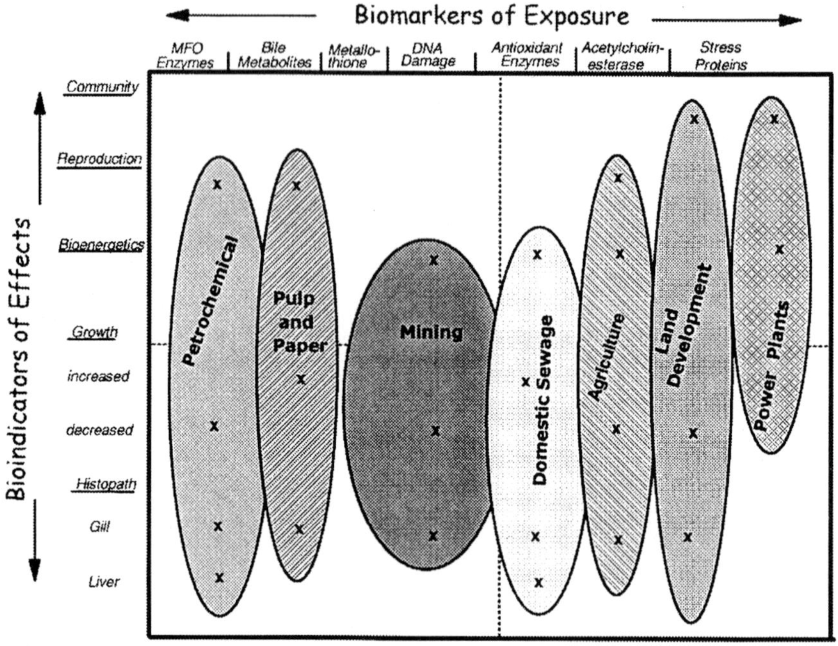

Fig. 2: Biomarker (exposure)-bioindicators (effects) profiles characteristic of each major type of anthropogenic activity which can be used to help identify major sources of stress in aquatic environments impacted by multiple stressors

lites are typically related to petrochemical and pulp and paper activities, even though a few other chlorinated organic compounds such as PCBs and insecticides are known to induce the P450 system. Metallothione-type protein compounds are usually induced only under heavy metal exposure (i.e. mining activities) while DNA damage is typically caused by carcinogenic and genotoxic compounds such as heavy metals and PAH-type compounds. Antioxidant enzymes such as superoxide dismutase and gutathione peroxidase are well-known indicators of exposure for several types of stressors including heavy metals, and even supersaturated oxygen levels in the environment can trigger the antioxidant defense system. Acetylcholinesterase (AChE) is very specific to organochlorine insecticides even though there is some evidence that levels of AChE may be mildly influenced by other environmental pollutants. There are no specific biomarkers of exposure for land-use and development activities and power plant operations because these two activities typically involve nonspecific stressors such as elevated temperature regimes and increased sediment and nutrient loading to aquatic systems in the case of land development. The stress proteins may be induced when the normal thermal tolerance of aquatic organisms is exceeded, but induction of these proteins is also relatively nonspecific to environmental stressors.

In several cases, bioindicator responses at the higher levels were similar for some of the major anthropogenic activities. This, of course, is not surprising, because biological responses at the population, community, and ecosystem levels are generally nonspecific to environmental stressors, and by the time stressor effects can be manifested at these ecological relevant levels, a variety of biotic and abiotic factors may have influenced or modified the overall stress response. For example, reproductive success of aquatic organisms was found to be impaired by petrochemical, pulp and paper, and agricultural activities. This is not to imply that the other anthropogenic activities in Fig. 2 cannot also affect the reproductive integrity of organisms, but just that reproductive competence was frequently found to be one of the more significant biological effects caused by these three activities. Also, different stressors may affect reproductive success in organisms by different mechanisms such as through hormonal imbalance, bioenergetic pathways, or directly through detrimental effects on eggs and larvae. Community level effects were frequently noted in the literature for activities related to land-use and development (logging, urban development, clear cutting). Environmental effects of power plant operation can be manifested either through direct pathways from siltation and temperature, or through indirect pathways which impair metabolic and feeding mechanisms and influence bioenergetic homeostasis of organisms. Growth was found to generally increase in organisms exposed to pulp and paper and domestic sewage discharges due to increased nutrient loading and increases in system-wide primary and secondary productivity. Conversely, decreased growth appears to be a major effect in systems perturbed by petrochemical, mining, agricultural, and land development activities. Decreased growth has been attributed directly to increases in metabolic demands when organisms are under environmental stress, and indirectly due to effects on the food chain which alter both the quality and quantity of available prey for higher-level consumers. Structural changes in tissues and organs, as determined by histopathological analysis, have consistently identified gill lesions as one of the most common individual-organism level responses to environmental stressors. All the anthropogenic activities shown in Fig. 1, except power plant operations, were commonly associated with gill pathologies (and thus metabolic and respiratory stress). Liver pathologies were also noted with domestic sewage and, in particular, petrochemical activities, where effluents from the latter are widely known to cause hepatic tumors.

Conclusions and Synthesis

The exposure-response profiles shown for the various anthropogenic activities in Fig. 2 were generated based on the major biomarkers of exposure and bioindicators of effects reported in the scientific literature. Each of these activities may have additional biomarkers of exposure and bioindicators of effects also associated with them, but for the purpose of generating these simplistic exposure-response profiles, only the principle biomarkers of exposure and bioindicators of effects reported for each activity were utilized in this presentation. The primary purpose of this exercise was to demonstrate the practical use of a diagnostic approach that can help identify sources of stress responsible for causing ecologically relevant responses in aquatic systems. Because it is difficult to establish causal relationships between environmental stressors and significant ecological endpoints, such an approach is valuable to help identify and diagnose sources of stress in environments which may be impacted by multiple stressors. For example, Chesapeake Bay is impacted by a number and variety of stressors including nonpoint source agricultural inputs, domestic and industrial sewage, and point-source inputs from Naval activities such as heavy metals and hydrocarbon (PAH) compounds. Various state and federal environmental laws require that the

Navy, for example, comply with these regulatory statutes in the form of environmetnal monitoring, assessment, and cleanup. Since the Navy is not the only source of environmental contaminants and other stressors present in Cheseapeake Bay, this method would help not only in identifying the types and sources of stress to the Bay ecosystem, but also in determining or separating out which of the observed effects at the population, community, or ecosystem level may be due to specific anthropogenic activities. By identifying the types and sources of environmental stressors, aquatic ecosystems can be more effectively protected and managed to maintain acceptable levels of environmental quality and ecosystem fitness.

References

Adams SM, Greeley MS, Ham KD, LeHew RF, Saylor CF (1996) Downstream gradients in bioindicator responses: point source contaminant effects on fish health. Can J Fish Aquat Sci 53:2177-2187

Adams SM (1990) Status and use of bioindicators for evaluating effects of chronic stress on fish. Am Fish Soc Symp 8:1-8

Adams SM, Greeley MS, Ryon MG (1999) Evaluating effects of stressors on fish health at multiple levels of biological organization: extrapolating from lower to higher levels. Hum Ecol Risk Assess (in press)

Alabaster JS, Lloyd R (1982) Finely divided solids. In: Alabaser JS, Lloyd R (eds) Water quality criteria for freshwater fish, 2nd edn. Butterworth, London, UK, pp 1-20

Andersson T, Forlin L, Hardig J, Larsson A (1988) Physiological disturbances in fish living in coastal water polluted with bleached kraft pulp mill effluents. Can J Fish Aquat Sci 45:1525-1536

Axelsson B, Norrgren L (1991) Parasite frequency and liver anomalies in three-spined stickleback, *Gasterosteus aculeatus* (L.), after long-term exposure to pulp mill effluents in marine mesocosms. Arch Environ Contam Toxicol 21:505-513

Bass ML, Berry CR, Heath AG (1977) Histopathological effects of intermittent chlorine exposure on bluegill (*Lepomis macrochirus*) and rainbow trout (*Salmo gairdneri*). Water Res 11:731-735

Baumann PC, Mac MJ, Smith SB, Harshbarger JC (1991) Tumor frequencies in walleye (*Stizostedion vitreum*) and brown bullhead (*Ictalurus nebulosus*) and sediment contaminants in tributaries of the Laurentian Great Lakes. Can J Fish Aquat Sci 48:1804-1810

Bergstedt LC, Bergersen EP (1997) Health and movements of fish in response to sediment sluicing in the Wind River, Wyoming. Can J Fish Aquat Sci 54:312-319

Bradley BP (1993) Are the stress proteins indicators of exposure or effect? Marine Environ Res 35:85-88

Cairns J (1981) Biological monitoring, part IV. Future needs. Water Res. 15:941-952

Coulliard CM, Hodson PV, Castonguay M (1997) Correlations between pathological changes and chemical contamination in american eels, *Anguilla rostrata*, from the St. Lawrence River. Can J Fish Aquat Sci 54:1916-1927

Coutant CC (1997) Thermal pollution in power plants. In: Encyclopedia of chemical technology, 4th edn, vol 23. John Wiley, New York, pp 963-984

Depledge M (1989) The rational basis for detection of the early effects of marine pollutants using physiological indicators. Ambio 18:301-392

DiGiulio RT (1992) Indices of oxidative stress as biomarkers for environmental contamination. pgs. 15-31. In: Mayes MA, Barron MG (eds) Aquatic toxicology and risk assessment. American Society for Testing and Materials, Philadelphia, ASTM 1124, vol 14

Doyotte A, Cossu C, Jacquin M, Babut M, Vasseur P (1997) Antioxidant enzymes, glutathione and lipid peroxidation as relevant biomarkers of experimental or field exposure in the gills and the digestive gland of the freshwater bivalve *Unio tumidus*. Aquat Toxicol 39:93-110

Ericson G, Lindesjoo E, Balk L (1998) DNA adducts and histopathological lesions in perch (*Perca fluviatilis*) and northern pike (*Esox lucius*) along a polycyclic aromatic hydrocarbon gradient on the Swedish coastline of the Baltic Sea. Can J Fish Aquat Sci 55:815-824

Farag AM, Stansbury MA, Hogstrand C, MacConnell E, Bergman HL (1995) The physiological impairment of free-ranging brown trout exposed to metals in the Clark Fork River, Montana. Can J Fish Aquat Sci 52:2038-2050

Huggett RJ, Kimerle RA, Mehrle PM, Bergman HL (eds) (1992) Biomarkers. Lewis, Boca Raton, Florida

Jimenez BD, Stegeman JJ (1990) Detoxification enzymes as indicators of environmental stress on fish. Am Fish Soc Symp 8:67-79

Kaur K, Dhawan A (1996) Effect of carbaryl on tissue composition, maturation, and breeding potential of *Cirrhina mrigala* (Ham.). Bull Environ Contam Toxicol 57:480-486

Lagadic L, Caquet T, Ramade F (1994) The role of biomarkers in environmental assessment (5). Invertebrate populations and communities. Ecotoxicology 3:193-208

Larsson A, Haux C, Sjobeck M (1985) Fish physiology and metal pollution: Results and experiences from laboratory and field studies. Ecotoxicol Environ Saf 9:250-281

Mallatt J (1985) Fish gill structural changes induced by toxicants and other irritants: a statistical review. Can J Fish Aquat Sci 42:630-648

McCarty LS, Munkittrick KR (1996) Environmental biomarkers in aquatic toxicology:fiction, fantasy, or functional? Human and Ecol. Risk Assess. 2: 268-274

Mitz SV, Giesy JP (1985) Sewage effluent biomonitoring. I. Survival, growth, and histopathological effects in channel catfish. Ecotoxicol Environ Saf 10:22-39

Moles A, Norcross BL (1998) Effects of oil-laden sediments on growth and health of juvenile flatfishes. Can J Fish Aquat Sci 55:605-610

Newcombe CP, MacDonald DD (1991) Effects of suspended sediments on aquatic ecosystems. N Am J Fish Manage 11:72-82

NRCC (National Research Council of Canada) (1985) The role of biochemical indicators in the assessment of ecosystem health: their development and validation. Publ NRCC 24371. National Research Council of Canada, Ottawa

Osborne LL, Iredale DR, Wrona FJ, Davies RW (1981) Effects of chlorinated sewage effluents on fish in the Sheep River, Alberta. Trans Amer Fish Soc 110:536-540

Owens JW (1991) The hazard assessment of pulp and paper effluents in the aquatic environment: a review. Environ Toxicol Chem 10:1511-1540

Power M (1997) Assessing the effects of environmental stressors on fish populations. Aquat Toxicol 39:151-169

Pyza E, Mak P, Kramarz P, Laskowski R (1997) Heat shock proteins (HSP70) as biomarkers in ecotoxicological studies. Ecotoxicol Environ Saf.38:244-251

Sanchez JC, Fossi MC, Focardi S (1997) Serum "B" esterases as a nondestructive biomarker for monitoring the exposure of reptiles to organophosphorus insecticides. Ecotoxicol Environ Saf 37:45-52

Sandstrom O (1996) In-situ assessments of the impact of pulp mill effluents on life-history variables in fish. In: Servos ME, Munkittrick KR, Carey JH, van der Kraak GJ (eds) Environmental fate and effects of pulp and paper mill effluents. St Lucie Press, Delray Beach, Florida, pp 449-457

Schlenk D, Perkins EJ, Hamilton G, Zhang YS, Layher W (1996) Correlation of hepatic biomarkers with whole animal and population-community metrics. Can J Fish Aquat Sci 53:2299-2309

Shugart LR, Bickham J, Jackim G, McMahon G, Ridley W, Stein J, Steinert S (1992) DNA alterations. In: Huggett RJ, Kimerle RA, Mehrle PM, Bergman HL (eds) Biomarkers. Lewis, Boca Raton, Florida, pp 125-153

Soimasuo R, Jokinen I, Kukkonen J, Petanen T, Ristola T, Oikari A (1995) Biomarker responses along a pollution gradient: effects of pulp and paper mill effluents on caged whitefish. Aquat Toxicol 31:329-345

Spies RB, Stegeman JJ, Hinton DE, Woodin B, Smolowitz R, Okihiro M, Shea D (1996) Biomarkers of hydrocarbon exposure and sublethal effects in embiotocid fishes from a natural petroleum seep in the Santa Barbara Channel. Aquat Toxicol 34:195-219

Stegeman JJ, Brouwer M, DiGiulio RT, Forlin L, Fowler BA, Sanders BM, Van Veld PA (1992) Molecular responses to environmental contamination: enzyme and protein synthesis as indicators of contaminant exposure and effect. In: Huggett RJ, Kimerle RA, Mehrle PM, Bergman HL (eds) Biomarkers. Lewis, Boca Raton, Florida, pp 235-335

Theodorakis CW, D'Surney SJ, Bickham JW, Lyne TB, Bradley BP, Hawkins WE, Farkas WL, McCarthy JF, Shugart LR (1992) Sequential expression of biomarkers in bluegill sunfish exposed to contaminated sediment. Ecotoxicology 1:45-73

Vetemaa M, Forlin L, Sandstrom O (1997) Chemical industry effluent impacts on reproduction and biochemistry in a North Sea population of viviparous blenny (*Zoarces viviparus*). J Aquat Ecosyst Stress Recovery 6:33-41

Wester PW, Vos JG (1994) Toxicological pathology in laboratory fish: an evaluation with two species and various environmental contaminants. Ecotoxicology 2:21-44

Wolfe DA (1996) Insights on the utility of biomarkers or environmental impact assessment and monitoring. Human Ecol Risk Assess 2:245-250

Woodward DF, Farag AM, Bergman HL, DeLonay AJ, Little EE, Smith CE, Barrows FT (1995) Metal-contaminated benthic invertebrates in the Clark Fork River, Montana: effects on age-0 brown trout and rainbow trout. Can J Fish Aquat Sci 52:1994-2004

Transport of Nitrogen, Phosphorus and Microorganisms from Manure into Surface- and Groundwater

MJ. Goss, A. Unc, S. Chen[1]

The quality of water resources is an important factor that must be considered in any analysis of sustainable agriculture or resource management. Most contaminants affecting water quality in rural areas can be grouped into one of three classes: simple inorganic ions, more complex organic molecules, and particulates. Contaminants of all three classes can result from animal manure. In particular, it potentially provides the source of nitrate, phosphate, toxic metals, and bacteria as well as materials that generate high biological oxygen demand.

The concentration of contaminants entering water can be different for each class of contaminant, as they are affected by different processes. Organic molecules are affected by vapour pressure, and solubility in water and soil organic matter, simple inorganic compounds are influenced by exchange and adsorption, whereas the transport of microorganisms are affected by surface charge and population dynamics.

The organic matter in manure also alters the physicochemical properties of the soil, thereby increasing the mobility of contaminants and resulting in losses from soil under conditions that might suggest that the materials would be less available to leaching and runoff.

Preferential flow paths can develop due to structures present in the surface soil or in the subsurface due to soil horizon boundaries. Preferential flow can carry water and contaminants to depth very rapidly. This reduces potential interactions occurring between the soil and contaminants in solution, and, again, can result in greater impact on water quality than might be expected.

Introduction

Goss et al. (1994) assessed the current knowledge related to the impacts of livestock manure on water quality in Ontario and concluded that contaminants derived from

1. Centre for Land and Water Stewardship, Richards Building, University of Guelph, Guelph, Ontario N1G 2W1, Canada

manure, such as phosphorus, nitrate and microorganisms, can severely impair the quality of water in nearby water resources. Indeed, there have been increasing numbers of reports of water resources being impacted by contaminants that are associated with manure. Strong public reaction to such events, coupled with concerns over odours, currently threatens the sustainability of livestock enterprises. An underlying factor has been that in intensive animal production systems, manure management has commonly been approached as a problem of waste disposal rather than the effective use of a by product which contains nutrients required for crop growth. The large organic carbon content of manure also can provide added benefit through improved structure and resilience of soil. Nevertheless, despite these benefits, manure management represents a net cost to nearly all intensive livestock operation. In part, this is because of the imbalance of constituent nutrients in manure compared with the ratio required by crops. An added problem on some farms is too little land for the optimal and safe application of the nutrients in manure for use in crop production.

This chapter reviews the processes involved in the transport of contaminants from manure to water resources, particularly through the unsaturated (vadose) zone and in surface runoff. We consider the importance of preferential flow paths in the transport of bacteria, and special features associated with manure being the main source of contaminants.

Contaminants and Their Transformations

Nitrogen and other nutrients tend to be voided in organic forms by animals, but some of the compounds are readily hydrolyzed to inorganic compounds. Thus, there are two potential sources of NO_3^- within manure: a mineral nitrogen (N) fraction, which is predominantly in the form of NH_4^+-N, and an organic-N fraction. Thus, both mineralization and nitrification processes are involved in determining the amount of NO_3^- present in the soil. N present in organic forms may be in complex molecules derived from the lignin in the feed, and in simpler compounds such as urea or ureide derivatives. It is these simpler compounds that are rapidly hydrolyzed by the enzyme urease and converted to ammonium ions. Animal feeding regimes can greatly influence the total amount of N excreted, and the proportion in the two key fractions (Haig and McBride 1996; de Lange 1996). The ammonium ions can undergo oxidative reactions to form nitrate and nitrite, but some will convert into dissolved ammonia and be subject to volatilization.

Once NO_3^- is formed in the soil by mineralization and nitrification, it may be subject to denitrification. The emission of gases such as N_2O, (the most important greenhouse gas generated in agriculture), NO, NO_2, and N_2 that results from denitrification is likely to be greater than that from soils receiving mineral nitrogen fertilizers (Thompson and Pain 1990). Paul and Zebarth (1993) found that denitrification rate was greater where manure was applied to the coarser of two sandy soils which had a shallow water table. Paul et al. (1993) showed that manure-amended soil produced N_2O and NO due to both nitrification and denitrification processes. Some of the carbon compounds present in liquid cattle manure are readily available to denitrifiers (Paul and Beauchamp 1989).

The proportion of phosphorus (P) that is in organic form is greater in solid than in liquid manure. P in liquid manure can occur as particulate P such as trimagnesium phosphate (Fordham and Schwertmann 1977), or as soluble components such as orthophosphates and low molecular weight organic phosphorus compounds. Leinweber (1997) found that the total P in dry poultry manure was less than in liquid swine manure but the proportion of soluble P was greater in poultry manure. Poultry manure tends to contain the largest concentration of total P (up to 0.92%) while cattle manure tends to contain the lowest level (Tietjen 1987). Swine slurry tended to have more than double the amount of P present in cattle slurry (Schweiger et al. 1989). The total P excreted per 100 kg live animal weight per year was also greatest for poultry (approximately 12.8 kg $year^{-1}$), followed by pigs (6.2 kg $year^{-1}$), beef cattle (4 kg $year^{-1}$), sheep (2.6 kg $year^{-1}$), and dairy cattle (2 kg $year^{-1}$) (Taiganides, 1987).

Manure is usually applied to soil as a solid material, usually containing a lot of high-carbon bedding material, or as a liquid with significant added water. Liquid manure is a complex mixture containing a number of solid components, all at or near equilibrium with each other, and reacting quite rapidly to any changes in solution composition. The distribution of components between phases can be expected to depend very closely on the condition imposed upon the system both before and after its application to the field (Fordham and Schwertmann 1977).

Toxic metals are also present in some animal manures. Copper sulphate is used as a feed additive in swine and poultry to promote weight gain and feed efficiency (O'Dell et al. 1979; Miller et al. 1991). Zinc is also used as a feed supplement for swine and poultry to reduce the excessive accumulation of Cu in liver, and to enhance general health and growth (O'Dell et al. 1979; Miller et al. 1991). Generally, zinc has been considered to be a safe mineral supplement, the animals tolerating high intakes of the element. It is excreted primarily in faeces (Miller *et al.* 1991). The efficiency with which pigs utilize copper is variable (20 to 40%), depending on the forms of Cu in feed and animal age, with excess Cu being eliminated mostly through biliary excretions (Miller et al. 1991).

As a result of the addition of metals to feed, manure from the animals is enriched with copper and zinc, leading to considerable amounts of these metals being applied to soil (Sutton et al. 1983). The total amount of Cu in fresh manure was found to be similar to or even slightly greater in poultry manure than in liquid swine manure (Japenga and Harmsen 1990), but on a dry matter basis the percentage was over six times greater in swine manure (Strauch 1987, Taiganides 1987). Liquid swine manure contained more soluble Cu than did poultry manure. The percentage of Cu in the liquid fraction of swine manure increases slightly with storage time (Japenga and Harmsen, 1990). Cattle excreted only trace amounts of these metals (Taiganides 1987). Cu and Zn are found in equal quantities within swine manure, and the level of Zn may be more than four times that present per unit of dry matter in poultry and cattle manure (Taiganides 1987).

Animal manure can be the source of pathogenic organisms such as bacteria, viruses, protozoa and helminthic worms (Strauch 1987). The amount of pathogenic bacteria is likely to be greater in swine and poultry manure than in cattle manure. Due to the greater mobility of bacteria in liquid phase compared to the solid phase, liquid manure tends to be more uniformly contaminated than the solid manure. Pathogens may be present in manure even if the animals present no symptoms, and a small number of infected animals can contaminate a whole source of manure (Strauch, 1988). Consequently, the greater the number of animals on a farm, the greater the likelihood of pathogens being present in the manure.

The bacterial groups present in greatest number in manure are faecal coliforms and streptococci (Table 1). While bacteria species from these two groups are always present in manure, *Salmonella* (another important group of pathogens) is present occasionally, mostly in swine and poultry manure.

Many organic compounds present in manure are readily subject to microbial breakdown in the soil, or if they enter adjacent surface water, and thereby establish a large oxygen demand. It is during the normal breakdown of organic compounds that nutrient ions are released, as well as CO_2.

Table 1. Examples of bacterial numbers in some animal manure (CFU ml^{-1} or g^{-1} fresh manure)

Manure type	Faecal coliforms	Faecal streptococci	*Salmonella* spp.	Reference
	4.3×10^3 to 1.3×10^5			Unc (1999)
Liquid swine manure	2.4×10^3	9.3×10^3	0	Weigel (1995)
	9.5×10^4 to 1.1×10^6 (*E. coli*)	7.2×10^4 to 4.5×10^5 (Streptococci-D)	0 to 1.5×10^3 (*S. infantis*)	Rüprich (1994)
	2.4×10^3	9.3×10^3	0	Weigel (1995)
Liquid cattle manure	4.5×10^2 to 1.5×10^6 (*E. coli*)	4.5×10^2 to 9.5×10^5 (Streptococci-D)	0	Rüprich (1994)
Dairy slurry	6.3×10^4 to 1.0×10^7 (Enterobacteria)			Östling and Lindgren (1991)
Solid beef manure	2.4×10^5	1.5×10^7	0	Weigel (1995)
	1.9×10^6 to 6.8×10^6			Unc (1999)
Solid dairy manure	2.0×10^5 to 1.0×10^7 (Enterobacteria)			Östling and Lindgren (1991)

Transport Processes

The variety of transport processes by which contaminants in manure move to water resources are reviewed in this section. For comprehensive reviews of solute transport see Addiscott and Wagenet (1985) and Jarvis et al. (1991).

Contaminants which originate in manure and affect water resources can be divided into three basic classes: simple inorganic ions (e.g. NH_4^+, NO_3^-, $H_2PO_4^-$), more complex organic molecules (e.g. phenolics, volatile fatty acids), and particulates (e.g. microorganisms). The concentration of inorganic ions is controlled by the equilibrium between solid and solution phases in soil water. This may involve the formation of sparingly soluble precipitates and adsorption reactions with soil particles. For organic molecules and some inorganic species such as NH_4^+, vapour pressure and solubility in water and in soil organic matter can determine the final concentration of contaminants in soil water. In contrast, particulates are generally affected by surface charge. In all cases, transport varies greatly according to soil structure, especially the size distribution and continuity of pores. For inorganic nitrogen compounds and bacteria, the soil is itself a source, and may also contain one or more types of sink.

Contaminant Characteristics Relevant to Their Transport

It is important to consider the characteristics of different contaminants that would affect transport. Nitrate travels in solution with infiltrating water. Plant uptake and de-nitrification remove nitrate from solution, but these processes are likely to be much less important when infiltration is rapid. Once nitrate enters a preferential flow path, there is little chance of it being removed from the solution. In contrast, NH_4^+ and $H_2PO_4^-$ can be removed from the water through interaction with the soil matrix.

Volatile organic compounds can also move as a vapour, so additional terms must be added to take account of the diffusive and dispersive fluxes in this phase. NH_3 can volatilize from NH_4^+ , and also move in the gaseous phase. In the case of manure, much of the ammonia nitrogen may be lost to the atmosphere before field application. Another factor is the variation in N deposition from the atmosphere. Goss and Goorahoo (1995) suggested that the considerable variation between monitoring sites observed in studies of N-deposition from the atmosphere (Barry et al. 1993) may be due to the impact of ammonia volatilization from livestock operations.

Transport of particulates such as microorganisms are transported as colloids. It is reasonable to assume that the concentration of a solute within the pore space is uniformly distributed, but that may not be true of colloids. Bacteria are much larger than nitrate ions, and their movement is more likely to be affected by the flow associated with the size of pores in which they are transported. They have variable surface charge which allows stronger adsorption of the bacteria to soil particles. Bacteria also have very large surface area to volume ratios that provide a large proportion of sites for adsorption. A third consideration with microorganisms is that their populations are dynamic.

Preferential Flow and Solute Transport

Soil pore characteristics are important for transport. However, the transport of contaminants in soils with strongly aggregated structure, or with large and continuous pores, is not well described by equations based on the CDE because preferential flow occurs (Wagenet 1990).

Preferential flow is the process whereby water and its constituents move by preferred pathways through a porous medium. It means that part of the matrix is effectively bypassed. The term preferential flow does not itself convey a mechanism for the process (Helling and Gish 1991), whereas the often-used term macropore flow implies transport through relatively large pores, channels, fissures, or other semicontinuous voids within the soil. Although there is no standardized definition for macropores, some pore classification has been proposed. Luxmoore (1981) suggested the classes of micro- meso- and macropore defined by equivalent pore diameters of < 10 µm, 10-1000 µm, and > 1000 µm, respectively. Skopp (1981) defined macroporosity as that pore space which provides preferential paths of flow so that mixing and transfer between such pores and the remaining pore space is limited. Some other classifications of soil pore size and their functions with respect to water movement or root penetration have been summarized by Helling and Gish (1991). Pore size and the corresponding capillary potential were given by Beven and Germann (1982). Macropores in soil may be developed by physical (shrink-swell, freeze-thaw, or tillage) or biological (earthworms, insects, soil fauna, or roots) processes. Continuous macropores can be formed in soil due to the activity of soil macrofauna, especially earthworms (Ehlers 1975). In soils with significant shrink-swell-behaviour, cracking may be important in the development of a preferential flow domain, and the extent of crack development is generally related to water extraction by roots. The channels created by roots can also dominate the transport process once the original roots have decayed (Barley 1954). Freeze-thaw cycles may also result in stress fractures. The size and direction of macropores can vary greatly.

A summary of the evidence of water and solute transport through macropores (Thomas and Phillips 1979) stimulated a reexamination of the *direct* (preferential flow) and *general* (matrix flow) drainage patterns initially observed by Lawes et al. (1882). A soil classification system has been proposed based on soil properties related to water and chemical transport (Quisenberry et al. 1993). Macropore flow has been well recognized as a significant component of water and solute transport in soils. Percolating water can bypass a large fraction of the soil matrix, thus moving deeper and with less displacement of the initial soil solution than would have been predicted by piston displacement (Beven and Germann 1982; Bouma 1981; Quisenberry and Phillips 1976, 1978; Quisenberry et al. 1993; Shirmohammadi et al. 1991). Watson and Luxmoore (1986) found that under ponded conditions, 73% of the flux was conducted through macropores (pore diam. > 0.1 cm). Furthermore, they estimated that 96% of the water was transmitted through only 0.32% of the soil volume. As much as 70 to 90% of applied chemicals may be moving preferentially through macropores (Ahuja et al. 1993). Quisenberry et al. (1994) found that 50% of applied water and Cl^- was transported through < 20% of the soil matrix.

Preferential flow may occur even in coarse-textured soil considered to be homogeneous (Steenhuis et al. 1990; Kung and Donohue 1991). Quisenberry and Phillips (1976) reported that macropore flow commenced at the tilled-untilled boundary in a tilled Maury silt loam. Similar results were observed on a Cecil sandy clay loam in the Piedmont of South Carolina (Hatfield 1988).

Jardine et al. (1990) investigated transport of Br through two distinct pore classes of an undisturbed pedon. They found that solutes were transported by convection and diffusion from small-pore to large-pore regions via hydraulic and concentration gradients, respectively, with small pores being a major source of the solute transported rapidly by large pores. A diffusion-based mechanism described by Luxmoore (1991), in which new water entering a soil gains the chemical attributes of old water, supported the results reported by Jardine et al. (1990). According to Luxmoore (1991), a large surface area of interaction combined with a short diffusion path between mesopore channels and micropores allows diffusion to be a significant contributor to chemical transport during preferential flow events.

Macropore flow can either increase or decrease the residence time of solutes in soil, depending on the location of solutes relative to macropores (Wild 1972; Edwards et al. 1994; Chen et al. 1997; 1999), although macropore flow has been known to be a major cause for the groundwater contamination and transport uncertainty. Leaching of nitrate added to the soil surface as fertilizer would be more rapid than that of nitrate formed from soil organic matter within the aggregates (Wild 1972 Goss et al 1987). Chen et al. (1999) examined the time required for solutes to become uniformly distributed at each depth after a leaching event, and the effect that degree of spatial distribution had on the subsequent vertical movement induced by a second leaching event in a well-structured clayey soil, under field conditions. Bromide tracer was practically uniformly distributed in the surface layer immediately following the first event, but the distribution became progressively more non-uniform as depth increased. The longer the delay in sampling following the first event, the more uniform was the lateral distribution of Br. The second leaching event caused Br^- to move deeper within the soil profile when it occurred the day following the first event than when it occurred later. Similarly, an initial small rain (5 to 10 mm) could move the solute into the soil matrix, thereby reducing the potential for transport in macropores during subsequent rainfall events (Shipitalo et al. 1990; Golabi et al. 1995).

The initial position of chemicals may affect their movement (Jardine et al. 1990; Shipitalo et al. 1990; Timlin et al. 1992; Golabi et al. 1995; Chen et al. 1997, 1999). For example, initial moisture content and rainfall intensity and duration may affect solute distribution and movement among small and large pores (Jardine et al. 1990). High-intensity rainfall or water application rate exceeds the soil infiltration rate, or when saturated-flow conditions occur, preferential flow can then be initiated. Soil morphology (Bouma 1981), clay films (Quisenberry et al. 1993; Chen et al. 1997), and surface condition (Phillips et al. 1989; Quisenberry et al. 1993) affect water and solute distribution or transport. Among these factors, those related to pedogenetic processes, such as soil morphology and structure, require longer time periods to show effects. Other factors such as moisture content, tillage, and cultural practice may be relatively transient in their effect. Understanding the effect of these factors on water and solute transport may ultimately lead to a more reliable prediction of transport processes in soil (Steenhuis et al. 1990; Ahuja et al. 1993), including the movement of contaminants.

Helling and Gish (1991) described some factors affecting the process of preferential flow including soil porosity, pore characteristics, structure, initial moisture content and soil management. Flow through tubes is proportional to the fourth power of their radii, therefore drainage will be much more rapid from large continuous macropores than from pores of smaller diameter. Mouldboard ploughing may destroy the continuity of pores between the plough layer and the deep horizons. Long-term no-tillage plots, on the other hand, often develop a high density of continuous, relatively large vertical channels (Goss et al. 1993a). Manure application may encourage the activity of earthworms, which may result in a greater continuity of macropores. Hence, there may be a faster breakthrough of contaminants than predicted (Munyankusi et al. 1994).

A relatively large water content might result in deeper movement of contaminants (Quisenberry and Phillips 1976), but the opposite effect has also been reported (White et al. 1986). Understanding the mechanism of bypass flow through convection and diffusion from regions with small pores to those with large pores may help to explain such differences.

Factors Important in Determining the Concentration of Contaminants in Point and Diffuse Sources

As indicated above, the concentration of the contaminant in the soil is a key factor in predicting its transport. Manure forms both point and diffuse sources of contamination.

Point Sources

The main point sources of manure are manure storages. The lignin-protein complexes and hemicellulose content between ruminants and non-ruminants will vary due to the ability of the former to break down cellulose, sugars and complex starches within the animal (MacLean et al. 1983). Thus the manure of non-ruminants will tend to be higher in cellulose than that of ruminants. Ruminants also release a large number of bacteria in their faeces. In addition to species and feed, the nitrogen concentrations in different manures depends on the health of the animals, the bedding content, and the amount of water added (Beauchamp 1983). Between one-quarter and one-third of manure C may be lost during normal storage periods (Patni and Jui 1987; Vanerp and Vandiyk 1992), and, in consequence, there is a relative increase in the concentration of nutrients.

The microbiological population in excreta undergoes considerable change during storage. Nodar et al. (1990) concluded that poultry excreta contained a high density of microorganisms, and was similar to cattle slurry in this respect. The number of microorganisms was an order of magnitude greater than those found in pig slurry. At the beginning of slurry storage, the population of viable organisms in most microbial groups abruptly declined (Nodar et al. 1992). Denitrifying and sulphate-reducing microbes, together with algae, increased during this time. Thereafter, the total population multiplied rapidly, becoming fivefold greater than the initial value after 14 weeks. The increase was mainly attributed to anaerobic bacteria (proteolytic, ammonific, amylolytic, anaerobic-cellulytic and anaerobic-nitrogen-fixing species); aerobic heterotrophic bacteria, actinomycetes and fungi showed little change.

In solid manure stores, pathogens close to the periphery of a pile may not be subject to the same temperature regime as those near the centre. Consequently, the former may survive and form the source for contamination when spread on the land (Sutton et al. 1983).

Diffuse Sources

After field application, manure forms a diffuse source. After field application, manure forms a diffuse source. The nutrients in the manure may be taken up by the crops to which they are applied, or they are available for transport. Nitrogen can be lost in gaseous form as ammonia, or it may be nitrified to nitrate, which is then subject to leaching or denitrification. Volatilization of NH_3 is a major reason for N loss during and immediately after manure application. Beauchamp et al. (1982) and Paul et al. (1990) found that the three most important variables which influence NH_3 volatilization appear to be temperature, soil pH and soil texture. The magnitude of gaseous losses of N is difficult to estimate. There is considerable uncertainty about the availability of manure N once incorporated in the soil. For example, Pratt et al. (1976) observed that there was a greater proportion of N that could not be accounted for at an application of 1750 kg N ha^{-1} than at 500 kg N ha^{-1}. Beauchamp (1986) found that about three-quarters of the ammonium-N fraction was as available as equivalent fertilizer N. In consequence, there are reports that groundwater contamination with NO_3^- is greater in areas where animal manures are applied regularly compared with areas receiving predominantly mineral-N fertilizer (Ritter and Chirnside 1987). Thomsen et al. (1993) suggested that nitrate leaching from land where manure slurry was applied may be greater than that from where mineral fertilizer was applied because no account was taken of the mineralization of the organic nitrogen in the slurry. In a study comparing the availability of N from liquid cattle manure (LCM), liquid poultry manure (LPM) and solid beef manure (SBM) to corn, Beauchamp found that the yield response to equal amounts of applied N was in the general order of LPM > LCM > SBM.

The experimental evidence shows that, compared with spring applications, manuring land in fall or winter results in lower recovery of applied nitrogen by the crops, and greater risk of leaching and denitrification (Thompson et al. 1987; Goss and Goorahoo 1995)

Phosphorus is not lost by volatilization, so the content in manure will not change provided there is no loss of bulk material. Increasing the amount of cattle manure applied to soils in Alberta over a period of 11 years increased the total phosphorus content of the soil and the available phosphorus (Chang et al. 1991). From agronomic and environmental considerations, one would want to apply only as much phosphorus, either as manure or fertilizer, as is required for the most economic crop production.

The availability of P from manure applied to the land is also subject to uncertainty, being a dynamic function of physical and chemical processes controlling both soluble P and bioavailable particulate P. Soluble P transport depends on desorption-dissolution reactions controlling P release from soil, fertilizer reaction products, vegetative cover and decaying plant residues (Sharpley et al. 1992). The behaviour of inorganic P applied to the soil as mineral or organic fertilizer is generally considered to be a function of the P sorption index. On the other hand, the labile organic P stays at more or less constant levels unless severely depleted during mineralization, after which a shift occurs from moderately labile and moderately resistant organic P, to labile organic P (Sharpley and Smith 1985). The P buffer power of a soil apparently depends on the availability of P sorption sites and their degree of saturation, and may not always be directly related to the ability of a soil to release P. The maximum quantity of P desorbed from soil is directly correlated to the P enrichment for that specific soil, and to the level of the native labile P content (Raven and Hossner 1993, Sharpley and Smith 1989).

The concentration of microorganisms in the manure at the time of application is an important determinant of the potential for their transport and eventual contamination of water resources. Manure may affect survival of bacteria as in the study by Östling and Lindgren (1991), where 20-40 times more indigenous *Bacillus* spores were present on manured crops than on unmanured crops, and these numbers remained constant with time to harvest. However, this was not the case for bacteria of manure origin, such as *Clostridium*, some coliforms and *E. coli*, all of which declined with time after manure application. The survival of any non-indigenous bacteria depends on several factors including soil pH, soil water content, organic matter content, soil texture, temperature, availability of nutrients, adsorption properties of the soil (soils containing clays with a large surface area can adsorb bacteria MacLean 1983) and biological interactions in the soil (Abu-Ashour et al. 1994b).

Free-living protozoa and nematodes are predators of bacteria in the soil, and *Bdellovibrio*, a soil bacterium, preys on other bacteria (Peterson and Ward 1989). Presence of these organisms may reduce or limit bacterial numbers; however, introduced bacteria may still be able to survive for an extended period after manure application. Chandler et al. (1981) found that on average 10% of faecal coliforms and faecal streptococci were still present in the soil 11 and 14 days, respectively, after application of pig manure.

Transport of Contaminants from Manure

N and P in runoff (surface and subsurface) are major environmental concerns related to agricultural activities. Concentrations of 0.3 mg inorganic N and 0.01 mg inorganic-P l^{-1} have been identified as levels above which excessive algal growth or eutrophication can occur (Nichols et al. 1994). In the case of NH_4^+-N, the tolerance limit may be as low as 0.1 mg N l^{-1} (Gangbazo et al. 1995). Runoff generally accounts for only a small portion of applied N compared with the leached portion (Burgoa et al. 1993). Blevins et al. (1996) found that after two growing seasons, less than 2% of fertilizer-N was lost to runoff whereas 30% had moved below 1 m in the soil. Agricultural land, particularly if it is under row crops, or is used for intensive animal production, is often associated with groundwater having NO_3^- concentrations near or above 10 mg l^{-1} (Goss et al. 1998; Toth and Fox 1998). In contrast, percolation from soil below forests and suburban areas generally has NO_3-N concentration < 1 mg l^{-1}.

Nitrogen

N in any manure solids left on the soil surface, or associated with fine particles that are readily moved during soil erosion, can be lost through surface runoff to a water course. The factors that determine N loss by erosion are: the amount of sediment moved, the N content of the soil moved, and the N content of the manure solids. The other material subject to loss to surface water is N dissolved in runoff water. This portion is usually small (Blevins et al. 1996; Meisinger and Randall 1991); however, it is variable, and depends on a number of factors, such as the degree of soil cover, source of N applied, application rate, timing and duration of the application. Surface conditions are also important, and are affected by slope, soil characteristics and land management. Finally, runoff is highly dependent on the intensity of rainfall after application. The largest losses occur if a soluble N source is applied to a bare soil surface, and a significant rainfall event occurs soon after application (Edwards and Daniel 1993; Sharpley 1997). In many cases, dissolved N is transported into the soil with the initial infiltration that precedes runoff (Meisinger and Randall 1991). Incorporating the N source such as manure dramatically reduces runoff losses. In most cases, runoff N losses are small, 3 kg N ha^{-1} annually or less (Meisinger and Randall 1991; Nichols et al. 1994; Blevins et al. 1996; Gascho et al. 1998). In the case of subsurface runoff through tile drains or as a result of a shallow hardpan, considerable amount of runoff may occur (Hubbard et al. 1991; Lowrance 1992; Goss et al. 1993b). Intensive rainfall shortly after fertilizer application generates the largest loss of NO_3^- in runoff (Hubbard and Sheridan 1983; Hubbard et al. 1991; and Lowrance 1992). Most of the runoff losses of NO_3^- N in the lower Southern Coastal Plain of the USA were from subsurface flow taking place in the top 30 cm of soil rather than from surface flow. Over a 10-year period, 20% of the N in the applied fertilizer was lost via surface and subsurface flow (Hubbard and Sheridan 1983). This was comparable with the loss in runoff reported by Edwards and Daniel (1993) for conditions of high rainfall intensity. Such results suggest that tile drainage systems can greatly reduce groundwater contamination at the expense of surface water contamination. However, not all drainage water may be intercepted by pipe drains, even during major flow events, so that groundwater contamination is still likely.

NO_3^- is the major N species lost by leaching. If economically optimum rates of N are applied to row crops, such as corn, NO_3-N losses by leaching from the root zone may be in excess of 10 mg l^{-1} (Jemison and Fox 1994; Toth and Fox 1998). The authors found that only when plants were visibly deficient in N was the flow-weighted NO_3 concentration below the drinking water standard. Sexton et al. (1996) studied optimum nitrogen and irrigation inputs for corn, and found that by applying urea-based fertilizer at 95% of that required for maximum yield, nitrate leaching could be reduced by 30 to 40%; and by using a variable deficit trigger for scheduling irrigation, nitrate leaching could be reduced by 50 to 55%. They also reported that at equivalent N rates, turkey manure produced equal or better crop yields than those following urea applications, but NO_3 leaching was equal to or less than urea. Dairy manure applied to a corn field resulted in similar or slightly smaller NO_3^- loading than agronomically equivalent rates of fertilizer N (Jokela 1992). In contrast, Nielsen and Jensen (1990) reported that NO_3^- losses from root zone in soils amended with liquid manure were greater than those from a similar soil to which the same amount of N had been applied as inorganic fertilizer. Jemison and Fox (1994) found very little difference in NO_3^- concentrations or mass of NO_3^- leached between non-manured and corn manured at economically optimum rate. The different results for the amount of NO_3^- leached following manure applications highlights the importance of N transformations, such as mineralization and denitrification, that influence the availability of nitrate in the soil.

Loss of N is affected greatly by soil water content (Randall and Iragavarapu 1995). Nitrate leaching tends to be greatest during the winter months, and may be minimized by applying manure during the late spring and early summer when crops can compete for NO_3^- with the smaller volume of water that moves downward through the vadose zone (Adams et al. 1994). Alfalfa crops included in the rotation resulted in a considerable reduction in the amount of NO_3 leaving a farm in leachate (Toth and Fox 1998). The effect is attributed to a longer period of evapotranspiration resulting in less drainage, and greater uptake and immobilization of N by the perennial crop. In a dry year when plant growth and N uptake are limited and percolation of soil water is negligible, mineralization continues to occur, and mineral N accumulates in soil profile, and it will eventually be subject to leaching when precipitation exceeds evapotranspiration. NO_3^- and Cl^- movement through undisturbed field soil was increased by large pores when the ions were applied in the infiltrating water and the soil profile was near field capacity. When soil water content was close to field capacity, the micropore space was filled with water. Application of more solution, such as liquid manure, tended to encourage flow in the macropore space, and hence resulted in deeper nitrate movement than when flow was limited to micropores.

Sharpley (1997) investigated N and P runoff on ten Oklahoma soils amended with poultry litter. Runoff N and P concentrations decreased with ten successive rains, starting 7 days after litter application. Increasing the time between litter application and rainfall from 1 to 35 days reduced total N from 7.54 to 2.34, NH_4-N from 5.53 to 0.11, dissolved P from 0.74 to 0.45, and bioavailable P from 0.99 to 0.65 mg l^{-1}; however, NO_3 concentrations were unaffected by rainfall frequency and timing.

Few studies exist that compare the potential for nitrate leaching of different manure types. Younie et al. (1996) found that nitrate leaching was higher where liquid cattle manure was the source of nitrogen than where solid beef manure was used.

Ritter et al. (1990) studied soil nitrate profiles under 16 sites, some of which received fertilizer N alone or in combination with either broiler manure or liquid swine manure. Although direct comparison of manure types was not made on the same site, N application rate was found to be the major determinant of N in the soil profile. It appears that manure from poultry, cattle or pig operations has the potential to contaminate groundwater if it is applied at excessive rates.

Liquid manure adversely affected tile drainage water quality when applied to the land following the farming guidelines current in Ontario (Dean and Foran 1992). These authors found that 75% of the manure-spreading events investigated resulted in water quality impairment. The difficulty of determining an acceptable rate of application of liquid manure due to the numerous factors which affect the contamination of water courses is apparent (Foran and Dean 1993).

Phosphorus

Phosphorus contribution to surface water in runoff from agricultural land was the major focus of the Soil and Water Environmental Enhancement Program (SWEEP). Studies under the Pollution from Land Use Reference Group (PLUARG) of the Great Lakes International Joint Commission, established during the 1970s, found that runoff from agricultural land was responsible for about 70% of the phosphorus reaching Lake Erie from the tributaries in Ontario (Miller et al. 1982). About 20% of this amount or 15% of the total was estimated to be due to direct inputs from livestock operations including runoff from storage areas and surface runoff from manure applied close to streams and not incorporated. The remainder was due largely to phosphorus associated with eroded sediment. Manure application would have two opposing effects on this latter contribution. It would increase the phosphorus content of the soil and hence the concentration on the eroded sediment. On the other hand, manure would tend to improve soil structural stability and hence reduce erosion. Where liquid or solid manure was not incorporated after application, the loss of phosphorus in surface run-off was greater from land under mouldboard cultivation than under no-till. Loss from no-till land was similar to that from land where the manure had been incorporated after application (King et al. 1994).

One aspect of phosphorus in runoff that is still poorly understood is the bioavailability of the different forms of phosphorus (Sharpley et al. 1992). While manure application may not increase the total phosphorus in runoff, it is more likely to increase the amount of bioavailable phosphorus. In the State of Delaware, continual land application of animal manures has resulted in an accumulation of P in the surface soil (Sharpley and Halvorson 1994). These authors concluded that there would be an increase in the bioavailability of P in runoff water from land following the application of manure because of the increased transport of low-density organic material together with the high solubility of manure P. The magnitude of the increase would be expected to vary, dependent on the density of the manure, the water and P content, for different animal sources.

P being a reactive ion, its enrichment generally decreases sharply with depth. Application of cattle feedlot waste resulted in increased proportion of available P in the first 30 cm, but with little increase below 50 cm (Campbell and Racz 1975). Decreasing enrichment or only slight enrichment with depth is not necessarily indicative of no leaching because the residence time of some drainage water in the subsoil may have been too short, perhaps because of preferential flow, to allow absorption of P on to soil particles (Johnston and Poulton 1997). Furthermore, some subsoils, e.g. sandy soils, may have no capacity to retain P. Leaching of P from manure may occur in both inorganic and organic forms (Campbell and Racz 1975; Eghball et al. 1996). Complexation of P with mobile organic compounds may favour the deep transport of P in organic forms even through layers with a great P adsorption capacity, such as carbonate soil layers. Experimental results from Eghball et al. (1996), showed that P from mineral fertilizer did not move under the carbonate layer (0.9 m) of soil even after 40 years of mineral P fertilization, while organic P from manure moved up to 1.8 m. The P movement in this soil was found to be unaffected by the P adsorption of the soil. Most importantly, P association with low molecular weight organic acids favoured increased mobility through both decreased adsorption and increased dissolution of P compounds leading to greater bioavailability (Bolan et al. 1994) and to an enhanced risk of leaching. Increasing labile, weakly bound P results in greater vulnerability of manure-treated soils to lose phosphorus by leaching (Stephenson and Chapman 1931, Robinson et al. 1995; Johnston and Poulton 1997). This results in deeper penetration of P compounds after manure application (Campbell and Racz 1975).

Leaching of P in water-soluble and particulate forms from soil is enhanced by the presence of tile drainage (Harrison 1987). Heckrath et al. (1997) found a critical concentration of soluble-P in the ploughed layer that if exceeded resulted in an enhanced contribution of P losses through tile drain in clay loam soils. Hergert et al. (1981) found P losses in field tile drain effluent to be increased where manure was applied compared to unfertilized control plots. Dils and Heathwaite (1997) found that subsurface transport of P may occur as water-soluble P and as particulate P for both undrained and tile-drained plots.

At a field scale, even when sorption capacities in the surface horizon are exceeded and the water-soluble P concentration becomes elevated, lower soil horizons may be able to sorb the leaching P and minimize the potential for water-soluble P movement to surface waters via drainage (Provin et al. 1995). Similarly immobilization of P on metal-oxide coatings within an aquifer can decrease as the P loading of soil increases (Walter et al. 1996). However, preferential flow can be an important factor in tile drain losses of particulate P (Gaynor and Findlay 1995, Heckrath et al. 1997). The importance of preferential flow for P transport was confirmed by the fact that P was leached from the soil despite the large adsorption potential of some subsoils (Thomas and Phillips 1979; Eghball et al. 1996). Stamm et al. 1998, observed that although the water-extractable P in the soil was concentrated in the uppermost layer of the profiles during most flow peaks, P concentrations in tile drain effluents strongly increased with increasing flow rates. Phosphorus was mainly transported as soluble-reactive and particulate P through preferential flow paths extending from the soil surface to the drains. On clayey soils in an intensive cropped area from Quebec, up to 50% of the P lost through tile drain effluent was in particulate forms, with less than 30% in soluble forms (Beauchemin et al. 1998).

In conclusion, although phosphorus leaching is considered less important than the leaching of nitrate, the specific combination of agricultural management practices, soil properties and climatic conditions can lead to significant losses of soluble and particulate P through leaching (Sims et al. 1998).

Toxic Metals

Transport of toxic metals through soil is mostly a function of pH and dissolved oxygen concentration. Accumulation of organic material close to the surface may possibly decrease the availability of Zn while increasing the solubility of Fe and Mn (Shuman 1988). Other studies show that the total dissolved Zn tends to remain constant, despite addition of organic material to soil via manure, with the Zn concentration shifting from light molecular weight organic particles to heavy molecular weight organic particles which tend to be adsorbed by the soil complexes (Del Castilho et al. 1993b). On the other hand, Cu interacts with low molecular dissolved organic carbon substances which are more likely to stay in solution, therefore increasing the percentage of mobile Cu (Köning et *al.* 1986; Del Castilho et *al.* 1993b)

Increases in the ionic concentration of the soil solution, as happens shortly after manure application to soils, may decrease the percentage of metallic ions attached to soil mineral and organic particles by increasing the competition for adsorption sites (Stevenson 1991).

Organic matter amendments are considered likely to lead to increases in the water-soluble forms of Zn rather than Cu, due to the direct effects of the organic matter and dissolved carbon, and to indirect effects on other soil properties (e.g. pH and redox status) caused by the addition of labile organic matter (Shuman 1991; Del Castilho et al. 1993a). Salomons and Forstner (1984) reported that the application of manure resulted in a decrease in soil pH which, even 2 months later, was an important factor in the dissolution of weakly bound toxic metals. Thus, even if the water soluble Cu content of fresh liquid swine manure is relatively small (Miller et al. 1986), application of manure to soil may mobilize some of the Cu and Zn already present from previous applications. Japenga et al. (1992) affirmed that the fate of toxic metals in manured soils is related to the fate of organic matter, with little difference due to the type of manure.

Bacteria

Bacterial transport is affected by soil pH. The application of cattle manure or pig slurry can result in a decrease in soil pH (Chang et al. 1991; Bernal et al. 1992). This will potentially alter bacterial transport due to an increase in the number of binding sites available for bacterial adsorption, and it may also affect bacterial survival. Cattle manure induced smaller changes in soil pH compared with those resulting from the application of pig manure.

Antecedent moisture content of the soil is another important factor in water movement, and consequently contaminant transport. Abu-Ashour et al. (1994a) conducted a series of experiments to determine factors influencing bacterial transport through soil. The findings of this study indicated that initial soil moisture was the critical variable in determining the extent of bacterial migration. In dry soil, none of the marked bacteria (biotracer) was detected below 87.5 mm. However, the biotracer travelled the full length of the soil columns (175 mm) when the soil was wet. Additional water applied after biotracer inoculation caused the biotracer to move deeper into the soil. The actual depth depended upon how close to saturation the soil became after the addition of a given volume of water. The water may have increased the soil water content sufficiently to allow the bacteria to move through the soil with percolating water. However, bacteria can be transported to depth even if the soil is not at saturation level (McMurry et al. 1998; Unc 1999).

A further factor influencing transport of bacteria after the application of animal manure from liquid storage is the high concentrations of salts that are present in manure. The salts may act as "bridges", allowing negatively charged bacteria to adsorb to negatively charged soil particles. High salt concentration can also decrease the thickness of the diffuse double layers around soil colloids, thereby allowing bacteria access to surfaces to which they can adhere. The addition of rainwater will dilute the salt concentration, thus increasing the thickness of the double layer, and may cause the elution of adsorbed bacteria (Tan et al. 1991). This will increase the number of bacteria that remain mobile in the soil solution and increase the risk to groundwater.

Harvey (1991) has shown that the transport of bacteria may be faster or slower than or similar to that of conservative tracers such as chloride or bromide. Bitton and Harvey (1992) concluded that bacteria move through soils and aquifers by several mechanisms, including continuous, discontinuous and chemotactic migration. Much of the modelling effort has treated transport as a continuous process, which assumes passive transport of bacteria. However, bacterial movement through the subsurface, especially over substantial distances, may be discontinuous because of processes that temporarily remove bacteria from solution. Bacteria are removed from the flowing water by straining or by reversible sorption on solid surfaces. They are then remobilized at some later time. Discontinuous transport creates an apparent retardation of the bacteria relative to conservative tracers. Retardation factors as large as 10 have been reported for bacterial populations travelling through porous aquifers (Matthess et al. 1988).

Bacteria may travel significantly faster than conservative tracers due to motility. Movement due to taxis (self-propulsion) is faster than that caused by random thermal (Brownian) motion. Jenneman et al. (1985) reported that motile bacteria penetrated Berea sandstone cores in the presence of a nutrient gradient up to eight times as fast as non-motile ones. Bacteria may also appear to travel faster than conservative tracers for other reasons. Bacterial transport is restricted to macropores, whereas conservative tracers diffuse into the soil matrix as well as into the larger pores. This may cause the average peak in bacterial concentrations to appear earlier than that of the conservative tracer. The bacteria are exploiting faster paths and can travel only during peak flow, whereas the average tracer concentration moving through the soil matrix and macropores would not peak until the majority had infiltrated through the soil matrix (Bitton and Harvey 1992). In addition to transport processes, the kinetics of population growth and decay must be considered in relation to the timing and numbers of organisms reaching a water resource. Microorganisms adsorbed to soil particles may survive longer than those in the liquid phase, as organic substrate and nutrients are more readily available to them (Sobsey 1983).

Runoff water from grazed and ungrazed grass pastures can contain large numbers of bacteria (Kirchmann 1994). Bacteria from poultry manure were not detected in runoff when the manure was applied to bare soil, but was present when the manure was applied to grassland (Giddens and Barnett 1980). More bacteria may be lost in overland flow from no-till land than from ploughed land within 24 h of manure application, but the rate of decline in the concentration of bacteria in the runoff water can also be greater (King et al. 1994).

The Ontario Farm Groundwater Quality Survey demonstrated the importance of preferential flow for bacterial transport, since they were found in wells more than 30 m deep (Goss et al. 1998). Preferential flow can thus facilitate the transport of contaminants to aquifers at depths that might be expected to remain unaffected by surface contaminants. This presents an important concern for modelling transport of bacteria. For example, when flow parameters in the theoretical model described by Corapcioglu and Haridas (1985) were taken to the permissible limits, the predicted extent of bacterial transport through unsaturated soil over 2 weeks was 0.2 m. But Smith et al. (1985) observed that *Escherichia coli* penetrated through a column of undisturbed soil to a depth of 0.3 m in 20 min, and Harvey et al. (1989) observed bacterial-sized microspheres transported through several meters of aquifer.

Conclusions

Groundwater quality has increasingly been the focus of attention, and this has also highlighted concerns over manure and tillage practices. In the context of agriculture the need is to address the transport of simple minerals, complex organic molecules and particulates, because there are examples of all three classes of contaminants in animal manures.

Preferential flow paths can have variable impacts on groundwater quality. Adverse effects result when contaminants are transported to greater depths at faster rates than would occur if travelling through the soil matrix. Physical and biological processes within the soil can lead to the formation of continuous macropores, which can provide pathways for preferential flow. However, most of the evidence presented suggests that contamination from animal manure is greatly increased in the presence of macropores.

References

Abu-Ashour J, Etches C, Joy DM, Lee H, Reaume CM, Shadford CB, Whiteley HR, Zelin S (1994a) Field experiment on bacterial contamination from liquid manure application, Final Report for RAC Project 547G. Ontario Ministry of Environment and Energy, Toronto, Ontario

Abu-Ashour J, Joy DM, Lee H, Whiteley HR, Zelin S (1994b) Transport of microorganisms through soil. Water Air Soil Pollut 75 (1/2):141-158

Adams PL, Daniel TC, Edwards DR, Nichols DJ, Pote DH, Scott HD (1994) Poultry litter and manure contributions to nitrate leaching through the vadose zone. Soil Sci Soc Am J 58 (4):1206-1211

Addiscott TM, Wagenet RJ (1985) Concepts of solute leaching in soils: a review of modelling approaches. J Soil Sci 36:411 - 424

Ahuja LR, Barnes BB, Rojas KW (1993) Characterization of macropore transport studied with the ARS root zone water quality model. Trans ASAE 36 (2):369-380

Barley KP (1954) Effects of root growth and decay on the permeability of a synthetic sandy loam. Soil Sci 78:205 - 210

Barry DAJ, Goorahoo D, Goss MJ (1993) Estimation of nitrate concentration in groundwater using a whole farm nitrogen budget. J Environ Qual 22:767-775

Beauchamp EG (1983) Response of corn to nitrogen in preplant and sidedress applications of liquid dairy cattle manure. Can J of Soil Sci 63 (2):377-386

Beauchamp EG, Kidd GE, Thurtell G (1982) Ammonia volatilization from liquid dairy cattle manure in the field. Can J of Soil Sci 62 (1):11-19

Beauchemin S, Simard RR, Cluis D (1998) Forms and concentration of phosphorus in drainage water of twenty-seven tile-drained soils. J Environ Qual 27 (3):721-728

Bernal MP, Roig A, Lax A, Navarro AF (1992) Effects of the application of pig slurry on some physico-chemical and physical properties of calcareous soils. Biores Technol 42:233-239

Beven K, Germann P (1982) Macropores and water flow in soils. Water Resour Res 18:1311-1325

Bitton G, Harvey RW (1992) Transport of pathogens through soil and aquifers. Environmental microbiology. Wiley-Liss, Toronto, pp 103-124

Blevins DW, Wilkison DH, Kelly BP, Silva SR (1996) Movement of nitrate fertilizer to glacial till and runoff from a claypan soil. J Environ Qual 25:584-593

Bolan NS, Naidu R, Mahimairaja S, Baskaran S (1994) Influence of low-molecular-weight organic acids on the solubilization of phosphates. Biol Fertil Soils 18 (4):311-319

Bouma J (1981) Soil morphology and preferential flow along macropores. Agric Water Manage 3:235-250

Burgoa B, Hubbard RK, Wauchope RD, Davis-Carter JG (1993) Simultaneous measurement of runoff and leaching losses of bromide and phosphate using tilted beds and simulated rainfall. Commun Soil Sci Plant Anal 24:2689-2699

Campbell LB, Racz GJ (1975) Organic and inorganic P content, movement and mineralization of P in soil beneath a feedlot. Can J Soil Sci 55:457-466

Chandler DS, Farran I, Craven JS (1981) Persistence and distribution of pollution indicator bacteria on land used for disposal of piggery effluent, Appl Environ Microbiol 42:453 - 460

Chang C, Sommerfeldt TG, Entz T (1991) Soil chemistry after eleven annual applications of cattle feedlot manure. J Environ Qual 20:475-480

Chen S, Franklin RE, Johnson AD (1997) Clay film effects on ion transport in soil. Soil Sci 162:91-96

Chen S, Franklin RE, Quisenberry V, Dang P (1999) The effect of preferential flow on the short and long-term spatial distribution of surface applied solutes in the structured soil. Geoderma (in press)

Corapcioglu MY, Haridas A (1985) Microbial transport in soils and groundwater: a numerical model. Adv Water Resour 8:188 - 200

de Lange CFM (1996) Animal and feed factors determining N and P excretion with pig manure. In: Goss MJ, Stonehouse DP, Giraldez JC (eds) Managing manure for dairy and swine. Towards developing a decision support system. SOS Publications, Fair Haven, New Jersey

Dean DM, Foran ME (1992) The effect of farm liquid waste application on tile drainage. J Soil Water Conserv 47 (5):368-369

Del Castilho P, Chardon WJ, Salomons W (1993a) Influence of cattle-manure slury application on the solubility of cadmium, copper and zinc in a manured acidic, loamy-sand soil. J Environ Qual 22:686-697

Del Castilho P, Dalenberg JW, Brunt K, Bruins AP (1993b) Dissolved organic matter, cadmium, copper and zinc in pig slurry- and soil solution-size exclusion chromatography fractions. Int J Environ Anal Chem 50:91-107

Dils RM, Heathwaite AL (1997) Phosphorus fractionation in grassland hill-slope hydrological pathways. In: Tunney H, Carton OT, Brookes PC, Johnston AE (eds) Phosphorus loss from soil to water.CAB International, Wallingford, pp 349-351

Edwards DR, Daniel TC (1993) Effects of poultry litter application rate and rainfall intensity on quality of runoff from fescue grass plots. J Environ Qual 22:361-365

Edwards DR, Benson VW, Williams JR, Daniel TC, Lemunyon J, Gilbert RG (1994) Use of the EPIC model to predict runoff transport of surface-applied inorganic fertilizer and poultry manure constituents. Trans ASAE 37:403-409

Eghball B, Binford GD, Baltensperger DD (1996) Phosphorus movement and adsorption in a soil receiving long-term manure and fertilizer application. J Environ Qual 25 (6):1339-1343

Ehlers W (1975) Observations on earthworm channels and infiltration on tilled and untilled loess soil, Soil Sci 119:242 - 249

Foran ME, Dean DM (1993) The land application of liquid manure and its effect on tile drain water and groundwater quality. In: Agricultural Research to Protect Water Quality. Proc Conf Febr 21-24, Minneapolis, Minnesota. Soil and water Conservation Society. Ankeney, Iowa, pp 279-281

Fordham AW, Schwertmann U (1977) Composition and reactions of liquid manure (gülle), with particular reference to phosphate: II Solid phase components. J Environ Qual 6 (2):136-139

Gangbazo G, Pesant AR, Barnett GM, Charuest JP, Cluis D (1995) Water contamination by ammonium nitrogen following the spreading of hog manure and mineral fertilizers. J Environ Qual 24 (3):420-425

Gascho GJ, Wauchope RD, Davis JG, Truman CC, Dowler JE, Hook HR, Sumner, Johnson AW (1998) Nitrate-nitrogen, soluble, and bioavailable phosphorus runoff from simulated rainfall after fertilizer application. Soil Sci Am J 62:1711-1718

Gaynor JD, Findley WI (1995) Soil phosphorus loss from conservation and conventional tillage in corn production. J Environ Qual 24:734-741

Giddens J, Barnett AP (1980) Soil loss and microbiological quality of runoff from land treated with poultry litter. J Environ Qual 9 (3):518-520

Golabi MH, Radcliffe DE, Hargrove WL, Tollner EW (1995) Macro effects in conventional tillage and no-tillage soils. J Soil Water Cons 50 (2):205-210

Goss MJ, Goorahoo D (1995) Nitrate contamination of groundwater: measurement and prediction. Fert Res 42:331-338

Goss MJ, Barry DAJ, Rudolph DL (1998) Groundwater contamination in Ontario farm wells and its association with agriculture. 1. Results from drinking water wells. J Contam Hydrol 32:267-293

Goss MJ, Colbourn P, Harris GL, Howse KR (1987) Leaching of nitrogen under autumn-sown crops and the effects of tillage. In: Jenkinson DS, Smith KA, (eds) Nitrogen efficiency in agricultural soils. EEC seminar. Edinburgh, September 1987. Elsevier, London, pp 269-282

Goss MJ, Curnoe WE, Beauchamp EG, Smith PS, Nunn BDC (1993a) Manure management to sustain water quality. Final Report on contract: National soil conservation program. Centre for Land and Water Stewardship, University of Guelph, Guelph, Ontario

Goss MJ, Howse KR, Lane PW, Christian DG, Harris GL (1993b) Losses of nitrate-nitrogen in water draining from under autumn crops established by direct drilling or mouldboard ploughing. J Soil Sci 44:35-48

Goss MJ, Ogilvie JR, Beauchamp EG, Stonehouse DP, Miller MH, Parris K (1994) Current state of the art of manure/nutrient management, Agriculture and Agri-Food Canada, COESA Report No: RES/MAN-001/94 prepared for Research Branch, London, Ontario

Haig PA, McBride BW (1996) Effect of animal and diet factors on dairy cattle manure quantity and composition. In: Goss MJ, Stonehouse DP, Giraldez JC (eds) Managing manure for dairy and swine. Towards developing a decision support system. SOS Publications, Fair Haven, New Jersey

Harrison AF (1987) Soil organic phosphorus: a review of world literature. CAB International, Wallingford

Harvey RW (1991) Parameters involved in modelling movement of bacteria in groundwater. In: Hurst CJ (ed) Modelling the environmental fate of microorganisms. American Society for Microbiology, Washington, DC, pp 89-114

Harvey RW, Garabedian SP (1991) Use of colloid filtration theory in modelling movement of bacteria through a contaminated sandy aquifer. Environ Sci Technol 25:178-185

Harvey RW, George LH, Smith RL, Leblanc DR (1989) Transport of microspheres and indigenous bacteria through a sandy aquifer: results of natural and forced-gradient tracer experiments. Environ Sci Technol 23:51-56

Hatfield WA (1988) Water and anion movement in a Typic Hapludults. Ph D Dissertation, Clemson University. Clemson, South Carolina

Heckrath G, Brookes PC, Poulton PR, Goulding KWT (1995) Phosphorus leaching from soils containing different phosphorus concentrations in the Broadbalk Experiment. J Environ Qual 24 (5):904-910

Heckrath G, Brookes PC, Poulton PR, Goulding KWT (1997) Phosphorus losses in drainage water from an arable silty clay loam. In: Tunney H, Carton OT, Brookes PC, and Johnston AE (eds) Phosphorus loss from soil to water.CAB International, Wallingford, pp 367-369

Helling CS, Gish TJ (1991) Physical and chemical processes affecting preferential flow. In: Gish TJ, Shrimohammadi A (eds) Preferential flow. American Society of Agricultural Engineers, St Joseph, Michigan, pp 77-86

Hergert GW, Bouldin DR, Klausner SD, Zwerman PJ (1981) Phosphorus concentration-water flow interactions in tile effluent from manured land. J Environ Qual 10 (3):338-344

Hubbard RK, Sheridan RG (1983) Water and nitrate losses from a small upland coastal plain watershed. J Environ Qual 12:291-295

Hubbard RK, Leonard RA, Johnson AW (1991) Nitrate transport on a sandy coastal plain soil underlain by plinthite. Trans ASAE 34 (3):802-808

Japenga J, Harmsen K (1990) Determination of mass balances and ionic balances in animal manure. Neth J Agric Sci 38:353-367

Japenga J, Dalenberg JW, Wiersma D, Scheltens SD, Hesterberg D, Salomons W (1992) Effect of liquid animal manure application on the solubilization of heavy metals from soil. Int J Environ Anal Chem 46:25-39

Jardine PM, Wilson GV, and Luxmoore RJ (1990) Unsaturated solute transport through a forest soil during rain storm events. Geoderma 46:103-118

Jarvis NJ, Jansson PE, Dik PE, Messing I (1991) Modelling water and solute transport in macroporous soil. I. Model description and sensitivity analysis. J Soil Sci 42 (1):59-70

Jemison JM, Fox RH (1994) Nitrate leaching from nitrogen B fertilized and manured corn measured with Zero-tension pan lysimeters. J Environ Qual 23:337-343

Jenneman GE, McInerney MJ, Knapp RM (1985) Microbial penetration through nutrient-saturated berea sandstone. Appli Environ Microbiol 50:383-391

Johnston AE, Poulton PR (1997) The downward movement and retention of phosphorus in agricultural soils. In: Tunney H, Carton OT, Brookes PC, Johnston AE (eds) Phosphorus loss from soil to water. CAB International, Wallingford pp 422-425

Jokela WE (1992) Nitrogen fertilizer and dairy manure effects on corn yield and soil nitrate. Soil Sci Soc Am J 56:148-154

King DJ, Watson GC, Wall GJ, Grant BA (1994) The effects of livestock manure application and management on surface water quality. Summary Technical Report GLWQP-AAFC Pest. Management Research Centre, London, ON, Canada, Agriculture and Agrifood Canada.

Kirchmann H (1994) Animal and municipal organic wastes and water quality. In: Lal R, Stewart BA (eds): Advances in soil science: soil processes and water quality. CRC Press, Boca Raton, Florida, pp 163-232

König N, Baccini P, Ulrich B (1986) Der Einfluß der natürlichen organischen Substanzen auf die Metallverteilung zwischen Boden und Bodenlösung in einem sauren Waldboden. Z Pflanzenernachr Bodenkde 149:68-82

Kung KJS, Donohue SV (1991) Improved solute-sampling protocol in a sandy vadose zone using ground-penetrating radar. Soil Sci Soc Am J 55 (6):1543-1545

Lawes JB, Gilbert JH, Warington R (1882) On the amount and composition of the rain and drainage-waters collected at Rothamsted. Williams Clowes, London

Leinweber P (1997) The concentrations and forms of phosphorus in manures and soils from the densely populated livestock area in north-west Germany. In: Tunney H, Carton OT, Brookes PC, Johnston AE (eds): Phosphorus loss from soil to water. CAB International, Wallingford, pp 425-427

Lowrance R (1992) Nitrogen outputs from a field-size agricultural watershed. J Environ Qual 21:602-607

Luxmoore RJ (1981) Micro-, meso-, and macroporosity of soil. Soil Sci Soc Am J 45:671-672

Luxmoore RJ (1991) On preferential flow and its measurement. In: Gish TJ, Shrimohammadi A (eds): Preferential flow. American Society of Agricultural Engineers, St Joseph, Michigan, pp 113-121

MacLean AJ (1983) Pathogens of animals in manure: environmental impact and public health. In: Farm animal manures in the Canadian environment. National Research Council of Canada Associate Committee on Scientific Criteria for Environmental Quality. Ottawa, Ontario, Canada, NRCC/No 189766, pp 103-107

MacLean AJ, Miller MH, Robinson JB (1983) The fertilizer potential of animal manures and environmental constraints of their use. In: Farm Animal Manures in Canadian Environment. Nation Research Council of Canada Associate Committee on Scientific Criteria for Environmental Quality. Ottawa, Ontario, Canada, NRCC/No. 189766, pp 21-51

Matthess G, Pekdeger A, Schroeter J (1988) Persistence and transport of bacteria and viruses in groundwater – a conceptual evaluation. J Contam Hydrol 2:171-188

McMurry SW, Coyne MS, Perfect E (1998) Fecal coliform transport through intact soil blocks amended with poultry manure. J Environ Qual 27 (1):86-92

Meisinger JJ, Randall GW (1991) Estimating nitrogen budgets for soil-crop system. In: Follet RF, Keeney DR, Cruse RM (eds) Managing nitrogen for groundwater quality and farm profitability. SSSA, Madison, Wisconsin, pp 85-124

Miller ER, Lei X, Ulley DE (1991) Trace elements in animal nutrition. In: Mortvedt JJ 2nd, Cox FR, Shuman LM, Welch RM (eds) Micronutrients in agriculture. Soil Science Society of America, Madison, Wisconsin, pp 601-616

Miller MH, Robinson JB, Coote DR, Spires AC, Draper DW (1982) Agriculture and water quality in the Canadian Great Lakes Basin: III. Phosphorus. J Environ Qual 11 (3):487-493

Miller WP, Martens DC, Zelazny LW, Kornegay ET (1986) Forms of solid phase copper in copper-enriched swine manure. J Environ Qual 15 (1):69-72

Minkara MY, Wilhoit JH, Wood CW, Yoon KS (1995) Nitrate monitoring and GLEAMS simulation for poultry litter application to pine seedlings. Trans ASAE 38:147-152

Munyankusi E, Gupta SC, Moncrief JF, Berry EC (1994) Earthworm macropores and preferential transport in a long-term manure applied Typic Hapludalf. J Environ Qual 23:733-784

Nichols DJ, Daniel TC, Edwards DR (1994) Nutrient runoff from pasture after incorporation of poultry litter of inorganic fertilizer. Soil Sci Soc Am J 58:1224-1228

Nielsen NE, Jensen HE (1990) Nitrate leaching from loamy soils as affected by crop rotation and nitrogen fertilizer application. Fert Res 26:197-207

Nodar R, Acea MJ, Carballas T (1990) Microbial composition of poultry excreta. Biol Wastes 33 (2): 95-105

Nodar R, Acea MJ, Carballas T (1992) Poultry slurry microbial population: composition and evolution during storage. Bioresour Technol 40 (1):29-34

O'Dell B, Miller ER, Miller WJ (1979) Literature review on copper and zinc in poultry, swine and ruminant nutrition. National Feed Ingredients Association, West Des Moines, Iowa

Östling CE, Lindgren SE (1991) Bacteria in manure and on manured and NPK fertilized silage crops. J Sci Food Agric 55:579-588

Parton WJ, Rasmussen PE (1994) Long-term effects of crop management in wheat-fallow: II. CENTURY model simulations. Soil Sci Soc Am J 58:530-536

Patni NK, Jui PY (1987) Changes in solids and carbon content of dairy-cattle slurry in farm tanks. Biol Wastes 20 (1):11-34

Paul JW, Beauchamp EG (1989) Effect of carbon constituents in manure on denitrification in soil. Can J Soil Sci 69:49-61

Paul JW, Zebarth BJ (1993) Nitrate leaching and denitrification following fall manure application. Int Summer Meet sponsored by The American Society of Agricultural Engineers, St Joseph, Michigan

Paul JW, Beauchamp EG, Whiteley HR, Sakupwanya JK (1990) Fate of manure nitrogen at the Arkell and Elora Research Stations 1988-1990. Report on Special Research Contract SR8710-SW001, Ontario Ministry of Agriculture and Food, Toronto, Ontario

Paul JW, Beauchamp EG, Zhang X (1993) Nitrous and nitric oxide emissions during nitrification and denitrification from manure-amended soil in the laboratory. Can J Soil Sci 73 (4):539-553

Peterson TC, Ward RC (1989) Development of a bacterial transport model for coarse soils. Water Resour Bull 25:349-357

Phillips RE, Quisenberry VL, Zeleznik JM, Dunn GH (1989) Mechanism of water entry into simulated macropore. Soil Sci Soc Am J 53:1629-1635

Pratt PF, Chirnside AEM, Scarborough RG (1976) A four-year field trial with animal manures. Hilgardia 44:99-125

Provin TL, Joern BC, Franzmeier DP, Sutton AL (1995) Phosphorus retention in selected Indiana soils using short-term sorbtion isotherms and long-term aerobic incubations. In: Steele K, (ed): Animal waste and land-water interface. CRC Press, Boca Raton, Florida, pp 35-42

Quisenberry VL, Phillips RE (1976) Percolation of surface-applied water in the field. Soil Sci Soc Am J 40:484-489

Quisenberry VL, Phillips RE (1978) Displacement of soil water by simulated rainfall. Soil Sci Soc Am J 42:675-679

Quisenberry VL, Smith BR, Phillips RE, Scott HD, Nortcliff S (1993) A soil classification system for describing water and chemical transport. Soil Sci 156 (5):306-315

Quisenberry VL, Phillips RE, Zeleznik JM (1994) Spatial distribution of water and chloride macropore flow in a well-structured soil. Soil Sci Soc Am J 58:1294-1300

Randall GW, Iragavarapu TK (1995) Impact of long-term tillage system for continuous corn on nitrate leaching to tile drainage. J Environ Qual 24:360-366

Raven KP, Hossner LR (1993) Phosphorus desorbtion quantity-intensity relationships in soil. Soil Sci Soc Am J 57:1501-1508

Ritter WF, Chirnside AEM (1987) Influence of agricultural practices on nitrates in the water table aquifer. Biol Wastes 19 (3):165-178

Ritter WF, Chirnside AEM, Scarborough RW (1990) Soil nitrate profiles under irrigation on Coastal Plain soils. J Irrig Drain Engin 116 (6):738-751

Robinson JS, Sharpley AN, Smith SJ (1995) The effect of animal manure applications on the forms of soil phosphorus. In: Steele K (ed). Animal waste and land-water interface. CRC Press, Boca Raton, Florida, pp 43-48

Rüprich A (1994) Felduntersuchungen zum Infiltrationsvermömgen und zur Lebensfähigkeit von Fäkalkeimen im Boden nach Gülledüngung. PhD Dissertation, University of Hohenheim, Hohenheim, Germany

Salomons W, Förstner U (1984) Metals in the hydrocycle. Springer berlin Heidelberg New York

Schweiger P, Binkele V, Traub R (1989) Nitrat im Grundwasser: Erhebungen und Untersuchungen zum Nitrataustrag in das Grundwasser bei unterschiedlicher Nutzung, Massnahmen zur Reduzierung und Verhalten von Nitrat im Untergrund. Stuttgart: E. Ulmer, Stuttgart, 84 pp

Sexton BT, Moncrief JF, Rosen CJ, Gupta SC, Cheng HH (1996) Optimizing nitrogen and irrigation inputs for core based on nitrate leaching and yield on a coarse-textured soil. J Environ Qual 25:982-992

Sharpley AN (1997) Rainfall frequency and nitrate and phosphorus runoff from soil amended with poultry litter. J Environ Qual 26:1127-1132

Sharpley AN, Halvorson AD (1994) The management of soil phosphorus availability and its impact on surface water quality. In: Lal R, Stewart BA (eds). Advances in soil science: soil processes and water quality. CRC Press, Boca Raton, Florida, pp 7-90

Sharpley AN, Smith SJ (1985) Fraction of inorganic and organic phosphorus in virgin and cultivated soils. Soil Sc Soc Am J 49:127-130

Sharpley AN, Smith SJ (1989) Mineralization and leaching of phosphorus from soil incubated with surface-applied and incorporated crop residue. J Environ Qual 18:101-105

Sharpley AN, Smith SJ, Jones OR, Berg WA, Coleman GA (1992) The transport of bioavailable phosphorus in agricultural runoff. J Environ Qual 21:30-35

Shipitalo MJ, Edwards WM, Dock WA, Owens LB (1990) Initial storm effects on macropore transport of surface-applied chemicals in no-till soil. Soil Sci Soc Am J 54:1530-1536

Shirmohammadi A, Gish TJ, Sadeghi A, Lehman DA (1991) Theoretical representation of flow through soils considering macropore effect. In: Gish TJ, Shirmohammadi A (eds). Preferential flow. American Society of Agricultural Engineers, St Joseph, Michigan, pp 233-243

Shirmohammadi A, Ulen B, Bergstrom LF, Knisel WG (1998) Simulation of nitrogen and phosphorus leaching in a structured soil using GLEAMS and a new submodel, APARTLE. Trans ASAE 41:353-360

Shuman LM (1988) Effect of organic matter on the distribution of manganese, copper, iron, and zinc in soil fractions. Soil Science 146 (3):192-198

Shuman LM (1991) Chemical forms of micronutrients in soil. In: Mortvedt JJ, Cox FR, Shuman LM, Welch RM (eds). Micronutrients in agriculture, 2nd edn. Soil Science Society of America, Madison, Wisconsin, pp 113-144

Sims JT, Simard RR, Joern BC (1998) Phosphorus loss in agricultural drainage: historical perspective and current research. J Environ Qual 27 (2):277-293

Skopp J (1981) Comment on "Micro-, meso-, and macroporosity of soil". Soil Sci Soc Am J 45:1244-1246

Smith MS, Thomas GW, White RE, Ritonga D (1985) Transport of Escherichia coli through intact and disturbed soil columns. J Environ Qual 14 (1):87-91

Sobsey MD (1983) Transport and fate of viruses in soils. In: Canter LW, Akin EW, Kreissl JF, McNabb JF (eds). Microbial health. Considerations of soil disposal of domestic wastewaters. Publication 600/9-83-017. US Environmental Protection Agency, Cincinatti, Ohio, pp 175-197

Stamm C, Fluhler H, Gachter R, Leuenberger J, Wunderli H (1998) Preferential transport of phosphorus in drained grassland soils. J Environ Qual 27 (3):515-522

Steenhuis TS, Parlange JY, Andreini MS (1990) A numerical model for preferential solute movement in structured soils. Geoderma 46:193-208.

Stephenson RE, Chapman HD (1931) Phosphate penetration in field soils. J Am Soc Agron 23 (10):759-770

Stevenson FJ (1991) Organic matter-micronutrient reactions in soil. In: Mortvedt JJ, Cox FR, Shuman LM, Welch RM (eds). Micronutrients in agriculture, 2nd edn. Soil Science Society of America, Madison, Wisconsin, pp 145-186

Strauch D (1987) Hygiene of animal waste management. In: Strauch D (ed) Animal production and environmental health. Amsterdam Elsevier, Amsterdam, pp 155-202

Strauch D (1988) Krankheitserreger in Fäkalien und ihre epidemiologhishe Bedeutung. Tierarztliche Praxis. Suppl. 321-27. p 12

Sutton AL, Nelson DW, Mayrose VB, Kelly DT (1983) Effect of copper levels in swine manure on corn and soil. J Environ Qual 12 (2):198-202

Taiganides EP (1987) Animal waste management and wastewater treatment. In: Strauch D (ed): Animal production and environmental health. Elsevier Amsterdam pp 91-154

Tan Y, Bond WJ, Rovira AD, Brisbane PG, Griffin DM (1991) Movement through soil of a biological control agent, Pseudomonas Fluorescens. Soil Biol Biochem 23:821-825

Thomas GW, and Phillips RE (1979) Consequences of water movement in macropores. J Environ Qual 8 (2):149-152

Thompson RB, Pain BF (1990) The significance of gaseous losses of nitrogen from livestock slurries applied to agricultural land. In: Merckx R, Vereecken H, Vlassak K, (eds): Fertilization and the Environment. Leuven University Press, Leuven, Belgium, pp 290-296

Thompson RB, Ryden JC, Lockyer DR (1987) Fate of nitrogen in cattle slurry following surface application or injection to grassland. J Soil Sci 38:689-700

Thomsen IK, Hansen JF, Kjellerup V, Christensen BT (1993) Effects of cropping system and rates of nitrogen in animal slurry and mineral fertilizer on nitrate leaching from a sandy loam. Soil Use Manage 9 (2):53-58

Tietjen C (1987) Influence of faecal wastes on soil, plant, surface water and ground water. In: Strauch D (ed): Animal production and environmental health. Elsevier, Amsterdam, pp 203-217

Timlin DJ, Heathman GC, Ahuja LR (1992) Solute leaching in crop row vs. inter-row zones. Soil Sc Soc Am J 56 (2):384-392

Toth JD, Fox RH (1998) Nitrate losses from a core-alfalfa rotation: lysimeter measurement of nitrate leaching. J Environ Qual 27:1027-1033

Unc A (1999) Transport of faecal bacteria from manure through the vadose zone. M Sc Thesis, University of Guelph, Ontario, Canada

Vanerp PJ, Vandiyk TA (1992) Fertilizer value of pig slurries processed by the promest procedure. Fert Res 32:61-70

Wagenet RJ (1990) Quantitative prediction of the leaching of organic and inorganic solutes in soil. Phil Trans R Soc Lond B 329:321-330

Walter DA, Rea BA, Stollenwerk KG, Savoie J (1996) Geochemical and hydrologic controls of phosphorus transport in a sewage-contaminated sand and gravel aquifer near Ashumet pond, Cape Cod, Massachusetts, United States geological survey. Watersupply paper 2463. United States government printing office, Washington

Watson KW, Luxmoore RJ (1986) Estimating macroporosity in a forest watershed by use of a tension infiltrometer. Soil Sci Am J 50:578-582

Weigel T (1995) Untersuchungen des Infiltrationsverhaltens von Mikroorganismen in Böden mittels Gruben- und Laborversuchen sowie eines selbst entwickelten Prototyps zur probennahme ohne Sekundärkontamination. PhD Dissertation, University of Hohenheim, Hohenheim, Germany

White RE, Dyson JS, Gerstl Z, Yaron B (1986) Leaching of herbicides through undisturbed cores of a structured clay soil. Soil Sci Soc Am J 50:277-283

Wild A (1972) Nitrate leaching under bare fallow at a site in northern Nigeria. J Soil Sci 23 (3):315-324

Younie MF, Burton DL, Kachanoski RG, Beauchamp EG, Gilham RW (1996) Impact of livestock manure and fertilizer application on nitrate contamination of groundwater. Final report for the Ontario Ministry of Environment and Energy. RAC 488G. Toronto, Ontario

Soil Biotests and Indicators

K. Becker-Van Slooten[1], J. Tarradellas[1]

This article gives a brief overview on soil biotests as ecotoxicological tools for the evaluation of the impact of substances on soil quality. Soil biotests or bioassays use biological indicators of bioaccumulation and effect. Toxicity tests evaluating the effect are described with some of their advantages and disadvantages. Their integration into the risk assessment of substances and of contaminated soils or sites is outlined.

Introduction and Definitions

Soil biotests are important ecotoxicological tools for the evaluation of the impact of substances on soil quality as well as for risk assessment of contaminated soils. A biotest or bioassay can be defined as a test using a biological system and involving the exposition of an organism to a test material and determining a response (USEPA 1998). This can be a toxicity test involving the determination of the effect of a material on a group of selected organisms, under defined conditions (Keddy et al. 1994), or a test of bioaccumulation. Two types of biological indicators are used in these test systems: indicators of bioaccumulation and indicators of effect. Bioindicators are therefore organisms which respond to an impact by their presence or absence, by a change of defined characteristics or activities, or by an increased content of a pollutant (Eijsackers 1983).

Fate and Bioaccumulation

The fate of pollutants in the soil is illustrated in Fig. 1. In this figure, pollutants may be volatilized and/or taken up by plants (A), solubilized in the water phase and therefore mobile (B), adsorbed and/or complexed with soil constituents (C), degraded by microorganisms (D), and degraded and/or bioaccumulated by organisms (E). Only a part of the pollutant will be available to organisms, and this bioavailability will depend on many factors such as type of soil and organism, physico-chemical properties of the pollutant, etc. Food-web relationships in the soil

1. DGR – GECOS – Ecotoxicologie, École Polytechnique Fédérale de Lausanne, 1015 Lausanne, Switzerland.

ecosystem are very complex. The bioaccumulative behavior of a persistent organic pollutant like PCB is illustrated in Fig. 2, where the concentrations increase along the food chain (= biomagnification), reaching concentrations which may cause sterility in predatory birds. Internationally standardized protocols for conducting bioaccumulation tests exist only for aquatic organisms (fish, clams), but many studies have been realized with terrestrial organisms including meso- and macrofauna (microarthropods, earthworms, isopods and mollusks) or plants. In laboratory tests, the risk of accumulation in the organism or in the food web can be evaluated, whereas analysis of *in situ* body burden is a helpful monitoring tool.

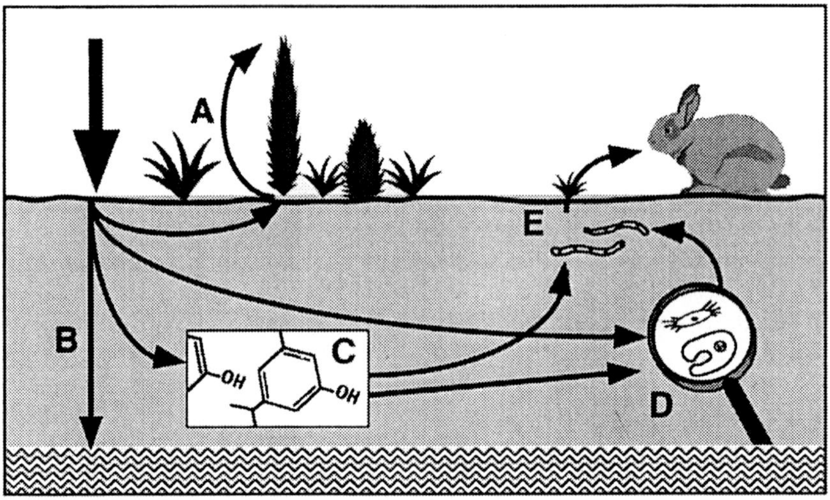

Fig. 1: Simplified scheme illustrating the fate of a substance in the soil (see text for explanation)

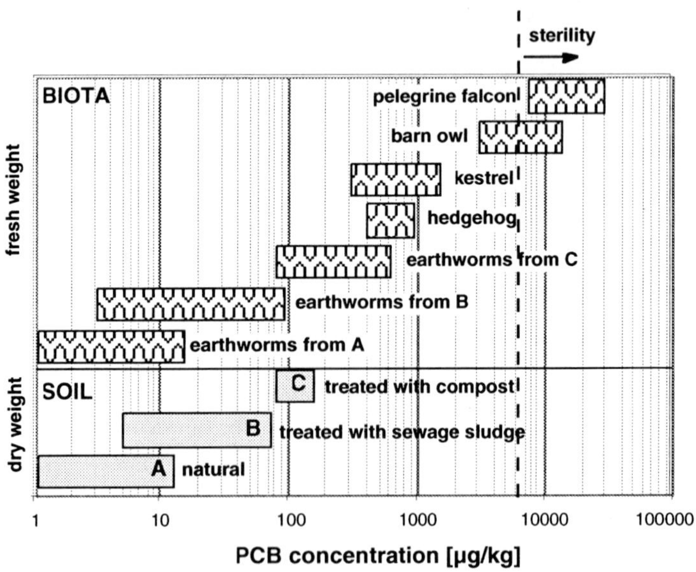

Fig. 2: Biomagnification of PCB in Swiss agro-ecosystems

Toxicity Tests

The "ideal" toxicity test would be at one and the same time sensitive, ecologically relevant, as well as very simple and inexpensive. Unfortunately, such an ideal bioassay does not exist. As illustrated in Fig. 3, relative response sensitivity decreases and ecosystem relevance increases with increasing complexity of the system and of the chosen toxicity endpoints.

On the organismal level, a certain number of assays on microorganisms, soil fauna, and plants have been standardized, in particular by OECD and ISO (International Standard Organization). These protocols are mentioned in Table 1, others are proposed by national institutions like American Standard for Testing and Materials (ASTM, USA), US Environmental Protection Agency (USEPA), Environment Canada, Association Française de Normalisation (AFNOR, France), and Deutsches Institut für Normung (DIN, Germany).

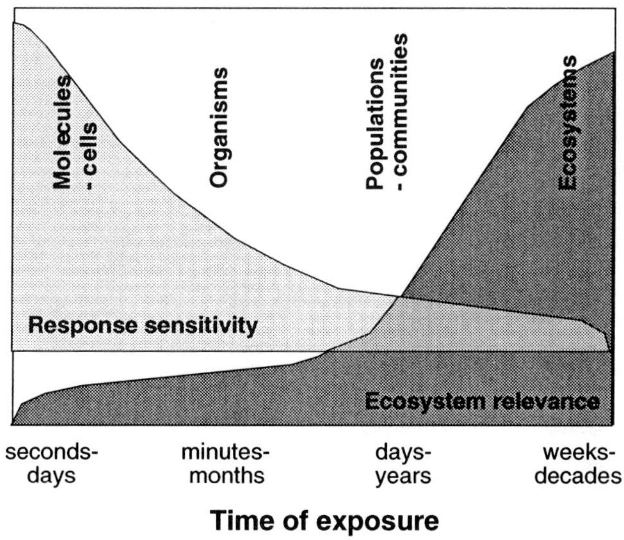

Fig. 3: Relative response sensitivity and ecosystem relevance of toxicity endpoints (Reprinted with permission from *Environmental Toxicology and Chemistry*, 1991. Assessing the toxicity of freshwater sediments, by G. A. Burton, Volume 10, Issue 12. Copyright Society of Environmental Toxicology and Chemistry (SETAC), Pensacola, FL, 1991).

Microbial tests involve processes which are essential for the functioning and fertility of a soil, as, for example, soil microbial biomass and soil respiration, nitrogen transformation, and enzyme activities. Despite difficulties in interpretation mainly due to the natural variability and to the influence of soil factors on the microbial activity and interaction with pollutants, the use of these parameters as biomonitors of soil pollution can be very useful (Pankhurst et al. 1998).

The endpoint measured in plant assays is seed germination, root elongation, as well as emergence and growth of seedlings. The tests on soil fauna concern acute (e.g., mortality) and chronic toxicity (e.g., reproduction), with contamination by food, substrate, or air. During reproduction tests, several parameters (e.g., number of cells, offspring, or eggs) can be followed. The results are expressed in effective concentrations, for example:

- EC 50: concentration causing an effect in 50% of the organisms.
- LC(D): lethal concentration (dose).
- NOEC: no observed effect concentration.
- LOEC: lowest observed effect concentration.

In fauna and plant tests, only one species is exposed at a time under controlled laboratory conditions. Generally, a battery of tests using species from different taxa and biological organization is applied to improve the ecological representativity and the extrapolation from the laboratory to the field. Keddy et al. (1994) have proposed usable and augmented batteries for soil quality assessment.

Table 1. Terrestrial toxicity tests

1. Microorganisms

Microbial respiration and mineralization of substrates

Determination of soil biomass (fumigation-incubation, fumigation extraction or substrate-induced respiration)

- OECD draft 217 (1999) Soil microorganisms: carbon transformation test
- ISO 14240-1 and 2 (1997) Soil quality – determination of soil microbial biomass
- "Litterbag tests"

Nitrogen transformation

- OECD draft 216 (1999) Soil microorganisms: nitrogen transformation test
- ISO/DIS 14238 (1995) Soil quality – determination of nitrogen mineralisation and nitrification in soils and the influence of chemicals on these processes
- Nitrogen fixation (e.g., *Rhizobium*)

Enzyme activities (e.g., dehydrogenase, hydrolytic enzymes)

Other methods (e.g. mycorhizial fungi, microbial communities, fatty acid analysis, DNA approaches)

2. Microfauna (< 0.02 mm)

Protozoa and Nematodes (promising, but no procedure from recognized standard organizations available)

3. Mesofauna (0.02-4 mm)

Oribatid mites, collembola, millipedes, isopods

- ISO/DIS 11267 (1998) Soil quality – inhibition of reprodution of Collembola (*Folsomia candida*) by soil pollutants

4. Macrofauna (> 4-80 mm)

Earthworms, enchytraeids, mollusks

- OECD 207 (1984) Guidelines for testing of chemicals: earthworm acute toxicity tests
- ISO 11268-1 and 2 (1994) Soil quality – effects of pollutants on earthworms (*Eisenia fetida*) (acute toxicity and reproduction)

5. Vascular plants

Many species from different taxa, like cabbage, wheat, lettuce, red clover, soybean, etc.

Endpoints: germination, shoot and root biomass, root elongation

- OECD 208 (1984) Guidelines for testing of chemicals: terrestrial plant, growth test

When conducting toxicity tests in the laboratory, the organisms can be exposed to artificially contaminated soil with chemicals (single or mixtures), or to soil sampled in the field. The tests can be realized directly with the soil or with soil extracts (water or solvents). In the second case, these extracts can be tested on aquatic organisms, such as bacteria, algae, or crustaceans. The selected approach depends on the aim of the assessment (Dechema 1995, Wetzel 1998):

- Evaluation of the living-space for soil communities: microbial and fauna tests.
- Evaluation of the living space for plant production: plant and root-symbiosis tests.
- Evaluation the retention capacity of soil and the risk for the groundwater: aquatic tests on soil leachates or eluates.

Different substrates can be used according to the approach. As mentioned above, tests can be conducted with natural soil, which are of course the most relevant for the environment, but which are very complex and difficult to standardize. Many test protocols refer to the artificial ISO soil (ISO 1994) containing sand, clay, and peat. Another possibility is to work with an inert substrate, which might be very helpful for screening or research purposes. The chosen substrate will have an important influence on the bioavailability of the substance(s).

The exposure routes in the soil ecosystem are the following, depending on the pollutant, the soil, and the organism:

- Contact to soil.
- Food.
- Pore water.
- Air (gas phase).

As the different tests may be used for different purposes, a good selection of these tools is a very important issue. Van Gestel et al. (1998) proposed the following criteria for selection for toxicity tests:

Selection of individual tests should take into account:

- Practical arguments (feasibility, cost effectiveness, rapidity).
- Acceptability of tests (standardization, reproducibility, statistical validity, GLP).
- Ecological significance (sensitivity, ecological realism, biological validity).

Set of assays (battery) should be representative of:

- Life-history strategies.
- Functional groups.
- Taxonomic groups.
- Exposure routes.
- Levels of biological organization.

More complex systems, like microcosms or *in situ* assays with appropriate indicators, enable the long term evaluation of effects on population and community level. These systems require expert interpretation and still need development to improve their predictivity (Sheppard 1997). Other organisms, like beneficial arthropods (e.g., honeybees), birds, and mammals, must, in certain cases, also be included in an impact assessment. Another promising field is the use of biomarkers, where a change in a biological response is related to an exposure to, or toxic effect of, environmental chemicals (Koeman et al. 1993). These responses can range from a molecular to a physiological level (in most cases biochemical responses) and they can constitute an early warning system of exposure and/or effect.

Risk Assessment

It is important to differentiate the risk assessment of chemicals and the risk assessment of soils or sites. Concerning the first, many approaches have been developed, especially in the field of pesticide evaluation. These approaches generally have a tiered structure (OECD 1995), based on a risk characterization scheme as shown in Fig. 4. Data on ecotoxicity and on expected environmental concentrations for the chemical are used to evaluate the exposure and the effect in order to determine the predicted environmental concentration (PEC) and the predicted noneffect concentration (PNEC). An assessment factor is applied for the determination of the PNEC – this factor depends on the relevance of the available ecotoxicity data (Table 2). If the PEC/PNEC ratio is higher than 1, an environmental hazard can be expected.

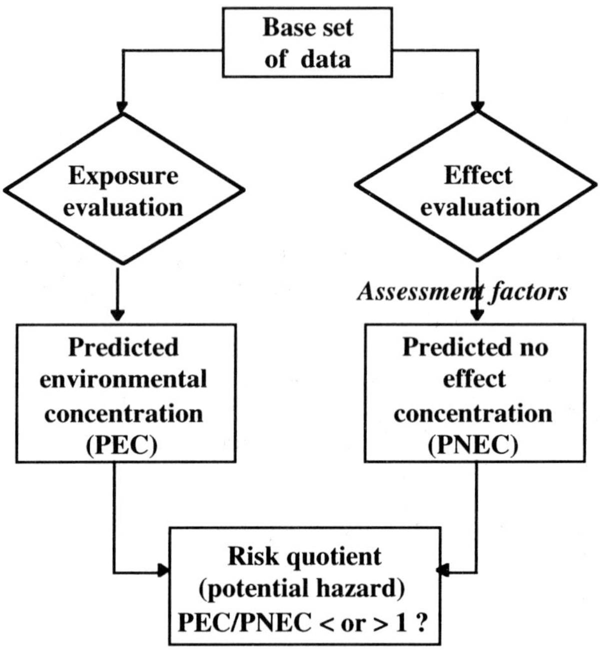

Fig. 4: Risk characterization (hazard assessment) of chemicals

Risk assessment always needs expert judgement in order to decide on the need for tests, to assess the validity of data, to interpret results, to contribute to risk management decisions, etc. (OECD 1995).

The ecological risk assessment (ERA) of contaminated soil is even more complex. Three general approaches can be distinguished (Ferguson et al. 1998).

1. Comparison of chemical data with soil screening or guideline values according to soil quality objectives (derived from toxicity data obtained in standardized tests).
2. Bioassays with solid material or soil extracts (supplement to chemical analysis).
3. Monitoring of biomarkers, bioaccumulation, indicator species, changes in community structure, etc.

Concerning the industrialized countries, the first approach is widely used and implemented and the second approach is common practice in some countries. The third approach, which is, of course, the most relevant, complex, and time – and money – consuming, is used only at a research level (Ferguson et al. 1998).

Table 2. Assessment factors for the determination of the PNEC of chemicals. (European Commission 1994)

Information available	Assessment factor
L(E)C50 short-term toxicity tests (e.g., plants, earthworms, or microorganisms)	1000
NOEC for one long-term toxicity test (e.g., plants)	100
NOEC for additional long-term toxicity tests of two trophic levels	50
NOEC for additional long-term toxicity tests for three species of three trophic levels	10
Field data/data of model ecosystems	Case by case

Advantages and Disadvantages

The advantages of bioassays are:

- Integration of bioavailability.
- Global response – integration of synergies and antagonisms.
- Early detection of ecotoxicological impact.
- Knowledge of toxicity mechanisms.
- Guidance for the development of nontoxic and/or more specific molecules.
- Promising tools for risk assessment.

The main problems encountered in the domain of soil ecotoxicology are the following:

- High complexity of soil system with phenomena like adsorption, complexation, interactions and with important natural fluctuations and variability.
- Choice of test substrate (artificial soil, natural soil, inert substrate?).
- Choice of organisms and endpoint.
- Difficulty to obtain a satisfying control or reference soil (for dilution and/or control organisms).
- Risk assessment for mixtures, adaptation and ecological recovery as well as multiple sources of stress are rarely taken into account.
- Extrapolation from laboratory to field remains difficult.

Conclusion

The important advantage of biotests is their global response to a toxic effect, taking into account bioavailability and interactions between different substances and between substances and soil parameters. This characteristic makes them indispensable for the control of soil pollution. There is, nevertheless, need for more research and development, and further validation is required, especially concerning the extrapolation from the laboratory to the field.

Further information about the topic can be found in the following books and reviews: Donker et al. (1994), Tarradellas et al. (1997), Texier et al. (1996), van Straalen and van Gestel (1993), and Verhoef and van Gestel (1995).

References

Burton G A (1991) Assessing the toxicity of freshwater sediments. Environ Toxicol Chem 10(12): 1585-1627

DECHEMA 1995. Biologische Testmethoden für Böden. Deutsche Gesellschaft für Chemisches Apparatewesen, Frankfurt, Germany, 46 pp

Donker M.H., Eijsackers H., Heimbach F. 1994. Ecotoxicology of soil organisms. Lewis, Boca Raton, Florida, 470 pp

Eijsackers H. 1983. Soil fauna and soil microflora as possible indicators of soil pollution. Environ Monit and Assess 3: 307-316

European Commission. 1994. Risk Assessment of existing substances. Technical guidance document. XI/919/94-EN. Brussels

Ferguson C., Darmendrail D., Freier K., Jensen B.K., Jensen J., Kasamas H., Urzelai A., Vegter J. 1998. Risk assessment for contaminated sites in Europe, vol 1: Scientific Basis. LQM Press, Nottingham, England. 165 pp

ISO 1994. Soil quality – effects of pollutants on earthworms (*Eisenia fetida*). International Standard Organization, ISO 11268, Geneva, Switzlerland

Keddy C., Greene J.C., Bonnell M.A. 1994. A review of whole organism bioassays for assessing the quality of soil, freshwater sediment, and freshwater in Canada. Environment Canada, Scientific Ser 198, 185 pp

Koeman J.H., Köhler-Günther A., Kurelec B., Rivière J.L., Versteeg D., Walker C.H. 1993. Applications and objectives of biomarker research. In: Peakall D.B. Shugart L.R (eds) Biomarkers. Springer, Berlin Hidelberg New-York, pp 1-13

OECD 1995. Report of the OECD workshop on environmental hazard/risk assessment. OECD Environment Monographs 105. Organisation for Economic Cooperation and Development, Paris

Pankhurst C.E., Rogers S.L., Gupta V.V.S.R. 1998. Microbial parameters for monitoring soil pollution. In: Lynch J.M., Wiseman A (eds) Environmental biomonitoring: the biotechnology ecotoxicology interface; Cambridge University Press, Cambridge, pp 46-69

Sheppard S.C. 1997. Toxicity testing using microcosms. In: Tarradellas J., Bitton G., Rossel D (eds) Soil ecotoxicology. Lewis, Boca Raton, Florida, pp 345-373

Tarradellas J., Bitton G., Rossel D. 1997. Soil ecotoxicology. Lewis, Boca Raton, Florida, 386 pp

Texier C., Cluzeau D., Cortet J., Gomot A. 1996. La faune indicateur de la qualité des sols, ADEME Editions - Série données et références, brochure 2588, 62 pp van Gestel C.A.M., Léon C.D., van Straalen N.M. 1998. Evaluation of soil fauna ecotoxicity tests regarding their use in risk assessment. In: Tarradellas J, Bitton G, Rossel D (eds) Soil ecotoxicology. Lewis, Boca Raton, Florida, pp 291-317

van Straalen N.M., van Gestel C.A.M. 1993. Soil invertebrates and micro-organisms. In: Calow P (ed) Handbook of ecotoxicology. Blackwell, Oxford, pp 251-277

USEPA 1998. Evaluation of dredged material proposed for discharge in waters of the U.S. – testing manual. United States Environmental Protection Agency, EPA 823-B-98-004

Verhoef H.A., van Gestel C.A.M. 1995. Methods to assess the effects of chemicals on soils. In: Linthurst R.A., Bourdeau P., Tardiff R.G. (eds) Methods to assess the effects of chemicals on ecosystems. Wiley, Chichester, pp 223-257

Wetzel A. 1998. Advances in biomonitoring: sensitivity and reliability in PAH-contaminated soil. In: Lynch J.M., Wiseman A (eds) Environmental biomonitoring: the biotechnology ecotoxicology interface. Cambridge University Press, Cambridge, pp 27-45

Part II
Nutrients

Ennio GALANTE

J.A. VAN VEEN
J.M. BAREA
D.F. BOESCH and R.B. BRINSFIELD
P. MOORE et al.
J. SCHEPERS et al.

Nutrients
Introductory Note

E. GALANTE[1]

"For the new millennium, the challenge to agriculture (agricultural systems) is to balance the priorities of food production with the needs for environmental protection and enhancement. It is recognised that there are many conflicting interests:

- higher yields need higher input of energy (mechanical and chemical) for plants and animals;
- in general, higher chemical inputs means that only a part of them are really absorbed by crop plants;
- the non-absorbed chemical fertilisers leach in the soil and pollute surface and ground water."

(Agriculture and the Environment - Challenges and Conflicts, Warwick, 14-16 April 1999).

These issues are not new. I wish here to recall some historical milestones within the past three decades.

Meadows and coworkers of MIT published *The Limits of Development* (1972), under the sponsorship of the Club of Rome. This study, which was a relevant example of research oriented to solve big problems, pointed out that it could be dangerous for developed nations to pursue the policy of continuing expansion of production and consumption. There were five main concerns: expanding population, production of food, fossil energy consumption, pollution and industrialisation. In conclusion, we must reflect that there are physical limits to all economical and social activities.

Some schools of economics criticised the MIT Report, considering it too catastrophic and non-realistic. It may be that Meadows' group did not at that time have this information about fossil energy reserve (oil and gas) that was soon after to be discovered. My opinion is that the MIT Report was of great value, because it suggested to politicians and economists a general analytical approach based on the study of interconnections between global phenomena and variables, and their feedback.

In the same period (1973) David Pimentel published in *Science* a paper (*Food Production and Energy Crisis*), demonstrating that in 25 years (1945 to 1970) the energetic parameters of maize production in USA were as follows:

1. Institute for Plant Biosynthesis, NRC, Via Bassini 15, 20133 Milano, Italy.

- consumption of nitrogen fertilisers increased by 16 times,
- consumption of phosphorous by 4.44 times,
- pesticides increased by 10 times,
- total inputs increased by 3.13 times,
- corn yield (output) increased only by 2.38 times,
- the ratio of kcal of output per kcal of inputs decreased by 20%.

Perhaps if they are to consider more recent data, one would find a worse ratio between output and input, due to a further increase in technical means.

In 1974 FAO held the World Food Conference and shortly after the National Academy of Science US started a study that was published in 1977: *World Food and Nutrition Study – The Potential Contribution of Research.* It was a great occasion for scientists to consider how their knowledge could be targeted to serious problems of society.

Two years later, the UNESCO Conference on Population alerted on the issue of demographic explosion.

The concept of sustainability of social, environmental and agricultural systems became gradually more and more popular.

In 1990 an international conference held at the University of Padua convened hundreds of scientists to discuss *Biotic Diversity in Agroecosystems.* David Pimentel was one of the organisers (see the five volumes of Proceedings).

In 1992 the Rio Conference on Environment and Development was held, and in 1997 the Rio + 5, in New York, which ended without any agreement on sustainable development. Meantime, the world population has been continuously growing at a rate of 1.4%/year. Millions of people have starved, particularly in tropical and sub-tropical arid regions, and many others suffer from malnutrition; but plant and animal productivity in the developed countries has increased significantly.

The international Conference on Managing Natural Resources for Sustainable Agricultural Production, which will be held in Delhi in February 2000, concludes its first announcement:

"...The goal of the conference would be to formulate performance-oriented action plans for the future by focusing on economically viable, socially acceptable, locally replicable, and geo-ecologically sustainable technologies of scientific resource use for potential agricultural production."

At the beginning of the third millennium, we are more than 6 billion humans on the planet. We need to be fed by an agriculture which will produce safe, high-quality food at an intensity sufficient to satisfy demand whilst reducing the impact on the environmental resources (either biological or physical). This impact is due to various causes, among them agricultural nutrients. NUTRIENTS are a relevant part of the strategy that has been called "Biological Resource Management for Sustainable Agricultural Systems". Therefore we expect that the distinguished speakers invited to report will try to set the point on the state of knowledge about nutrients and environment.

Nitrogen: Recent Developments in Related Microbial Processes

JA. VAN VEEN[1]

Nitrogen is a key nutrient for the production of crops on earth. More than any other, it is the growth-limiting nutrient for plant and therefore, it has been added to arable land as fertiliser in enormous quantities, all over the world. Due to its (bio)chemical "flexibility" and its mobility, nitrogen is easily lost to the environment, where it has become a major polluting element in the atmosphere, water and soil. Thus, sound management of the nitrogen resources is a prerequisite for sustainable agriculture and for a clean environment. This calls for understanding of the fate of nitrogen in plant/soil systems. Here, three examples of recent developments in research on nitrogen will briefly be discussed, with emphasis on microbiological processes. These cases are the modelling of nitrogen flow through the soil food web, the use of molecular biological techniques for the identification of ammonium-oxidising bacteria and the mechanisms of host recognition by symbiotic nitrogen-fixing bacteria. They also exemplify progress that has been made in the science on terrestrial ecosystems during the period of the present OECD programme on Biological Resource Management, progress that is needed to develop future strategies for proper management of the biological nitrogen resources.

Introduction

Nitrogen is one of the key elements for life on earth. In contrast to other life-essential elements, nitrogen is present in vast quantities (80% N_2 in the atmosphere), but only a small fraction of the world's nitrogen occurs in biologically available forms. Prior to modern industrial development, nitrogen became available for life either through the process of biological nitrogen fixation or by physicochemical transformations, e.g. during lightning. The development of the Haber-Bosch process of industrial fixation of atmospheric nitrogen created a third major input of life-available nitrogen.

In the processes of rendering nitrogen available for life, as well as in the subsequent transformation processes, microorganisms play vital roles. The fixation of

1. Netherlands Institute of Ecology, Centre for Terrestrial Ecology, P.O. Box 40, 6666 ZG Heteren, The Netherlands

atmospheric nitrogen by the activity of nitrogen-fixing bacteria is the major input of nitrogen in the global N-cycle. Moreover, the transformation of nitrogen into inorganic and organic forms, as well as in the redox processes involving different forms of inorganic nitrogen, are mainly the result of microbially mediated processes.

Nitrogen is essential for crop production, and ever-increasing nitrogen inputs have been necessary to meet the food demands of the world's growing population. However, nitrogen is also a mobile element, which is easily lost from crop production sites into the surrounding environment, where it may cause substantial damage when present in large quantities. So, efficient use of nitrogen is required for sustainable agricultural production and with minimal environmental damage.

Until the past decade, biological nitrogen fixation was by far the largest input of atmospheric nitrogen in the global N-cycle. Estimates of approximately 100 Tg N per annum have been mentioned (Vitousek 1994). Since the 1980s, a dramatic increase in the amount of anthropogenically fixed nitrogen, AFN, has occurred (Fig 1). AFN includes industrially fixed nitrogen as well as nitrogen fixed by leguminous crops used as green manure. Today, it is estimated that AFN exceeds naturally fixed nitrogen as the major source of life-available nitrogen.

Only occurring more recently, the emission of nitrogen oxide, N_2O, into the atmosphere has also increased considerably (Fig 1). N_2O is one of the three main greenhouse gases, next to CO_2 and CH_4, held responsible for the global warming.

The amount of AFN used for crop production depends on the demand for food, the development stage of agricultural crop production and on the efficacy of the management of the N resources.

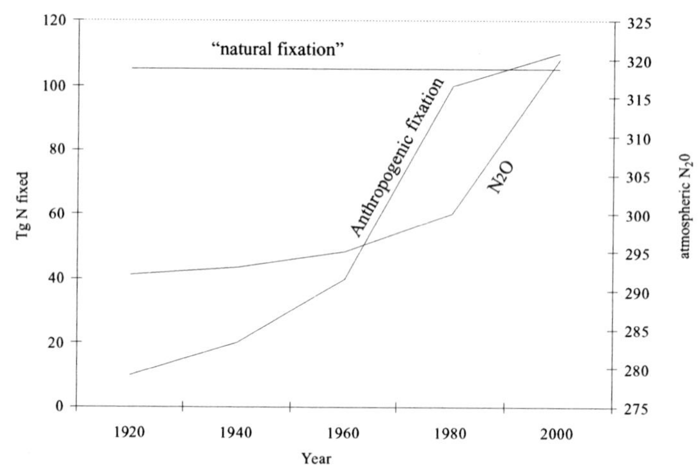

Fig. 1: Global N-fixation in relation to increase of atmospheric N_2O concentration. (After Vitousek 1994)

As shown in Table 1, the amount of N used is largest in Asia, Europe and North America, whereas the amounts used in Africa and South America are rather low. If one uses the N_2O emission from livestock as an indicator for the efficiency of N use, Africa and South America show remarkably poor relative efficiencies as compared to the other continents (Table 2).

Table 1. Estimated annual application of fertilizer and manure N, in 1990. (After Smith et al. 1997)

Region	Amount (Tg)	% of total
Africa	2,1	2,7
North and central America	13,1	16,9
South America	1,7	2,2
Asia + Oceania	38,2	49,4
Europe	13,6	17,6
Former USSR	8,7	11,2
Total	77,4	100

Clearly, efficient management of the available N resources is urgently needed for both sustaining and increasing the agricultural crop production necessary to meet the increasing demand for food and reducing N losses into the environment. Cole et al. (1996) suggested a number of measures to be integrated into agricultural practices, which could be taken to reduce the emission of N into the environment (Table 3). They estimated that, by applying all these measures, the annual N_2O emission into the atmosphere could be reduced by one-third.

Table 2. Estimated total N_2O emission from cattle dung and urine. (After Oenema et al. 1997)

Continent	N_2O emission ($Tg\text{-}year^{-1}$)
Africa	0,3
N. America	0,1
S. America	0,3
Asia + Oceania	0,7
Europe	0,1
Total	1,5

Table 3. Practices to improve N fertiliser efficiency and expected reduction in N2O output (After Cole et al. 1996)

Agricultural practice	Estimated reduction in N_2O output ($Tg\ N\text{-}year^{-1}$)
Match N supply and crop demand	0.24
Tighten N flow cycles	0.14
Use advanced fertiliser techniques (incl. nitrification inhibitors)	0.15
Optimise tillage, irrigation and drainage	0.15
Total	0.68

Efficient management of available nitrogen resources requires:

1. Proper regulation of the availability of nitrogen for crop uptake.
2. Reduction of losses of nitrogen to the surrounding environment.
3. Increase in the input of nitrogen by biological fixation.

These three cases, which will be worked out in this chapter, exemplify recent progress in our understanding of some of the key processes of each of the items listed above.

With regard to (1), the development of a new generation of mathematical simulation models to describe the turnover of nitrogen in soil has enabled the prediction of the impact of agricultural practices on the community of microorganisms and soil animals. This is necessary to develop crop production systems in which the availability of nitrogen fits the demand of the crop.

Losses of nitrogen to the environment, (2), may occur by several processes, including microbial transformation of ammonium and nitrate. In these processes, nitrifying bacteria play a key role. Molecular biological techniques have tremendously increased the possibilities of identifying and detecting microorganisms in soil. Such techniques have greatly increased our knowledge of the bacteria involved in the oxidation of ammonium, which is one of the major sources of N_2O. This knowledge is a prerequisite for future strategies to reduce the emission of N_2O into the atmosphere.

The fixation of atmospheric nitrogen, (3) by the bacteria that live in symbiosis with leguminous plants is the most effective way of rendering atmospheric nitrogen available for life. The symbiosis is created through the intermediary role of specific compounds excreted by the bacteria and the plants with are used as recognition signals. Modern biochemical techniques have facilitated the identification of some of these compounds, which may help to optimise the symbiotic relationship between the nitrogen-fixing bacteria and their host plants.

Mathematical Modelling of Nitrogen Flow Through Soil Food Webs

According to the estimates of Cole et al. (1996); Table 3, the greatest reduction in the emission of nitrogen oxides into the atmosphere may be achieved by synchronising nitrogen fertiliser application with the demand of the crop. This requires a proper understanding of the processes that regulate the availability of nitrogen in soil. Nitrogen availability is determined by the continuously and simultaneously occurring processes of mineralisation and immobilisation. Because of the complexity of these processes, mathematical models have been used since the 1950s to quantify the rates of these processes in soil. (e.g. Kirkham and Bartholomew 1955). The use of models greatly expanded as better calculation procedures, computers and suitable simulation languages became available in the 1970s (e.g. Frissel and Van Veen 1981).

The first models were mainly based on the description of biochemical transformation of the different nitrogen compounds. The second generation of models included some of the agents responsible for the different transformations, such as nitrifiers and the total microbial biomass as the transforming agent of organic into inorganic nitrogen forms and vice versa (e.g. Verberne et al. 1990).

The third generation of models considers the turnover of nitrogen as the result of food-web interactions in soil (De Ruiter et al. 1993, 1995). Microbes are considered to be the primary decomposers of organic materials, thereby mineralising or immobilising nitrogen, depending on the C/N ratio of the decomposed material. Micro-organisms, i.e. bacteria and fungi, are food to predators like protozoa, nematodes and collembola, which, subsequently, are a source of food for higher organisms like mites. During predation, waste products are produced including nitrogen compounds, which are directly or indirectly available for plant uptake.

These models provide the detailed description of the processes underlying nitrogen transformations, that is necessary to predict the seasonal dynamics in the turnover of nitrogen and its availability for plant uptake (Fig. 2).

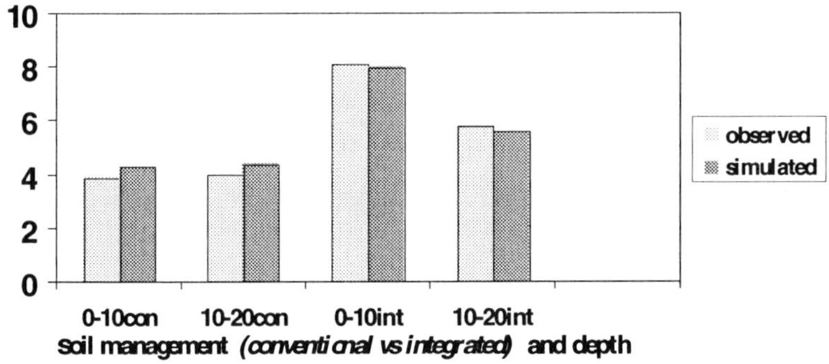

Fig. 2: Simulated food web model of De Ruiter et al. (1993) and observed N mineralization (kg N-ha^{-1}-year^{-1}-cm^{-1})

The dynamics of each of the populations and groups of organisms which contribute to the flow of energy and nitrogen through the system might differently be affected by environmental conditions of climate and soil. So, it is important to have insight into the relative contribution of each of the participating groups of organisms (Table 4).

Table 4. Contribution of soil organisms to the mineralization of nitrogen in soils under conventional and integrated practices (kg N-ha^{-1}-year^{-1}-layer^{-1})

Organisms	Conventional practice	Integrated practice
Microbes	92,3	114,1
Protozoa	23,1	29,0
Nematodes	8,2	11,5
Microarthropodes	0,6	0,3
Larger fauna	1,3	1,5
Total	125,5	156,4

Moreover, these models enable the prediction of the effect of agricultural practices on the food web and thus on the interactions that largely determine the functioning of the ecosystem. Table 5 shows the effects of the (simulated) inactivation or killing of certain groups of the food web on the flow of energy through the system.

In this case, energy, often expressed as C equivalents, is linearly related to the flow of nitrogen. Effects such as those simulated in Table 5 might be exerted, for instance, by the application of pesticides or the input of other toxic compounds.

Table 5. Effect of group removal on relative reduction (%). In below-ground biomass. (After Moore et al. 1993)

Site	Central Plains (USA)	Lovinkhoeve (NL)	Horshoe Band (USA)
Protozoa	12 (4)[a]	36 (4)	72 (4)
Nematodes			
Bacteriovores	14 (5)	10 (5)	1 (5)
Fungivores	1 (6)	1 (6)	1 (8)
Predators	8 (23)	12 (19)	- (-)

[a] In brackets ratio between biomass reduction and own biomass.

Use of Molecular Biological Techniques to Identify and Detect Nitrifying Bacteria

In order to understand the processes which cause loss of nitrogen in the environment, it is necessary to know the mechanisms of these processes. Several processes of the nitrogen cycle produce forms of nitrogen which may cause serious damage to the environment. One of those processes is nitrification. Table 6 shows the negative influences that the process may exert on the environment.

Table 6. Negative environmental effects of nitrification

Production of protons	Acidification of soils Deterioration of buildings
Production of nitrate	Pollution of (ground) water Eutrophication of surface water
Production of N_2O/NO	Greenhouse effect Degradation of ozone layer

Nitrification is the process by which ammonium is oxidised to nitrite, which is subsequently further oxidised to nitrate. The process is largely the result of the activity of specialised groups of chemoautotrophic bacteria. Until recently, our knowledge of the occurrence of these bacteria was largely based on information obtained from cultivated bacteria. With regard to the bacteria involved in the oxidation of ammonia, Belser (1979) mentioned that seven species belonging to five genera of bacteria were involved in the oxidation of ammonium in soil and water. The relative lack of information available for ammonium-oxidising bacteria and other nitrifying bacteria can, at least in part, be attributed to the difficulties involved with their poor culture isolation and their slow growth rates

One of the greatest developments in ecology, and particularly in microbial ecology in the past decade, is the application and development of methods based upon nucleic acid (i.e. DNA and RNA) analysis. At present, a large variety of methods have become available for the research of microbes in natural environments (e.g. Van Elsas et al. 1997). This has broadened our knowledge of the role of microorganisms in the functioning of ecosystems, and, at the same time, has raised many new questions on topics of which we were not even aware before these methods became available.

Using a PCR-based analysis of 16S rDNA directly extracted from soil, Kowalchuk et al. (1998) showed that the diversity of organisms in soil capable of oxidising ammonium is much larger than could be envisaged on the basis of cultivation-based techniques (Fig. 3). Similar molecular studies suggested the specialised nature of particular nitrifier groups, and it is our understanding of these specialised physiologies in the heterogeneous environment that are the key to future attempts to minimise negative environmental impacts.

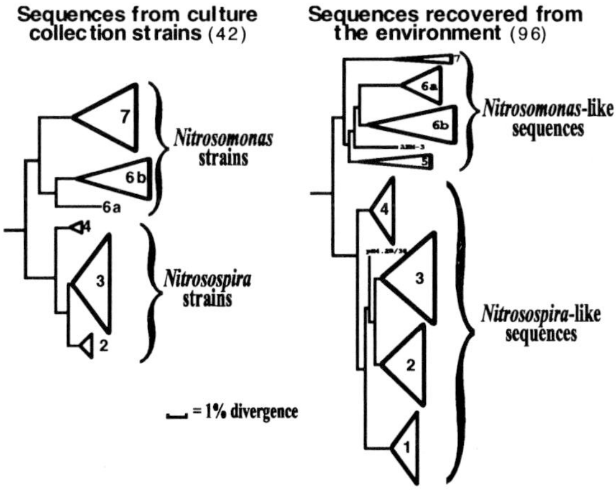

Fig. 3: Comparison of the number of ammonium-oxidizing bacterial species as determined after isolation and after direct extraction of DNA followed by PCR-DGGE-analysis

Mechanisms of Host Recognition by Symbiotic N_2-Fixing Bacteria

Fixation of N_2 by symbiotic bacteria, such as *Rhizobium*, *Bradyrhizobium* and others, is the most effective way of rendering atmospheric nitrogen available for life. The amounts of nitrogen that can, potentially, be fixed by the different symbioses range from 10 to 300 kg N $ha^{-1}year^{-1}$. (Table 7). These are much higher values than the amounts that are estimated to be fixed by non-symbiotic nitrogen fixation. According to Peoples and Craswell (1992), symbiotic nitrogen fixation provides 80% of the biologically fixed nitrogen on land.

Biological nitrogen fixation is a specific process that involves bacteria that are either in symbiosis with a host plant or not. The symbiosis between the bacterium and its host plant is effected by the formation of nodules. In these nodules a differentiated form of the bacterium is able to fix nitrogen, which can be used by the plant. The symbiosis can be rather specific including only one bacterial species and one host plant, but certain nitrogen-fixing bacteria can form a symbiosis with several hosts (Table 8).

The host-specific aspect of the symbiosis is very pronounced and has led to the classification of bacteria according to their group of host plants.

The determination of the host-specific nodulation involves at least a two-step signal exchange. In the first step, flavonoids excreted by the plant induce the transcription of bacterial nodulation genes. In the second step the bacterium produces lipo-oligosaccharide signals by means of the nodulation genes. The role of these lipo-oligosaccharide signals is related to the induction of the nodule meristem, which is a most important phenotypic process for the establishment of the bacterium-plant symbiosis.

The backbone of the signal molecule of all rhizobia consists of a β-1.4-linked *N*-acetyl-D-glucosamine varying in length between three and five sugar units (Spaink 1995). The substitutions to the chitin molecule are often rhizobium strain-specific.

Table 7. Diversity of symbiotic N_2-fixing associations. (After Vance 1996)

Plant type	Genus	Microbe	Range of N_2-fixation rate (kg N ha^{-1}-$year^{-1}$)
Leguminosae	*Pisum, Glycine, Medicago,* (etc)	*Rhizobium, Bradyrhizobium, Azorhizobium*	10-350
Ulmaceae	*Prasponia*	*Bradyrhizobium*	20-70
Betulaceae	*Alnus*	*Frankia*	15-300
Casuarinaceae	*Casuarina*	(Actinomycete)	10-50
Eleagnaceae	*Eleagnus*	(Actinomycete)	ND
Rosaceae	*Rubus*	(Actinomycete)	ND
Pteridophytes	*Azolla*	*Anabaena*	40-120
Cycads	*Ceratozamia*	*Nostoc*	19-60
Lichens	*Collema*	*Nostoc*	ND

Table 8. Species of the family of Rhizobiaceae and their respective host plants. (After Dénarié et al. 1992)

Bacterial species	Host plant
Rhizobium leguminosarum	
biovar *viciae*	*Pisum, Vicia, Lathyrus*
biovar *trifolii*	*Trifolium*
biovar *phaseoli*	*Phaseolus*
Rhizobium meliloti	*Melilotus, Medicago*
Rhizobium loti	*Lotus*
Rhizobium etli	*Phaseolus*
Rhizobium tropici	*Phaseolus, Leucaena*
Rhizobium sp. NGR 234	*Various tropical legumes*
Bradyrhizobium japonicum	*Glycine*
Bradyrhizobium elkanii	*Glycine*

The possibilities of manipulating the biological nitrogen fixation process have been extensively studied in the past decade, because of its great economic and environmental importance. The mechanism of nitrogen fixation involves regulation of the nitrogenase enzyme complex. This enzyme complex requires specific conditions, particularly related to the redox conditions of its environment. This hampers the possibilities for manipulation. For instance, until now, it has not been possible to obtain the enzyme operating in plants outside the location of the symbiosis. Given the possibilities provided by DNA-transformation technologies, many people are, of course, interested in introducing the genes responsible for nitrogen fixation into plants. As this dream is still far from reality, other ways of maximising the use of biological nitrogen fixation should be taken into consideration. One of the possibilities is to broaden the host range of rhizobia and to increase the efficacy of inoculated rhizobium strains in competition with the native bacterial community and of their interaction with the host plant. Here, the improvement of our knowledge on the mechanisms of the initial interactions between rhizobia bacteria and their host plant is indispensable.

Conclusions

- Use of anthropogenically fixed nitrogen has increased enormously, as has its loss into the environment. Given the continuously increasing demand for food, this calls for proper management of the biological nitrogen resources, as nitrogen is the most commonly deficient nutrient for crop production.
- Proper management of biological nitrogen resources should include the regulation of nitrogen availability for plant uptake, reduction of losses of nitrogen to the atmosphere, surface water and groundwater and the optimisation of the use of biological nitrogen fixation.
- Progress during the period of the OECD Program in molecular biology, biochemistry and mathematical modelling has considerably increased our understanding of the microbiology of the N cycle. This includes aspects concerning the synchronisation of soil nitrogen availability and plant uptake, identification and detection of microorganisms responsible for emission of nitrogen into the environment and insight into the mechanisms of recognition of symbiotic nitrogen fixing bacteria and their host plant.

References

Belser LW (1979) Population ecology of nitrifying bacteria. Annu Rev Microbiol 33:309-333

Cole V, Cerri C, Minami K, Mosier A, Rosenberg N (1996) Agricultural options for mitigation of greenhouse gas emissions. In: Watson RT, Zinyowera MC, Moss RH, Dokken DJ (eds) Climate change 1995. Impacts, adaptations, and mitigation of climate change: scientific-technical analysis. Cambridge Univ Press, Cambridge, pp 745-771

Dénarié J, Debellé F, Rosenberg C (1992) Signalling and host range variation in nodulation. Annu Rev Microbiol 46:497-531

De Ruiter PC, Moore JC, Bloem J, Zwart KB, Bouwman LA, Hassink J, De Vos JA, Marinissen JCY, Didden WAM, Lebbink G, Brussaard L (1993) Simulation of nitrogen dynamics in the belowground food webs of two winter-wheat fields. J Appl Ecol 30:95-106

De Ruiter PC, Neutel AM, Moore JC (1995) Energetics, patterns of interaction strengths, and stability in real ecosystems. Science 269:1257-1260

Frissel MJ, van Veen JA (1981) Simulation of nitrogen behavior in soil-plant systems. Pudoc, Wageningen, 277 pp

Kirkham D, Bartholomew WV (1955) Equations for following nutrient transformations in soil, utilizing tracer data: II. Soil Sci Soc Am Proc 19:189-192

Kowalchuk GA, Bodelier PLE, Heilig GHJ, Stephen JR, Laanbroek HJ (1998) Community analysis of ammonia-oxidizing bacteria, in relation to oxygen availability in soils and root-oxygenated sediments, using PCR, DGGE and oligonucleotide probe hybridisation. FEMS Microbiol Ecol 27:339-350

Moore JC, de Ruiter PC, Hunt HW (1993) Soil invertebrate/microinvertebrate interactions: disproportionate effects of species on food web structure and function. Vet Parasit 48:247-260

Oenema O, Velthof GL, Yamulki S, Jarvis SC (1997) Nitrous oxide emissions from grazed grassland. Soil Use Manage 13:288-295

Peoples MB, Craswell ET (1992) Biological nitrogen fixation: investments, expectations, and actual contributions to agriculture. Plant Soil 141:13-39

Smith KA, McTaggart IP, Tsuruta H (1997) Emissions of N_2O and NO associated with nitrogen fertilisation in intensive agriculture, and the potential for mitigation. Soil Use Manage 13:296-304

Spaink HP (1995) The molecular basis of infection and nodulation by Rhizobia: the ins and outs of sympathogenesis. Annu Rev Phytopathol 33:345-368

Vance CP (1996) Root-bacteria interactions: symbiotic nitrogen fixation. In: Waisel Y et al. (eds) Plant roots; the hidden half. Marcel Dekker, New York, pp 723-755

Van Elsas JD, Trevors JT, Wellington EMH (1997) Modern soil microbiology. Marcel Dekker,New York, 683 pp

Van Veen JA, Frissel MJ (1981) Simulation model of the behavior of nitrogen in soil. In: Frissel MJ, van Veen JA (eds) Simulation of nitrogen behavior in soil-plant systems. Pudoc, Wageningen, pp 126-144

Verberne EJL, Hassink J, de Willigen P, Groot JJR, van Veen JA (1990) Modelling organic matter dynamics in different soils. Neth J Agric Sci. 38:221-238

Vitousek PM (1994) Beyond the global warming: ecology and global change. Ecology 75:1861-1876

Rhizosphere and Mycorrhiza of Field Crops

JM. Barea[1]

The rhizosphere is the zone of influence of plant roots on the associated microbiota and soil constituents. The supply of photosynthates, as substrates to soil microbiota, is a key fact in rhizosphere formation. Developing root-soil interfaces create dynamic microenvironments where microorganisms, plant roots and soil components interact. Because some rhizosphere microorganisms, which include saprophytes and mutualistic symbionts, play fundamental roles to benefit plant growth and health and soil physicochemical properties microbial colonisation of root environments has a substantial impact on plant fitness and soil quality.

Mycorrhizal fungi and nitrogen (N_2)-fixing bacteria, mutualistic symbionts, are particularly important. Arbuscular mycorrhiza are is the mycorrhizal type formed by crop species of agronomic interest. The responsible mycorrhizal fungi, after the biotrophic colonization of root cortex, develop an external mycelium which is a bridge connecting the root with the surrounding soil microhabitats. Such mycorrhizal (fungal-root) symbiosis is critical for nutrient cycling in soil-plant systems. Mycorrhiza also improve plant health through increased protection against biotic and abiotic stresses. In cooperation with other soil organisms, the external mycorrhizal mycelium forms water-stable aggregates necessary for good soil tilth. Mycorrhiza establishment is known to change mineral nutrient composition, hormonal balance, C allocation patterns, and other aspects of plant physiology. Changes in the chemical composition of root exudates in mycorrhizal plants, in addition to physical modifications induced by the mycorrhizal soil mycelium on the environment surrounding the roots, affect both quantitatively and qualitatively the microbial populations in the rhizosphere and develop the so-called mycorrhizosphere. Biotechnological and molecular management of rhizosphere/mycorrhizosphere interactions are key factors for a sustainable plant productivity.

Introduction

Sustainability, either in natural ecosystems or in agroecosystems, is dependent on a biological balance in the soil, which is mainly governed by the activities of microbial com-

1. Departemento de Microbiología del Suelo y Sistemas Simbióticos, Estación Experimental del Zaidín, CSIC, Prof. Albareda 1, 18008 Granada, Spain

munities. Some of such activities can be managed as a natural resource tool (Bethlenfalvay and Linderman 1992; Barea and Jeffries 1995). Many of the soil-borne microbes are bound to the surface of soil particles or found in soil aggregates, while others interact specifically with the plant root system (Glick 1995); actually, a large number of microorganisms are living in the soil-plant interfaces to develop what is known as the rhizosphere (Lynch 1990; Azcón-Aguilar and Barea 1992; Linderman 1992).

Soil microbial dynamics largely govern ecosystem functioning (Kennedy and Smith 1995) through a number of activities, carried out mainly by rhizosphere microbiota constituents, which are known to enhance soil and plant quality (Bethlenfalvay and Schüepp 1994). The functions involved include improvement of plant establishment, increased availability of plant nutrients, enhacement of nutrient uptake, protection against cultural and environmental stresses, improvement of soil structure, etc. (Barea et al. 1997). Certainly, plant health and productivity depend on soil quality which, in turn, is dependent on the diversity and effectiveness of its microbiota (Bethlenfalvay and Schüepp 1994).

Rhizosphere

The rhizosphere was recognized as the zone of influence of plant roots on the associated microbiota and soil constituents. Key facts in rhizosphere formation and function are the supply of nutrients to plants, including those derived from microbial activities, and the supply of plant photosynthates as substrates for the root-associated microbiota. Developing root-soil interfaces create dynamic microenvironments where microorganisms, plant roots and soil components interact to carry out activities known to greatly influence plant health and soil quality (Lynch 1990).

The division or compartmentalization of the rhizosphere has been a controversial topic. As Kennedy (1998) suggests, a simple solution is to divide this area into rhizosphere, rhizoplane, and root itself. The rhizosphere is the zone of soil influenced by roots which release, mainly as root exudates, carbon-containing substrates that affect microbial activity. The rhizoplane is the surface of a plant root, but including strongly adhered soil particles. Certain microorganisms can colonize the area within the root to develop activities involved in plant growth promotion and plant protection (Kloepper 1994; Chanway 1996). These microbes, so-called endophytes, can be considered as colonizing the root itself rather than a portion of the rhizosphere.

The rhizophere is accepted to extend from 1 to 4-5 mm from the root surface and the rhizoplane with adhering soil can be up to 2 mm in width; thus, from a methodological point of view, some overlapping can be found to distinguish between the rhizoplane and the rhizosphere (Kennedy 1998). Microbial colonization of the rhizoplane and/or the root tissues, which is known as root colonization, while the colonization of the adjacent volume of soil under root influence is known as rhizosphere colonization (Kloepper et al. 1991).

Exudates and other forms of carbon compounds are being supplied by the plant to microbial populations, as either signals or growth substrates (Werner 1998), thereby microorganisms are stimulated to grow around plant roots. Figures 1 and 2 illustrate the development and function of rhizosphere and the key concepts related to the rhizosphere effect. Two main groups of microorganisms can be distinguished: saprophytes and symbionts. Both of them comprise detrimental, neutral, and beneficial bacteria and fungi. Detrimental microbes include major plant pathogens, and minor parasitizing and nonparasitizing, deleterous rhizosphere organisms, either bacteria or fungi, (Weller and Thomashow 1994; Nehl et al. 1996). Beneficial microorganisms are known to play fundamental roles in agroecosistem and natural ecosystem sustainability, and some of them can be used as inoculants to benefit plant growth and health (Barea et al. 1997). A key concept, mainly referred to rhizosphere bacteria, concerns a subset of the total rhizosphere bacterial community, which is termed rhizobacteria (Kloepper 1994, 1996) known to display a rather specific ability for root colonization. The beneficial root colonist rhizosphere bacteria, the so-called plant growth-promoting rhizobacteria (PGPR), carry out many important ecosystem processes, such as those involved in the biological control of plant pathogens, nutrient cycling and/or seedling establishment (Kloepper et al. 1991; Lugtenberg et al. 1991; Haas et al. 1991; O'Gara et al. 1994; Weller and Thomashow, 1994; Schippers et al. 1995; Glick, 1995; Broek and Vanderleyden 1995; Bashan and Holguin 1998). Recently, Bashan and Holguin (1998) proposed to divide PGPR into two groups to specify if they act as biocontrol agents or not. Mycorrhizal fungi and nitrogen (N_2)-fixing bacteria are relevant members of mutualistic symbionts. Figure 3 summarizes the main types and activities of beneficial rhizosphere microorganisms.

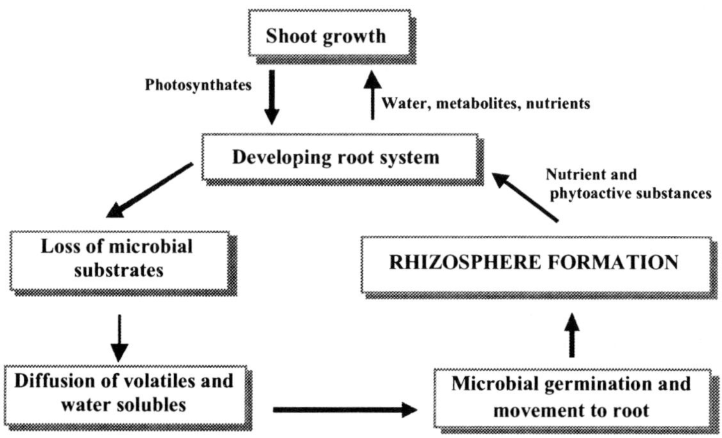

Fig. 1: Rhizosphere formation and functions

Mycorrhiza

The roots of most plant species live associated in symbiosis with certain soil fungi by establishing what are known as mycorrhiza. The mycorrhizal fungi, after the biotrophic colonization of root cortex, develop an external mycelium which is a bridge connecting the root with the sorrounding soil microhabitats. Such mycorrhizal (fungal-root) symbiosis is critical in nutrient cycling in soil-plant systems. In cooperation with other soil organisms, the external mycorrhizal mycelium forms water-stable aggregates necessary for good soil tilth. Mycorrhizas also improve plant health through increased protection against biotic and abiotic stresses, thus with possible applications in biocontrol and in bioremediation (Bethlenfalvay and Linderman 1992; Barea and Jeffries 1995; Jeffries and Barea 1999).

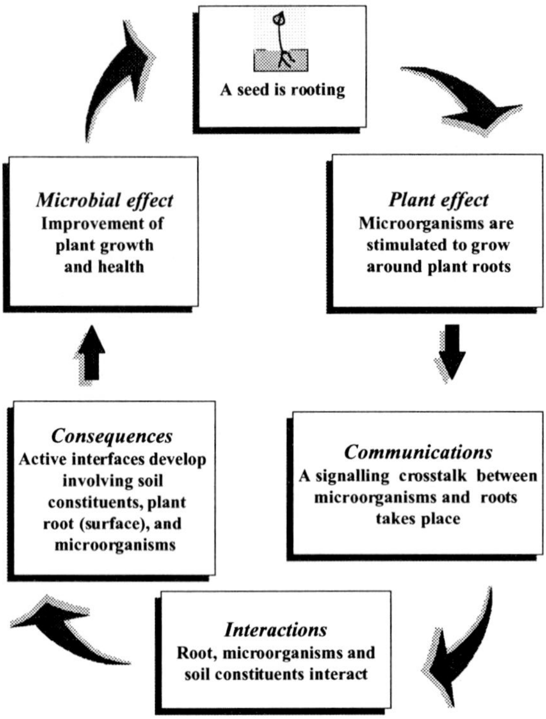

Fig. 2: The rhizosphere effect

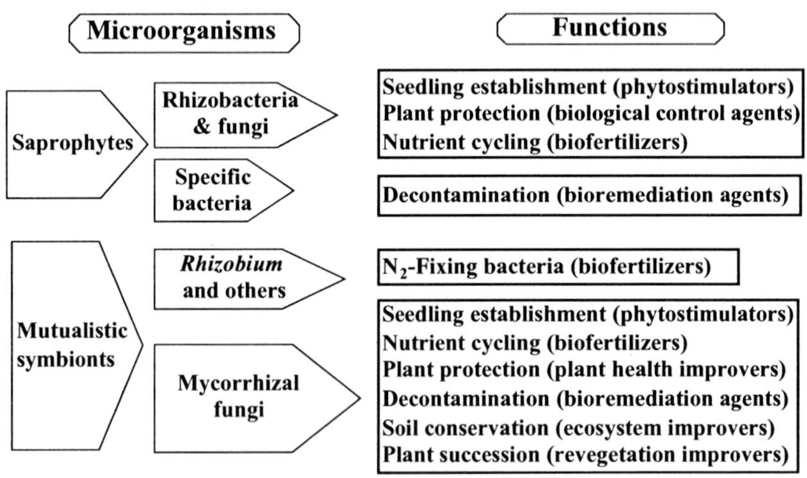

Fig. 3: Beneficial microrganisms in the soil-plant systems

The mycorrhizal associations are usually considered mutualistic symbioses because of the highly interdependent and mutually beneficial relationship established between both partners, where the host plant receives mineral nutrients via fungal mycelium (mycotrophism), while the heterotrophic fungus obtains carbon compounds from the host's photosynthesis (Harley and Smith 1983; Azcón-Aguilar and Bago 1994).

Mycorrhiza can be found in almost any kind of soil. All but a few vascular plant species (those belonging mainly to the families Cruciferae, Chenopodiaceae, Cyperaceae, Caryophyllaceae, and Juncaceae) are able to form mycorrhiza. The universality of this symbiosis implies a great diversity in the taxonomic features of the fungi and the plants involved. There are, in fact, considerable differences in the morphology of the mycorrhizal types, and this is reflected in the resulting physiological relationships (Azcón-Aguilar and Bago 1994; Smith and Read 1997). Five types of mycorrhiza can be recognized, whose structural and functional features have been detailed in several publications (Barea 1991; Azcón-Aguilar and Barea 1997; Smith and Read 1997); thus only a brief consideration will be given here to differentiate these groups.

About 3% of the higher plants, mainly forest trees in the Fagaceae, Betulaceae, Pinaceae, Eucalyptus, and some woody legumes, form ectomycorrhiza. The fungi involved are mostly higher Basidiomycetes and Ascomycetes which colonize the cortical root tissues. A lack of intracellular penetrations is characteristic. In general, the fungus develops a sheath or mantle around the feeder roots. Three other types of mycorrhiza can be grouped as endomycorrhiza, in which the fungus can colonize the root cortex intracellularly. One of these is restricted to some species in the Ericaceae (ericoid mycorrhiza), the second to the Orchidaceae (orchid mycorrhiza), and the third, the arbuscular mycorrhizas, which is by far the most widespread type. There is a fifth group, the ectendomycorrhiza, composed of plant species in families other than Ericaceae but in the Ericales, and in the Monotropaceae. They have sheaths and produce intracellular penetrations (arbutoid and monotropoid mycorrhiza).

Most of the major plant families are able to form mycorrhiza naturally, the arbuscular mycorrhizal (AM) associations being the commonest mycorrhizal type involved in agricultural systems (Barea et al. 1993). Since AM symbiosis can benefit plant growth and health, there is an increasing interest in ascertaining their effectiveness in particular plant production systems and, consequently, in manipulating them, when feasible, so that they can be incorporated into production practices. Evidence is accumulating to show that indigenous and/or introduced AM fungi can benefit either field-sown and plantation crops, or transplantable horticultural crops. A number of diverse plant crops of economic interest are related to AM activity. These include annual crops, such as cereals and legumes, in which the natural endophytes normally present in arable soils can operate. In these cases, however, the biotechnological approaches required for the appropriate management of AM potential are rather complex. For other plant production systems, however, the application of mycorrhizal biotechnology is feasible. These include vegetable crops, temperate fruit trees or shrubs, tropical plantation crops, ornamentals, spices, etc.; in summary, horticultural (including fruitculture) plant crops.

Many high-value ornamental and edible horticultural crops form arbuscular mycorrhiza. The plant species involved usually display a considerable degree of mycotrophy and their optimal development is, therefore, dependent on early AM symbiosis establishment. This is particularly relevant when the seedlings are produced in potting mixtures or in disinfected nursery beds, where AM propagules are present in low numbers, or even absent, such as in micropropagation of plant material (Lovato et al. 1995, 1996).

During the process of mycorrhiza formation, in which the plant "accepts" the fungal colonization without any significant rejection reaction, a series of root-fungus interactions give way to the integration of both organisms. The establishment of the symbiosis must be the result of a continuous molecular "dialogue" between plant and fungus, as exerted through the exchange of both recognition and acceptance signals. The result of this dialogue will finally depend on the genome expression of both partners (Gianinazzi-Pearson et al. 1996).

Both the fungus and the AM symbiosis are distributed worldwide. The fungi belong to the class Zigomicotina, order Glomales (Rosendahl et al. 1994; Morton et al. 1995; Walker 1995). About 150 species in the only six genera which are able to form AM associations (Acaulospora, Entrophospora, Gigaspora, Glomus, Sclerocystis and Scutellospora) have been systematized; none of these fungi has yet been successfully cultured axenically (Azcón-Aguilar et al. 1998). Molecular biology approaches are now being developed to explore the biodiversity and characterization of AM fungi (Redecker et al. 1997; van Tuinen et al. 1998).

As Fig. 3 summarizes, the AM symbiosis influences several aspects of plant physiology, such as mineral nutrition, plant development, and plant protection. The primary effect of AM symbiosis is to increase the supply of mineral nutrients to the plant, particularly those whose ionic forms have a poor mobility rate, or those which are present in low concentration in the soil solution. This mainly concerns phosphate, ammonium, zinc and copper (Barea 1991). It has also been recently recognized that AM colonization affects a wide range of morphological parameters in developing root systems (Atkinson et al. 1994; Berta et al. 1995), a greater root branching being the most commonly described effect. Another activity of arbuscular mycorrhizas concerns plant protection from biotic stress. However, it is important to point out that the enhancement of root resistence/tolerance to pathogen attack in AM plants is not exerted with the same effectivity by all AM fungi, is not applicable to all pathogens, and is not expressed in all substrata or in all environmental conditions. Nevertheless, there are examples demonstrating that prior colonization of selected AMF protect plants against pathogenic fungi, such as *Phytophthora, Gaeumanomyces, Fusarium, Thielaviopsis, Pythium, Rhizoctonia, Sclerotium, Verticillium*, Aphanomyces, etc., or nematodes such as *Rotylenchus, Pratylenchus, Meloidogyne*, etc. (Linderman 1994; Azcón-Aguilar and Barea 1996) The mechanisms that have been suggested to explain the protective action of AM symbiosis include the improvement of plant nutrition, damage compensation, competition for photosynthates or for colonization/infection sites, induction of changes in the morphology/anatomy of the root system, induction of changes in mycorrhizosphere populations and activation of plant defence mechanisms.

Mycorrhizal fungi also enable plants to cope with abiotic stress by means of alleviating nutrient deficiencies, improving drought tolerance, overcoming the detrimental effects of salinity, enhancing tolerance to pollution and improving the adaptation of sterile micropropagated plants to unsterile substrata and to field conditions (Barea et al. 1993; Azcón-Aguilar and Barea 1997).

Mycorrhizosphere and Other Rhizosphere Interactions Involving Mycorrhiza

Mycorrhiza establishment is known to change the mineral nutrient composition, hormonal balance, C allocation patterns, and other aspects of plant physiology (Azcón-Aguilar and Bago 1994; Smith et al. 1994). Besides, the nutrient transport attributes of mycorrhizal fungi, when active, and the release of materials, when senescing, represent a supply of nutrients in plant rhizosphere (Harley and Smith 1983). Thus, the AM symbiotic status affects the chemical composition of root exudates while the development of a mycorrhizal soil mycelium introduces physical modifications into the environment surrounding the roots. These changes affect both quantitatively and qualitatively the microbial populations in either the rhizosphere or the rhizoplane (Azcón-Aguilar and Barea 1992; Linderman 1992) and develop the so-called mycorrhizosphere (Linderman 1992). The terms hyphosphere or mycosphere are also used. A typical function of the AM soil mycelium is to serve as a carbon source to microbial communities, even outside the limit of the proper rhizosphere, and it results in an important contribution of the mycorrhizosphere to improve soil quality (Bethlenfalvay and Schüepp 1994). Figure 4 summarizes the interactions and activities in the mycorrhizosphere microhabitats.

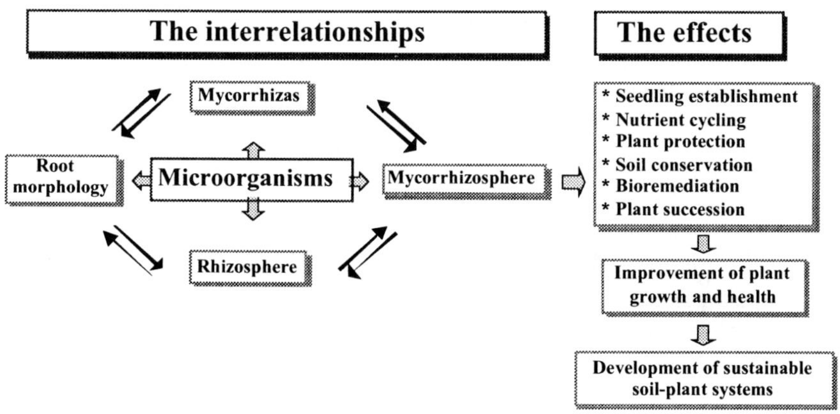

Fig. 4: Micorrhizosphere development and functions

Specific types of organisms are able to establish relationships with mycorrhizal fungi which affect plant growth and health, and also soil quality. These interactions are recorded in Fig. 5, as summarized from my own recent publications (Barea et al. 1996, 1997, 1998; Barea 1997; Requena et al. 1997; Toro et al. 1997, 1998).

Manipulating Rhizosphere/Mycorrhizosphere Interactions

Current public concerns about the side-effect of agrochemicals are attracting interest to research areas such as those involving biological balance in soil, microbial diversity, microbial dynamics in soil, and rhizosphere interactions (Altieri 1994). In fact, an increasing demand for low-input agriculture has resulted in greater interest in the manipulation and use of beneficial soil microorganims that increase soil fertility or improve plant nutrition and health. It is obvious that reducing the use of chemicals and energy in agriculture will lead to a more economical and sustainable production while minimizing pollution (Jeffries and Barea 1999).

Type of microorganism	Results of the interaction
N_2-fixing bacteria (*Rhizobium, Frankia..*)	N_2 fixation, N-cycling, N "transfer"
Phosphate solubilizers (*Bacillus, Pseudomonas..*)	P cycling, use of rock and organic phosphates as an alternative P source
Plant hormone producers (*Azospirillum*, PGPR..)	Rooting and establisment of seedlings
Agents of biological control of plant diseases (*Trichoderma, Gliocadium, Pseudomonas, Bacillus..*)	Increased resistence/ tolerance to microorganisms causing root diseases
Microorganisms related to formation of stable aggregates	Improvement of soil quality
Microfauna components	Fragmentary information

Fig. 5: Interactions of arbuscular mycorrhizas with beneficial soil microorganisms

Consequently, some biotechnological inputs have been proposed concerning rhizosphere technology. These include aspects such as the use of microbial inoculants and of transgenic plants (rhizospheres). With regard to microbial inoculants, the key issues refer to: inoculum technology, inoculum registration, bioethical and risk assessment considerations (particularly for inocula based on genetically modified microorganisms), quality control, technology transfer, commercialization, etc. Concerning inoculant technology, topics that must be considered include microbial selection and characterization, competitiveness, inocula production (substrates) and formulation, delivery (field release), and inoculation rate, the establishment of inoculated microorganism, and the fate and behaviour of inoculants after application (monitoring population dynamics, effect on target microorganisms and microbial processes, effect on plant growth and health, etc.), and others (Barea and Olivares 1998).

Mycorrhizal technology deserves some particular comments. As obligate symbionts, the use of AM fungi in plant biotechnology differs from that of other beneficial soil microorganisms. Thus, specific procedures are required to multiply AM fungi and to produce high-quality inocula. These are being developed in cooperation with companies involved in the R&D activities, through the COST Action 8.38. This is a pan-European network that has been established to study managing arbuscular mycorrhizal for improving soil quality and plant health in agriculture. A working group devoted to mycorrhizal technology has been created within the framework of such a COST Action, where research efforts are being addressed to:

1. Inoculum technology. This mainly concerns the development of appropriate inoculum products compatible with those used for other beneficial soil microorganisms, particularly the production of second-generation inocula, obtained by mixing AM fungi with other microbial inocula.
2. Inoculum registration. The working group will provide the guidelines for inoculum registration and use, considering ecological, biosafety, and bioethic concerns that fit specific European standards.

3. Quality control. Specific protocols for quality control of AM fungal inoculum will be developed and standarized for application.
4. Technology transfer. Scaling up of production and use of AM fungal inoculum to made it economically feasible for SME, by means of concerted R & D field experiments.
5. Case studies application. These will be carried out in the areas of horticulture, fruit production, revegetation of desertified ecosystems, and in bioremediation programmes for contaminated, in general degraded, areas.

Acknowledgments. This study was supported by the EC Biotechnology Programme, IMPACT Project BIO4 CT96-0027 and the TMR Network ERB FMR XCT 960039.

References

Altieri MA (1994) Sustainable agriculture. Encycl Agricult Sci 4:239-247

Atkinson S, Berta G, Hooker JE (1994) Impact of mycorrhizal colonisation on root architecture, root longevity and the formation of growth regulators. In: Gianinazzi S, Schüepp H (eds). Impact of arbuscular mycorrhizas on sustainable agriculture and natural ecosystems. ALS, Birkhäuser, Basel, pp 89-99

Azcón-Aguilar C, Bago B (1994) Physiological characteristics of the host plant promoting an undisturbed functioning of the mycorrhizal symbiosis. In: Gianinazzi S, Schüepp H (eds). Impact of arbuscular mycorrhizas on sustainable agriculture and natural ecosystems. ALS, Birkhäuser, Basel, pp 47-60

Azcón-Aguilar C, Barea JM (1992) Interactions between mycorrhizal fungi and other rhizosphere microorganisms. In: Allen MJ (eds). Mycorrhizal functioning. An integrative plant-fungal process. Routledge, Chapman & Hall, New York, pp 163-198

Azcón-Aguilar C, Barea JM (1996) Arbuscular mycorrhizas and biological control of soil-borne plant pathogens. An overview of the mechanisms involved. Mycorrhiza 6:457-464

Azcón-Aguilar C, Barea JM (1997) Applying mycorrhiza biotechnology to horticulture: significance and potentials. Sci Hortic 68:1-24

Azcón-Aguilar C, Bago B, Barea JM (1998) Saprophytic growth of arbuscular-mycorrhizal fungi. In: Hock B, Varma A (eds). Mycorrhiza: structure, function, molecular biology and biotechnology. Springer, Berlin, Heidelberg, New-York, pp 391-408

Barea JM (1991) Vesicular-arbuscular mycorrhizae as modifiers of soil fertility. In: Stewart BA (ed). Advances in soil science. Springer, Berlin, Heidelberg, New York, pp 1-40

Barea JM (1997) Mycorrhiza/bacteria interactions on plant growth promotion. In: Ogoshi A, Kobayashi L, Homma Y, Kodama F, Kondon N, Akino S (eds). Plant growth-promoting rhizobacteria, present status and future prospects. OCDE, Paris, pp 150-158

Barea JM, Jeffries P (1995) Arbuscular mycorrhizas in sustainable soil plant systems. In: Hock B, Varma A (eds). Mycorrhiza: structure function, molecular biology and biotechnology. Springer Berlin, Heidelberg, New York, pp 521-559

Barea JM, Olivares J (1998) Manejo de las propiedades biológicas del suelo. In: Jiménez Díaz R, Lamo de Espinosa R (eds). Agricultura sostenible. Editorial Mundi Prensa, Madrid, pp. 173-193

Barea JM, Azcón R, Azcón-Aguilar C (1993) Mycorrhiza and crops. In: Tommerup I (ed). Advances in plant pathology, vol 9. Mycorrhiza: A Synthesis. Academic Press, London, pp 167-189

Barea JM, Tobar RM, Azcón-Aguilar C (1996) Effect of a genetically-modified *Rhizobium meliloti* inoculant on the development of arbuscular mycorrhizas, root morphology, nutrient uptake and biomass accumulation in *Medicago sativa* L. New Phytol 134: 361-369

Barea JM, Azcón-Aguilar C, Azcón R (1997) Interactions between mycorrhizal fungi and rhizosphere microorganisms within the context of sustainable soil-plant systems. In: Gange AC, Brown VK (eds). Multitrophic interactions in terrestrial systems. Blackwell, Oxford, pp. 65-77

Barea JM, Andrade G, Bianciotto V, Dowling D, Lohrke S, Bonfante P, O'Gara F, Azcón-Aguilar C (1998) Impact on arbuscular mycorrhiza formation of *Pseudomonas* strains used as inoculants for the biocontrol of soil-borne plant fungal pathogens. Appl Environ Microbiol 64:2304-2307

Bashan Y, Holguin G (1998) Proposal for the division of plant growth-promoting rhizobacteria into two classifications: biocontrol-PGPB (plant growth-promoting bacteria) and PGPB. Soil Biol Biochem 30:1225-1228

Berta G, Trotta A, Fusconi A, Hooker JE, Munro M, Arkinson D, Giovannetti M, Morini S, Fortuna P, Tisserant B, Gianinazzi-Pearson V, Gianinazzi S (1995) Arbuscular mycorrhizal induced changes to plant growth and root system morphology in Prunus cerasifera. Tree Physiol 15:281-293

Bethlenfalvay GJ, Linderman RG (1992) Mycorrhizae in sustainable agriculture. ASA Spec publ 54, Madison, Wisconsin, 124 pp

Bethlenfalvay GJ, Schüepp H (1994) Arbuscular mycorrhizas and agrosystem stability. In: Gianinazzi S, Schüepp H (eds). Impact of arbuscular mycorrhizas on sustainable agriculture and natural ecosystems. Birkhäuser, Basel, pp 117-131

Broek AV, Vanderleyden J (1995) Genetics of the *Azospirillum*-plant root association. Crit Rev Plant Sci 14:445-466

Chanway CP (1996) Endophytes: they're not just fungi! Can J Bot 74:321-322

Gianinazzi-Pearson V, Dumas-Gaudot E, Gollotte A, Tahirialaoui A, Gianinazzi S (1996) Cellular and molecular defence-related roots responses to invasion by arbuscular mycorrhizal fungi. New Phytol 133:45-57

Glick BR (1995) The enhancement of plant growth by free-living bacteria. Can J Microbiol 41:109-117

Haas D, Keel C, Laville J, Maurhofer M, Oberliansli T, Schnider U, Voisard C, Wüthrich B, Defago G (1991) Secondary metabolites of *Pseudomonas fluorescens* strain CHA0 involved in the suppresion of root diseases. In: Hennecke H, Verma DPS (eds). Advances in molecular genetics of plant-microbe interactions. Kluwer, Dordrecht, pp 450-456

Harley JL, Smith SE (1983) Mycorrhizal symbiosis. Academic Press, New York, 483 pp

Jeffries P, Barea JM (1999) Arbuscular mycorrhiza a key component of sustainable plant-soil ecosystem. The mycota (in press)

Kennedy AC (1998) The rhizosphere and Spermosphere. In: Sylvia DM, Fuhrmann JJ, Hartel PG, Zuberer DA (eds). Principles and applications of soil microbiology. Prentice Hall, Upper Saddle River, New Jersey, pp 389-407

Kennedy AC, Smith KL (1995) Soil microbial diversity and the sustainability of agricultural soils. Plant Soil 170:75-86

Kloepper JW (1994) Plant growth-promoting rhizobacteria (other systems) In: Okon Y (ed). *Azospirillum*/plant associations. CRC Press, Boca Raton, Florida, pp 111-118

Kloepper JW (1996) Host specificity in microbe-microbe interactions. BioScience 46:406-409

Kloepper JW, Zablotowick RM, Tipping EM, Lifshitz R (1991) Plant growth promotion mediated by bacterial rhizosphere colonizers. In: Keister DL, Cregan PB (eds). The rhizosphere and plant growth. Kluwer Dordrecht, pp. 315-326

Linderman RG (1992) Vesicular-arbuscular mycorrhizae and soil microbial interactions. In: Bethlenfalvay GJ, Linderman RG (eds). Mycorrhizae in sustainable agriculture. ASA Spec Publ, Madison, Wisconsin, pp. 45-70

Linderman RG (1994) Role of VAM fungi in biocontrol. In: Pfleger FL, Linderman RG (eds). Mycorrhizae and plant health. APS Press, St Paul, pp. 1-26

Lovato PE, Schüepp H, Trouvelot A, Gianinazzi S (1995) Application of arbuscular mycorrhizal fungi (AMF) in orchard and ornamental plants. In: Varma A, Hock B (eds). Mycorrhiza structure, function, molecular biology and biotechnology. Springer Berlin, Heidelberg, New York, pp 521-559

Lovato PE, Gianinazzi-Pearson V, Trouvelot A, Gianinazzi S (1996). The state of art of mycorrhizas and micropropagation. Adv Hortic Sci 10:46-52

Lugtenberg BJJ, Weger de LA, Bennett JW (1991) Microbial stimulation of plant growth and protection from disease. Curr Opin Microbiol 2:457-464

Lynch JM (1990). The rhizosphere. John Wiley, New York, 462 pp

Morton JB, Franke M, Bentivenga SP (1995) Developmental foundations for morphological diversity among endomycorrhizal fungi in Glomales. In: Varma A, Hock B (eds). Mycorrhiza structure, function, molecular biology and biotechnology. Springer Berlin, Heidelberg, New York, pp 669-683

Nehl DB, Allen SJ, Brown JF (1996) Deleterious rhizosphere bacteria: an integrating perspective. Appl Soil Ecol 5:1-20

O'Gara F, Dowling DN, Boesten B (1994) Molecular ecology of rhizosphere microorganisms. VCH, Weinheim, Germany, 173 pp

Redecker D, Thierfelder H, Walker C, Werner D (1997) Restriction analysis of PCR-amplified internal transcribed spacers of ribosomal DNA as a tool for species identification in different genera of the order Glomales. Appl Environ Microbiol 63:1756-1761

Requena N, Jimenez I, Toro M, Barea JM (1997) Interactions between plant-growth- promoting rhizobacteria (PGPR), arbuscular mycorrhizal fungi and Rhizobium spp. in the rhizosphere of *Anthyllis cytisoides*, a model legume for revegetation in Mediterranean semi-arid ecosystems. New Phytol 136:667-677

Rosendahl S, Dodd JC, Walker C (1994) Taxonomy and phylogeny of the Glomales. In: Gianinazzi S, Schüepp H (eds). Impact of arbuscular mycorrhizas on sustainable agriculture and natural ecosystems. ALS, Birkhäuser, Basel, pp 1-12

Schippers B, Scheffer RJ, Lugtenberg BJJ, Weisbeek PJ (1995) Biocoating of seeds with plant growth-promoting rhizobacteria to improve plant establishment. Outlook Agric 24:179-185

Smith SE, Read DJ (1997) Mycorrhizal symbiosis. Academic Press, San Diego, 605 pp

Smith SE, Gianinazzi-Pearson V, Koide R, Cairney JWG (1994) Nutrient transport in mycorrhizas: structure, physiology and consequences for efficiency of the symbiosis. In: Robson AD, Abbott LK, Malajczuk N (eds). Management of mycorrhizas in agriculture, horticulture and forestry. Kluwer, Dordrecht, pp 103-113

Toro M, Azcón R, Barea JM (1997) Improvement of arbuscular mycorrhizal development by inoculation with phosphate-solubilizing rhizobacteria to improve rock phosphate bioavailability (32P) and nutrient cycling. Appl Environ Microbiol 63:4408-4412

Toro M, Azcón R, Barea JM (1998) The use of isotopic dilution techniques to evaluate the interactive effects of *Rhizobium* genotype, mycorrhizal fungi, phosphate-solubizing rhizobacteria and rock phosphate on nitrogen and phosphorus acquisition by Medicago sativa. New Phytol 138:265-273

Tuinen Van D, Jacquot E, Zhao B, Gollotte A, Gianinazzi-Pearson V (1998) Characterization of root colonization profiles by a microcosm community of arbuscular mycorrhizal fungi using 25S rDNA-targeted nested PCR. Mol Ecol 7:879-887

Walker C (1995) AM or VAM: What's in a word? In: Varma A, Hock B (eds). Mycorrhiza structure function, molecular biology and biotechnology. Springer Berlin, Heidelberg, New York, pp 25-26

Weller DM, Thomashow LS (1994) Current challanges in introducing beneficial microorganisms into the rhizosphere. In: O'Gara F, Dowling DN, Boesten B (eds). Molecular ecology of rhizosphere microorganisms biotechnology and the release of GMOs. VCH, Weinheim, Germany, pp 1-18

Werner D (1998) Organic signals between plants and microorganisms. In: Pinton R, Varanini Z, Nannipieri P (eds). The rhizosphere: biochemistry and organic substances at the soil-plant interfaces. Marcel Dekker, New York (in press)

Coastal Eutrophication and Agriculture: Contributions and Solutions

DF. Boesch[1], RB. Brinsfield[2]

Many coastal waters of developed nations have experienced widespread and rapid eutrophication (the increase in supply of organic matter) during the last half of the 20th century. This has resulted in increased phytoplankton production, decreased water clarity, often-severe depletion of dissolved oxygen in bottom waters, loss of seagrasses and, in some cases, declines or changes in the quality of fisheries production. Large-scale changes resulting from eutrophication have been documented for continental shelf waters in the Gulf of Mexico, Mediterranean, Black and North Seas, relatively confined seas such as the Baltic and Seto Inland Sea, large bays such as the Chesapeake Bay and Long Island Sound and numerous smaller estuaries and lagoons. These trends are closely tied to the increased use of chemical fertilizers in agriculture, human population growth, and increasing atmospheric deposition of nitrogen resulting from fossil fuel combustion. Although atmospheric and human waste sources are significant in some heavily populated areas, agricultural inputs of phosphorus and nitrogen are the largest source of nutrients driving the increased production of organic matter in most extensively affected areas, including coastal waters receiving drainage from large river basins with extensive agriculture (e.g., Mississippi, Po, and Danube Rivers).

Agricultural inputs of nutrients are driven not only by applications of chemical fertilizers, but also by animal wastes, irrigation, drainage, and the conversion of wetlands and riparian zones (important sinks for nutrients) for agricultural land uses. More efficient agronomic practices, use of crop rotation and cover crops, and avoiding the over-application of manure can result in reductions in nutrient losses by 20 to 30%. Reconfiguration of agricultural landscapes through reconstruction of strategically placed wetlands, riparian forests and flood plains can trap a similar fraction of the remaining nutrient losses, such that total reductions of 50% may be feasible without devastating economic impacts and with numerous local benefits to environmental quality. Efforts to restore large coastal ecosystems such as the Baltic Sea,

1. University Center for Environmental Science, P.O. Box 775, Cambridge, Maryland 21613, USA
2. Wye Research and Education Center, College of Agriculture and Natural Resources, University of Maryland, College Park, Queenstown, Maryland 21658, USA

northern Gulf of Mexico, and Chesapeake Bay through commitments to reduce nutrient loading have been underway or are beginning. They represent substantial challenges in working across political jurisdictions and across scientific disciplines and internalizing the external environmental costs of food production.

Introduction

Eutrophication – the increasing supply of organic matter (Nixon 1985) – is one of the most serious threats to coastal marine ecosystems around the world (GESAMP 1990; NRC 1994). However, the most substantial and widespread eutrophication has resulted from increased delivery of plant nutrients, not organic wastes, to coastal waters as a result of human activities. This differs from the typical lay person's notion of pollution, namely the direct discharge of a toxic chemical. On one hand, inputs of nitrogen, phosphorus, and other nutrients are essential to their high productivity. On the other hand, oversupply by nutrients may cause excessive algal growth. Such over enrichment results in diminished light penetration and shading of bottom-dwelling plants, harmful or noxious blooms of algae, altered food chains, and depletion of the dissolved oxygen as the organic matter decomposes.

Agriculture has played an important role in the delivery of nutrients to coastal waters since the land was first cleared and cultivated. Virtually any land-use activity adds nutrient; changes the capacity of the native plants, microbes, and soils to retain nutrients; and increases water and sediment runoff. However, the increased use of chemical fertilizers and nitrogen-fixing crops, the expansion of areas in cultivation, higher yields, and intensification of animal production have greatly increased nutrient loss from agricultural activities into streams, rivers, and eventually the sea.

Coastal eutrophication has grown to such proportions and severity that coastal constituencies and governments are demanding control on the sources of nutrient inputs, not only within the coastal zone but far upstream. Various national and international agencies are aggressively promoting voluntary practices for reduction of nonpoint sources, including those in agriculture, and, in some cases, implementing mandatory requirements. At the same time, agriculture is facing the challenge of feeding a growing world population, the economic pressures of new global markets, and competition with growing cities for productive land.

This chapter provides an overview perspective of the current state of knowledge regarding the causes and consequences of coastal eutrophication, the role of agriculture, and the new approaches that could be taken to reduce nutrient losses from agriculture. We also discuss efforts underway to control nutrient inputs from large, multijurisdictional watersheds and the institutional and scientific challenges they present.

Coastal Eutrophication: A Late 20th Century Phenomenon

Cultural Eutrophication

The environmental effects on coastal ecosystems of the addition of large quantities of organic matter from sewage or industrial discharges such as pulp plants have been recognized and studied for some time. Such concentrated inputs of organic matter can result in depletion of dissolved oxygen due to the biochemical oxygen demand of the wastes and often cause a buildup of organic matter in sediments, dramatically affecting bottom-dwelling (benthic) organisms (Pearson and Rosenberg 1978). While these effects may be severe and require long-term recovery, they are generally of limited spatial extent.

While the effects of inorganic nutrients on eutrophication of lakes have been recognized and studied for some time, recognition of the extent of coastal eutrophication from land-based nutrient inputs and research on the subject is relatively recent, with rapid growth in the numbers of papers published only since 1985 (Nixon 1995). There are three explanations for this. First, the confined nature of lakes and low turnover of their water volume makes them particularly susceptible to nutrient eutrophication (Smith 1998). Second, the loading of nutrients to the coastal zone has dramatically increased since the 1950s and 1960s (Howarth et al. 1996). Finally, scientists have only recently developed the tools to distinguish the effects of eutrophication in highly variable coastal ecosystems and to reconstruct historical changes.

The effects of overenrichment by nutrients are more insidious than those due to organic discharges. They are not concentrated at the point of input, but are spread over broader areas as nutrients stimulate the production of particulate organic material that is subsequently transported and mineralized. Moreover, an atom of carbon contained in organic matter is practically lost from the system once it is oxidized, while an atom of nitrogen or phosphorus may be recycled several times during residence in the coastal ecosystem, each time facilitating the production of more organic carbon (Ryther and Dunstan 1971).

The delivery of nitrogen and phosphorus to coastal waters has increased multifold due to human activities during the 20th century. In creating N and P budgets for the North Atlantic Ocean, Howarth et al. (1996) estimated that nitrogen fluxes from the continents had increased from 2 to 20 times from the preindustrial period. The increase in N and P delivery to coastal waters can be attributed to the growth, concentration, and consumption of human populations and the dramatic increase in the use of chemical fertilizers and combustion of fossil fuels during the last half of the century. Fossil fuel combustion results in the release of oxides of nitrogen to the atmosphere, which return to the Earth's surface in acidic wet deposition or adsorbed onto particles. Agricultural and atmospheric deposition sources vary among regions of the North Atlantic basin as a result of differences in population density, fuel consumption, and agricultural activity.

The extent and consequences of eutrophication in the developing world are less well understood. Coastal ecosystems in these regions have been as extensively studied as those in developed regions. Nutrient limitation and the effects of eutrophication in tropical waters are, in particular, poorly known. Nutrient inputs per capita are lower in the developing world because of the lower consumption of fertilizers and fuels. However, as population growth takes place primarily in the developing world, urbanization intensifies, and consumption of fertilizers and fuels grows to support demands of development, serious risks are posed for the worsening eutrophication outside of North America, Europe, and Japan (Nixon 1995).

Although phosphorus and other micronutrients may stimulate algal production in coastal waters, particularly in brackish waters, nitrogen is the limiting nutrient for most coastal marine ecosystems. Algal production in marine ecosystems has been shown to be nitrogen-limited because nitrogen is continuously lost to the atmosphere due to denitrification, while phosphorus is relatively conserved. Moreover, nitrogen sources are more diverse and rapidly growing, and are typically harder to control. Coastal eutrophication is but one outcome of what Vitousek et al. (1997) have described as a global nitrogen problem that also affects air quality, buildup of a green house gas (N_2O), acidic deposition, long-term soil fertility, and terrestrial plant biodiversity. This "global change" has happened at a much faster rate than the buildup of CO_2 in the atmosphere. Vitousek et al. estimated that on a global basis inputs of fixed nitrogen into the biosphere are presently at least equal to the amount of natural nitrogen fixation. The amount of nitrogen industrially fixed has doubled in just 10 years, while it took over 150 years to witness the same rate of change in CO_2 release.

Coastal Eutrophication in the Developed World

As the indicators of eutrophication of coastal waters have become better known, investigations and assessments have shown that a large proportion of the bays, lagoons, and estuaries in North America, Europe, and Japan are suffering some impairment from nutrient overenrichment (Figure 1). For example, in a recent survey of available data and expert knowledge, Bricker et al. (1999) found high levels of eutrophic conditions in 48 of the 139 estuaries in the United States and moderate problems exhibited by an additional 36 estuaries. Estuaries, lagoons, and harbors show similar signs of eutrophication in northwestern Europe (Borum 1996), the Mediterranean Sea (Duarte 1995), Japan (Goda 1991; Nakata 1995) and Australia (Shepard et al. 1989).

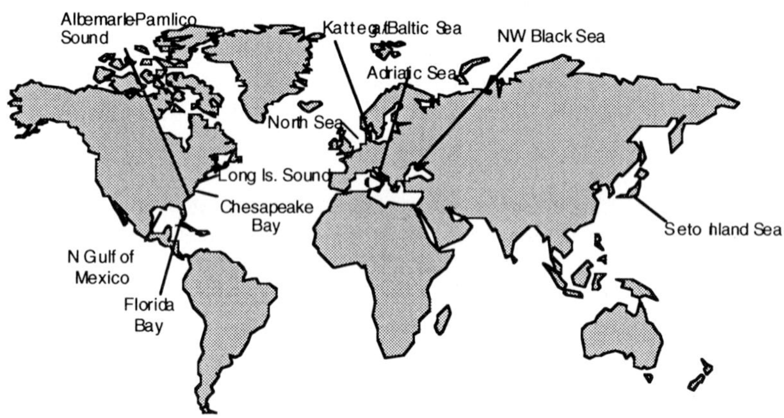

Fig. 1: Location of the large coastal ecosystems experiencing eutrophication that are discussed in this chapter

Perhaps more surprising is the documentation over the past 20 years of nutrient enrichment from human activities in large semienclosed seas and open continental shelf waters, where large volumes or active exchange of water provide greater dilution of nutrient inputs. Among other effects of this large-scale enrichment is the occurrence of extensive hypoxia – dissolved oxygen conditions too low to support active marine animals, including most fish and crustaceans, typically $< 2 mg \cdot l^{-1}$ – in bottom waters. These regions of seasonal hypoxia in some cases cover thousands of square kilometers, with the most extensive hypoxia found in the Baltic, Black and Adriatic Seas and the northern Gulf of Mexico (Diaz and Rosenberg 1995). These regions receive runoff from large rivers draining the continents and have modest tides for mixing waters. As a result, the water masses are often seasonally stratified, restricting the resupply of oxygen to cooler and saltier bottom waters sufficient to meet the demands of decomposition of organic matter from the nutrient-enriched surface waters, causing depletion or elimination of dissolved oxygen in bottom waters.

Evidence is also building up in many affected areas that eutrophication has increased dramatically in the 20th century. The best records are for European waters such as the Baltic Sea (Larsson et al. 1985; Jansson 1997; Jansson and Dahlberg 1999), the Kattegat (Jørgensen and Richardson 1996), and the Adriatic Sea (Malone et al. 1999), where there is a long history of scientific observation to document the changes taking place. For example, in the northern Adriatic Sea analysis of oxygen data beginning in 1911 shows there was a significant shift toward supersaturated oxygen conditions in surface waters (reflecting high rates of photosynthesis) and hypoxic bottom waters between 1966 and 1972 (Justić et al. 1987; Justić 1991). This was coincident with more than doubling of the average concentration of N and P in the discharge of the Po River, the largest source of land-based inputs (Justić et al. 1994). In the Baltic Sea declining oxygen concentrations in the deep basins (for which records began in the 1930s) were observed beginning in the 1960s and continued to the present (Diaz and Rosenberg 1995), while in the Kattegat (for which observations began in 1912) worsening hypoxia and progressive mortalities of benthic organisms began in 1980 (Jørgensen and Richardson 1996).

In other parts of the world, including North America, the history of careful scientific observation is not as long. However, the use of biological and biogeochemical paleoindicators in sediments allows the reconstruction of the history of eutrophication. In the Chesapeake Bay – the largest estuary in the United States (4400 km^2) into which drains a 166 000 km^2 watershed – records of sedimentation, diatoms and pollen chronicle the first signs of eutrophication associated with land clearing by European colonists, which brought increased loads of sediments and nutrients to the estuary beginning in the late 18th century (Cooper and Brush 1991; Cooper 1995). Based on geochemical indicators, periodic hypoxia began to occur in the deep waters of the bay; however, seasonal anoxia intensified during the 1950 to 1970 time period. On the continental shelf off the mouth of the Mississippi and Atchafalaya Rivers, the first systematic observations of hypoxia in bottom waters began in 1985 and continued to document the occurrence of hypoxia over approximately half of the year and over an area from 9000 to 18000 km^2 in mid-summer. Although there is some suggestion in the paleoindicators of a signal of land clearing in the Mississippi basin in the 19th century, the most dramatic changes in planktonic production and hypoxic stress on benthos were observed during the late 1950s and 1960s (Rabalais et al. 1996).

Anoxia and hypoxia are extreme modifications resulting from eutrophication. Other effects may precede serious hypoxia. Moreover, increased algal biomass, decreased light penetration and altered benthic habitats may result in smaller coastal bays and estuaries not susceptible to hypoxia because of the lack of water mass stratification. While the timing of changes noted in these different parts of the world varies somewhat, it is clear that significantly accelerated eutrophication took place during the 1960s and 1970s. For example, algal blooms and hypoxia on the northwestern shelf of the Black Sea expanded dramatically (Tolmazin 1985), while nitrate and phosphate concentrations in the Danube River increased by up to tenfold (van Bennekom ans Salomons 1981) and nutrient concentrations in the Denstr River doubled (Balkas et al. 1990). Increases in nutrient loading were coincident with the rapid increase in fertilizer use and crop production in North America and Europe. In fact, Turner and Rabalais (1991) demonstrated the close coherence between the concentration of nitrate in the Mississippi River discharge, which nearly tripled (Goolsby et al. 1999), and the use of chemical fertilizers in the United States. However, the role of agricultural activities in coastal eutrophication is complex and will be explored later.

Effects of Eutrophication

While the processes resulting in oxygen depletion of bottom waters from eutrophication are now reasonably well understood, the consequences of the these and other ecosystem changes induced to fisheries production and overall health of coastal ecosystems are poorly known.

Nutrient inputs to coastal ecosystems, whether from runoff or upwelling from the deep ocean, are essential to maintaining these highly productive ecosystems. Primary production in marine ecosystems is directly proportional to the supply of nutrients (Nixon et al. 1986); consequently, increase in the nutrient supply invariably leads to increase in plant production somewhere in the receiving system, particularly by phytoplankton. For example, Richardson and Heilmann (1995) calculated that annual primary production in the southern Kattegat had increased by at least a factor of two between the periods 1954-1960 (when ^{14}C methods of measuring production were first used) and 1984-1993. While some of this increased production is consumed by zooplankton grazers and may thus support higher trophic level production, under eutrophic conditions a larger proportion of the organic production typically sinks (passively or through biodeposition) to bottom waters and the seabed, where its decomposition consumes oxygen.

In addition to affecting the rate of plant production, nutrient enrichment also affects the standing crop biomass (and thus light penetration) and the qualitative composition of the phytoplankton. The sediment record in the Chesapeake Bay shows an increase in small planktonic diatoms and decrease in benthic diatoms as the Bay transitioned from a clear water system with extensive benthic primary production to a more turbid system dominated by smaller phytoplankton (Cooper 1995). Such shifts to smaller, fast-growing phytoplankters (particularly flagellates and nannophytoplankton) under eutrophic conditions are thought to alter planktonic food webs in favor of small protist consumers and microbial recyclers – the so-called microbial loop. This not only shunts production from the food chains supporting fish larvae and other higher consumers, but results in more rapid recycling of nutrients, which can then stimulate more production.

Some of the phytoplankters favored by eutrophication bloom to nuisance proportions or produce toxins that kill other organisms and may even affect human health. For example, in the Neuse River estuary, a part of the extensive, shallow Albemarle-Pamlico system in North Carolina, USA, large agricultural and other inputs of nutrients stimulate blooms of cyanobacteria and dinoflagellates that cause water discoloration and nuisance conditions as well as depletion of oxygen (Pinckney et al. 1998). Nutrient enrichment has also been suggested to contribute to outbreaks of the mainly heterotrophic dinoflagellate *Pfiesteria piscicida* (Burkholder and Glasgow 1997), an organism that produces toxins that can kill fish and result in impairment of short-term memory and other cognitive functions in humans (Grattan et al. 1998). Because of risks to human health, these and other outbreaks of toxic *Pfiesteria* along the US east coast have raised public fear and stimulated regulatory and legislative action to control nutrient inputs to estuarine waters. There is international concern about the apparent increased frequency of a variety of harmful algal blooms around the world (Anderson 1995). While it is not clear that eutrophication is responsible for this general increase in harmful algal blooms – and certain that it is not for some types of algae – some toxic blooms in European and Japanese waters have been linked with nutrient overenrichment (Smayda and Shimizu 1991). Much yet needs to be learned regarding the relationships between harmful algae and nutrient enrichment. For example, the form of nutrients (e.g., organic N versus nitrate) may be important. Already, though, concern about risks to human health is a new driving force shaping policies for the control of nutrient pollution both in North America and Europe.

Organisms living at the seabed may be affected by eutrophication in several ways. Increased nutrient availability may favor certain epibenthic algae over others. In the Baltic, for example, annual filamentous algae are favored over coarse perennial algal species (Jansson 1997). Shading by dense phytoplankton and epiphytic growth may keep sufficient light from reaching seagrasses (Duarte 1995) or corals causing their demise. Increased organic matter deposited on the seabed provides an enriched food resource for some deposit feeders living in the sediments, while increasing physiological stress (due to low oxygen or high sulfide levels) in others. As a result, the density, biomass, and composition of benthic communities are frequently greatly altered (Diaz and Rosenberg 1995). Finally, hypoxia or anoxia may extend into the water column, stressing or killing animals living on or swimming near the bottom. In addition to influencing the abundance and composition of the benthic biota, eutrophication may affect rates of sedimentation, decomposition of detrital material, bioturbation of sediments, nutrient regeneration and material exchange across the sediment-water interface (Heip 1995).

Experience from several regions of the world indicates that modest eutrophication can increase fishery harvests, particularly of pelagic species, as a result of food chain stimulation (Caddy 1993; Houde and Rutherford 1993). On the other hand, when seasonal or permanent hypoxia begins to occur, catches of benthic feeding fish decline rapidly. If oxygen deficiencies and other manifestations of hypereutrophy continue, even pelagic populations may be adversely affected. Perhaps the most extreme case of devastation of fisheries by eutrophication is the northwestern Black Sea, where numerous stocks of benthic and pelagic species have collapsed (Mee 1992; Caddy 1993). Of the 26 commercial species fished in the 1960s, only 6 still support a fishery.

The interactions between fishing pressures and eutrophication make the relationship of enrichment to fishery yields difficult to interpret. Fishing effort and, consequently, catches have in some cases increased along with eutrophication, perhaps giving the false impression that fish populations have increased or at least not been affected by nutrient enrichment. On the other hand, in some areas, overfishing has reduced stocks while eutrophication proceeded, potentially lending the false impression that nutrient over-enrichment was responsible.

Finally, eutrophication may have manifold effects on the overall "health" of a coastal ecosystem in the sense of its vigor, organization, and resilience (Costanza 1992). Vigor embodies the throughput or productivity of an ecosystem. Organization represents not only its species diversity, but also the degree of connectedness of the constituent species. Resilience refers to an ecosystem's ability to maintain structure and patterns of behavior in the face of disturbance. A healthy ecosystem, then, is one that is active, maintains its biological organization over time and is resilient to stress. The gross productivity of the Chesapeake Bay, for example, has increased as a result of two centuries of cultural eutrophication (Boesch, et al. 1999), but this increase results from rapidly growing, small organisms at the expense of larger, more long-lived organisms that are more valuable to humans. From this perspective, the health of the Chesapeake Bay has deteriorated as a result of nutrient overenrichment, concomitant reduction of light availability, and loss of habitats that produce complexity. This has resulted in an ecosystem that is a less vigorous producer of valuable fish and shellfish, less diverse and well organized, and more susceptible to and slower to recover from disturbance (Ulanowicz, 1997).

Contributions from Agriculture

Nutrient delivery to coastal waters is affected by agricultural activities through a number of interacting activities. Most attention has been given to the expanded use of inorganic fertilizers in agriculture, for indeed it is this form of anthrogenically fixed nitrogen that has grown most dramatically since the 1960s (Vitousek et al. 1997). While the use of industrially fixed nitrogen is a key driving force for agricultural inputs of nitrogen to coastal waters, the sources and pathways are diverse and interacting and, thus, must be considered in control strategies.

Chemical fertilizer use in Europe and North America grew rapidly after the 1950s, but has leveled off since the 1980s. Worldwide, however, fertilizer use has continued to grow in meet increased demand in developing countries (Nixon 1995). Nitrogen and phosphorus in chemical fertilizers applied to fields may be transported to coastal waters in a number of ways, including movement of unassimilated materials into groundwater or surface waters, transport in the air of volatilized (NH_3) or dust-associated materials, and the mineralization of plant residues. Nitrogen and phosphorus behave differently in these regards, with nitrogen as nitrate being highly soluble and prone to groundwater intrusion, and phosphorus being less soluble and highly particle-reactive. The proportion of the N or P applied to fields that is lost to aquatic systems depends on the degree to which the application rate meets or exceeds the crop's requirements (overapplication results in proportionately greater losses), tillage practices, soil type and underlying geology, drainage, and climate conditions.

Although excluded from some budgets of new nitrogen inputs and agricultural source controls, nitrogen-fixing crops contribute nitrogen to terrestrial ecosystems far beyond the rate of nitrogen fixation in forests and grasslands. Jordan and Weller (1996) estimated that new N inputs by nitrogen-fixing crops such as soybeans and alfalfa are more than two-thirds of that due to inorganic fertilizers for the entire United States. The decomposition of the unharvested plants releases nitrogen, which may be lost to surface or groundwaters (Angle 1990), although Goolsby et al. (1999) concluded that the extensive soybean cultivation in the Mississippi River basin contributed little of the nitrogen exported by the river.

Considerable losses of N and P to coastal waters may result from the handling, storage and application of animal wastes to agricultural land. The trend of intensification of poultry and livestock production concentrates animal wastes, thus increasing the likelihood of these losses. Nutrients may be lost due to volatilization of wastes, discharge from waste holding facilities, and the application of manure as a soil fertilizer.

Atmospheric return of NH_3 volatilized from fertilizers or animal wastes is an important route for N input into coastal waters in some regions. For example, atmospheric deposition has been shown to account for over 30% of the nitrogen supplied to North Carolina estuarine waters (Paerl 1995) and is similarly important in regions of the Baltic and North sea, including The Netherlands, Belgium, and Denmark (Asman and Larsen 1996). Monitoring of waste streams and receiving waters may underestimate the importance of atmospheric transport routes of N delivery.

Artificial drainage of agricultural lands and alterations of the hydrology and plant communities of field-edge and downstream riparian zones may also greatly increase the delivery of nutrients to the coast. Soils in many highly fertile farmlands, such as the corn belt of the upper Mississippi River valley or low coastal lands or flood plains of the are naturally hydric and are able to grow crops only because they are artificially drained. This has three major effects on nitrogen export: (a) large reservoirs of organic matter in the soil are oxidized, releasing nutrients through mineralization; (b) soluble nitrate is more readily leached into drainage tiles or ditches; and (c) denitrification, which occurs in anoxic, wet soils, is reduced. In fields in the upper Mississippi basin that have intense tile drainage loss of nitrogen from mineralizing soil organics may be greater than that from chemical fertilizers (Cambardela et al. 1999).

The complex sources, sinks, and transport and transformation processes operating on nutrients make it difficult to account for the origins and causes of coastal eutrophication downstream. Through a variety of mass balance and statistical modeling approaches, Goolsby et al. (1999) have developed a nitrogen mass balance for the Mississippi River Basin. Fertilizer accounts for 54% of the new nitrogen inputs, N fixation 34%, and atmospheric deposition of nitrate (from fossil fuel combustion) the remainder. Recycled nitrogen (from people, animals, the atmosphere, and mineralized soil) is estimated to equal nearly 85% of the new nitrogen inputs, with inputs of N from soil mineralization about equal to that from fertilizer. Discharge from the river into the Gulf of Mexico accounts for only 7% of the total nutrient budget (or 13% of the new nutrients), slightly less than the amount of nitrogen lost to the atmosphere due to denitrification. In aggregate, though, agricultural sources account for an overwhelming proportion of the nitrogen budget of the Mississippi basin (Figure 2).

The relative contributions of agricultural sources to nutrient delivery to coastal waters vary widely, however, among the regions experiencing eutrophication (Fig. 2). The Mississippi River basin is heavily dominated by agricultural sources. Municipal and industrial point sources contribute only a very small percent of the N or P that reaches the Gulf of Mexico. Although agricultural sources are significant both for the Chesapeake Bay (with its grain fields and meadows supplying intense animal production), the northern Adriatic (which receives the effluent from the agricultural Po River valley) and Denmark, their more concentrated human populations contribute relatively more wastes via point sources and combustion by-products. Moreover, some eutrophied coastal waters, such as Long Island Sound (New York) and the Seto Inland Sea (Japan), receive little input of nutrients from agriculture but large inputs from outfalls and the atmosphere from nearby major population centers.

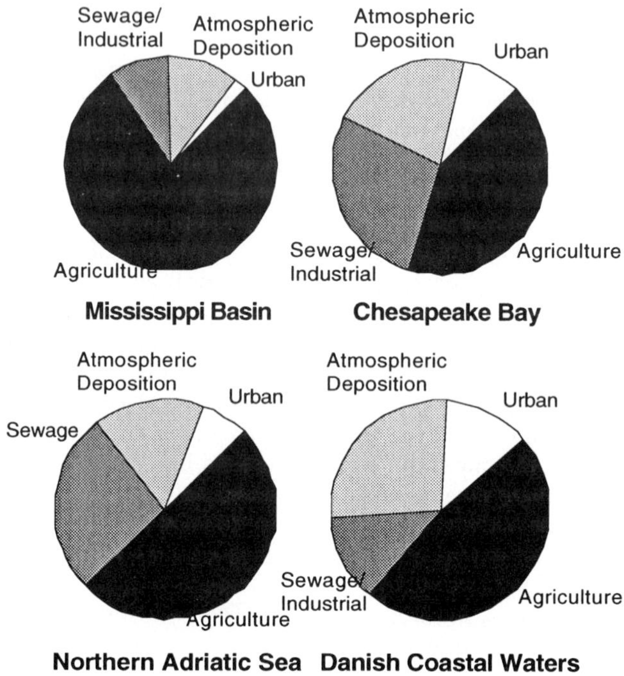

Fig. 2: The relative contribution of agricultural sources in comparison to other anthropogenic nitrogen loads varies among eutrophic coastal regions (Boesch et al., 1999; Goolsby et al. 1999; Seagle et al. 1999; Kronvang et al. 1993)

Approaches to Reduce Nutrient Losses

Introduction

Mitigating the environmental impacts of food and fiber production on our natural resources represents one of the greatest challenges facing agriculture as we enter the next millennium. Finding solutions that are environmentally sound and economically feasible heighten the challenge. Presently, these challenges are particularly acute for the livestock and poultry industries. The general trend from small-scale, diverse family farms to large integrated containment facilities located on limited land area further exacerbates these problems and makes economically viable solutions ever more challenging. Also challenging is the management of nitrogen losses from extensive grain production resulting from artificial drainage or seepage into groundwaters. Control of these losses requires not only more efficient application of fertilizers, but also the large-scale restoration of wet landscapes that act as sinks for nutrients.

Due to differing mobility in the soil, groundwater, and surface waters, approaches to the reduction of nitrogen and phosphorus losses are considered separately below. N controls focus on the movement of nitrate (NO_3) through the soil profile to groundwater, while P generally moves as either soluble or particulate P in surface water runoff.

Nitrogen

Nitrate contamination of groundwater has been documented in most regions of the United States and Europe, where N fertilizers, including sewage sludge and animal manure, are applied. Elevated N in groundwater not only poses a human health concern due to contaminated drinking water, but also contaminates surface waters due to lateral transport of groundwater into streams.

Most of the regions that have experienced increasing delivery of nitrogen to coastal waters are associated with shallow aquifers, coarse-textured soils, excessive manure-N application rates, dense animal populations, overapplication of fertilizer, highly mineralizing soils, or hydrologic conditions conducive to leaching, e.g., excess rainfall, heavy irrigation, artificial drainage, limited crop uptake, and limited denitrification losses (Sharpley et al. 1997).

With increasing attention to protection of groundwater, future attempts to improve water quality should focus first on source control; specifically on preventing nitrate from entering subsurface flow paths. Developing strategies to reduce nitrate loss to groundwater requires information about the temporal and spatial patterns of nitrate leaching in agricultural systems. Rates of N leaching have been correlated with N fertilization rates in various systems (Baker and Johnson 1981; Mitsch et al. 1999; Sharpley et al. 1997). N leaching rates where animal manure or sewage sludge is applied based on crop N requirements are consistently higher than where crop N needs are met with inorganic N sources. In all the above scenarios, N leaching to groundwater occurs as a result of elevated concentrations of NO_3 in the soil profile during groundwater recharge. Therefore, strategies that minimize pore water NO_3 concentrations at the onset of groundwater recharge are keys to reducing rates of NO_3 losses from agricultural systems. For most regions of the US and Europe, evapotranspiration rates exceed average rates of precipitation during the growing season (May through September). Therefore, the primary factor dictating N leaching rates is the availability of NO_3 in the soil profile following fall crop harvest.

A multistate commitment for the reduction of eutrophication in the Chesapeake Bay made in 1987 called for a 40% reduction of controllable N and P inputs by the year 2000 (Boesch et al. 1999). Each state has implemented a variety of programs to reduce agricultural nutrient sources. Maryland's program consisted of a volunteer N-based nutrient plan that relied on matching N inputs to crop demand by fertilizing for realistic yields, splitting N applications and using a pre-sidedress nitrogen test (PSNT) to determine additional mid-season N requirements. As of 1999, N-based nutrient management plans had been written and implemented for approximately 400000 hectares. Unfortunately, N concentrations in groundwater, streams, and rivers in areas dominated by agricultural inputs have not decreased. Uncertainty in weather, droughts, disease, insect pressure, and fall mineralization of the organic pool following crop harvest limit the long-term potential for significant reductions in N losses based on matching N inputs with crop demands. Furthermore, research suggests that N losses to groundwater following soybean harvest (for which there is no N application) can be as great as following crops for which N is applied.

Reducing fertilizer application rates, optimum timing of fertilizer application, crop rotation, and the use of winter cereal cover crops (Figure 3A) could result in approximately a 20% reduction in the discharge of N to groundwater, streams and rivers (Staver and Brinsfield 1994; Mitch et al. 1999). If further reductions in N inputs to crops are required, crop yields would very likely be reduced below the economic optimum. N losses from fields increase substantially as the crop requirements are approached and exceeded. Consequently, agronomic prescriptions that closely match the amount and timing of crop requirements provide significant environmental benefits, but with modest cost savings in terms of reduced fertilizer use and increased costs of testing. It is often cheaper to avoid risks of lowered yields and somewhat overapply fertilizers. Precision agricultural practices (Schepers, this volume) offers benefits, in addition to fertilizer savings, of increased yields. In addition, in the United States there is active consideration of other steps that would reduce application of nutrients in excess of crop requirements, such as insurance coverage that would compensate farmers if adherence to a rigid prescription reduced crop yields.

In addition to reductions from on-farm practices, other techniques, including the creation and restoration of wetlands and riparian ecosystems between farmland and streams and rivers and downstream, offer real opportunities for additional significant N reductions from agriculture. For example, research by several investigators (Jørgensen 1994; Lowrance et al. 1997) has shown significant N reductions using combinations of grass and tree buffer zones between farm fields and streams. Although significant N reductions are possible, optimum position of the riparian buffer or wetland in the landscape is critical (Staver and Brinsfield 1996; Mitsch et al. 1999). Mitch et al. estimate that creating or restoring 21,000 to 52,000 km^2 of wetlands in the Mississippi River basin (0.7 to 1.8% of the basin) would result in a 20 to 50% reduction in N loading to the Gulf of Mexico. While ecosystem restoration strategies have tremendous potential, adequately compensated volunteer programs, including permanent easements, must be available to the landowner.

Similar struggles to balance food production and water quality are occurring in Europe. In Germany, where rising NO_3 concentrations in groundwater are the most immediate concern, use of commercial N fertilizer grew about four fold from 1951 to 1990, while the estimated N content of farm produce increased just over two fold (van der Ploeg et al. 1997). Thus, N-use efficiency decreased and the surplus available to contaminate goundwater and surface waters increased greatly. The European Union nations are required by the Union's 1991 Nitrate Directive to have a voluntary code of good agricultural practices (Council of European Communities 1991). They are also required to identify vulnerable zones where additional action must be taken to reduce N losses to acceptable levels. More recently, a team of six European research institutions began evaluating economic alternatives for reducing N losses from agriculture. The so-called Nitrotax study is coordinated by the Centre for Agriculture and Environment in The Netherlands. Researchers are using a combination of farm models, literature reviews and expert judgment to better understand the implications of economic systems for N control. One of the goals is to form a "European Community" which seeks to merge individual countrie's agricultural programs into a unified effort to develop fair N policies that stabilize farm incomes and simultaneously meet regional water quality goals.

Phosphorus

While N losses to the environment from agriculture are widespread in most developed countries, most P losses occur in areas where livestock and poultry are highly concentrated, sewage sludge is applied or where there are high rates of soil erosion. Unlike N, which is highly dependent on dynamic soil microbial processes, P levels in soils as well as in manure and sewage sludge are easier to quantify and track. Furthermore, soil P levels change slowly, allowing for development of mass balances within fields, farms, and watersheds.

In the context of ecosystem protection, strategies for P reductions have focused mainly on educational and incentive programs to encourage implementation of practices designed primarily for reducing the movement of sediment from cropland (reduced tillage, grassed waterways, retention ponds, terraces, and contours). This strategy has been successful for reducing sediment and sediment-bound P losses on cropland where erosion rates are high. However, research suggests that these strategies for soils with high P levels not predisposed to high erosion rates may be less effective.

Soil P levels are, in general, increasing throughout Europe and the U.S., particularly where livestock and poultry manure or sewage sludge is applied to cropland. Crop nutritional requirements and grain removal rates indicate a N:P supply ratio of approximately 6:1 on a weight basis, which is twice the ratio of plant-available N:P in manure or sewage sludge. Consequently, repeated applications of animal manure or sewage sludge to cropland, based on N crop requirements, results in unused P and increasing soil P levels. For example, in the states of Maryland and Delaware, many soils have soil P levels sufficient to meet crop P requirements for several years without additional P fertilization (Sims 1987, 1993). Problems associated with high soil P levels are exacerbated because they are often located near sensitive freshwater bodies in which P enrichment causes excessive algal growth.

On the Delmarva Peninsula, an intensive poultry producing region lying between the Atlantic Ocean and the Chesapeake Bay, repeated application of poultry litter has been increasing topsoil P inventories by an average of 10 kgha^{-1} annually beyond P removal rates required for the optimum production of crops (Kenneth Staver, pers comm). From the perspective of the total regional P budget, approximately one-half of the total P imported to the Maryland portion of the peninsula in the form of fertilizer, sewage sludge, grain, and dietary supplements for poultry remains as a residual in the soils. Similar studies in Denmark, Norway, Finland and Sweden suggests that the main reason for high P losses from cropland is the net input of approximately 20 kg-ha^{-1} annually for the past several decades (Svendsen and Kronvang 1991). Iserman (1990) reported even higher P surpluses for Germany and the Netherlands.

The potential for P loss from these soils is worsened by high water tables, tile drainage, and soils with low water-holding capacity. Breeuwsma et al. (1989) estimated that for a watershed where over 80% of the soils were P-saturated, leaching losses of 2.5 kg-ha^{-1} and runoff losses of 0.3 kg-ha^{-1} of total P were observed annually. For soil types not predisposed to high erosion rates, reductions in surface runoff of P will require better management of P levels in the uppermost soil horizon. However, one of the most effective best management practices for erosion control (no-till) actually increases P levels in the top soil horizons and under certain conditions can increase dissolved P losses. In a study by Staver and Brinsfield (1994), total P losses from fields under no-till for many years was twice as high on an annual basis compared to conventional till (Figure 3B). When toxic *Pfiesteria piscicida* outbreaks occurred in subestuaries of the Chesapeake Bay along the Delmarva Peninsula in 1997, the combination of highly P-enriched soils in the region, emerging evidence regarding loss of dissolved P from enriched soils, and high levels of P in the estuarine waters provided the impetus for legislation in Maryland mandating management of manure applications based both on N and P crop requirements.

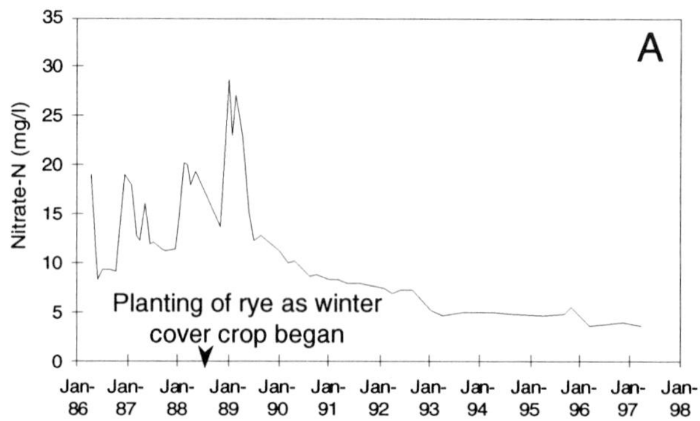

Fig. 3A: Experimental field plots on the coastal plain near the Chesapeake Bay have demonstrated unanticipated losses of nutrients under conventional nutrient management: A. Fall cereal cover crops reduced the losses of nitrate to groundwater in a field under conventional nutrient management (Staver and Brinsfield, 1996, 1998)

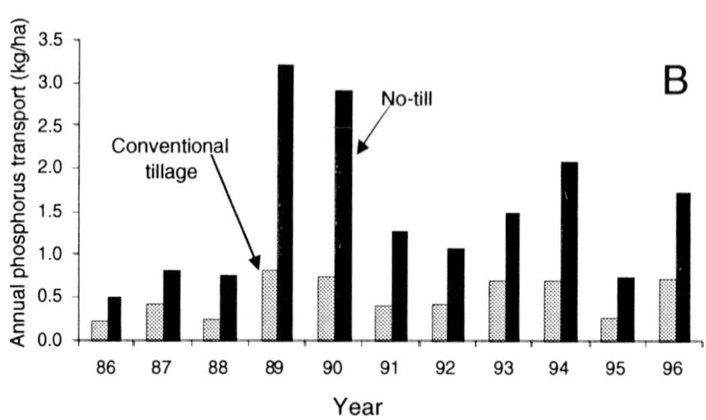

Fig. 3B: No-till field management resulted in increased phosphorus concentrations in surface soils, with increased losses of dissolved P in runoff compared to tilled fields (Staver and Brinsfield, 1994, updated)

While strategies which reduce sediment-bound P transport are important in the long term, better management of P inputs on a watershed basis will be required. This is especially true for managing manure in regions with concentrated animal production. A more holistic systems approach is needed. Efforts should include alternative uses for manure including burning, pelletizing, and composting and further research on feed, nutrition, and manure handling and storage. Nutrient composition of manure can be altered by changing diet composition to provide a better balance in nutrient output. Greater feed efficiency is generally economically desirable and can reduce the quantity of manure. Recent poultry diet research by Sims et al. (1999) has shown that the combination of the enzyme phytate and low phytic acid corn reduced total P in poultry manure by 50% and soluble P by 84% when compared to traditional diets. If the P in manure could be reduced by 50% below current levels, then the N:P ratio of the manure would be closely balanced with crop N:P requirements, thereby reducing the buildup of soil P following repeated applications, based on crop N requirements.

Integrated Strategies

Managing nutrients (both N and P) in farm fields, feedlots and the environment to achieve significant reductions in nutrient inputs from agriculture requires integrated strategies. The problems related to phosphorus accumulation in soils managed solely for nitrogen and the recognition of distant eutrophication driven by nitrogen losses from fields under P-based management demonstrate the need to manage both nutrients in an integrated manner. Furthermore, some nutrient losses from agricultural activities, whether they originate from fertilizers, mineralization of soil organic matter, or animal wastes are inevitable. More efficient agronomic practices, moderation of excessive soil draining, and improved animal waste handling and disposal could result in reductions of nutrient loadings to coastal waters of 20 to 30%. Also, nutrient traps or sinks could be restored and enhanced in watersheds under intense agricultural production in order to capture some of the nutrients inevitably lost downstream. Expanded restoration and creation of wetlands and riparian forests and reopening flood plains could trap another 20-30% of the nutrients. Thus, reduction of nutrient inputs from agricultural activities at the present level of intensity of 50% or more is potentially feasible. However, this will require substantial technical and financial assistance and greatly increased governmental and private efforts to restore and manage critical nutrient-trap habitats (Keeney and DeLuca 1993; Mitsch et al. 1999). In the process, the esthetics, wildlife, and fisheries production, and biodiversity of "overdeveloped" agricultural landscapes could be enhanced.

In an economic assessment of the costs of agricultural nutrient controls in the Mississippi River Basin, Doering et al. (1999) concluded that N losses could be reduced by 20% through a combination of fertilizer restrictions and wetland restoration with minimal net economic impact and no loss of production. Based on this model, reductions in production would only become significant when N loss restrictions reach 30% and higher, but the effects on aggregated producer net returns would be partially offset by increased commodity prices. In practice, though, it is the returns to individual producers rather than the aggregate that determine the economic viability of agriculture. The inequities are further compounded because it is more effective for reducing downstream nutrient loads in downstream nutrient loads to target nutrient use restrictions and transformation of agricultural lands to wetlands geographically rather than to apply these efforts uniformly throughout the catchment basin. These conditions challenge society to find ways to internalize the heretofore ignored external environmental costs within agricultural commodity prices and to provide targeted technical assistance, subsidies and incentives to achieve environmental benefits while mitigating inequities.

Large Watershed Management

The mounting pressures to control nutrient inputs to coastal waters, including those from inputs from agriculture, will only grow in the future. Although major gains in environmental quality have been realized by reducing organic and toxic loading from municipal and industrial waste discharge, advanced wastewater treatment to remove nutrients (especially N) has been accomplished in only a few regions. However, reduction of sources of P (elimination of phosphate-based detergents) and chemical and biological treatment to reduce both P and N are being shown to be feasible and successful. While significant reductions in N and P from wastewater treatment plants have occurred, the application of sludge generated by these processes may result in losses of these nutrients from croplands on which the sludge is applied. There is even promise that atmospheric inputs of N will be reduced due to air quality regulations requiring greater reduction of nitrous oxide emissions from both stationary sources and motor vehicles.

Major challenges still confront the reduction of agricultural nonpoint sources of nutrients. These sources are numerous and diffuse. Agriculture is economically and socially important but is undergoing serious conflicts between traditional operations and global economic forces. Consequently, production agriculture is generally marginally profitable, at best, and can ill afford increased costs for environmental controls. The market place does not easily allow the transfer of these external environmental costs to the consumers, who desire cheap food prices and have many choices. The impacts of agricultural nutrient pollution are often far removed (thousands of kilometers in some cases) from their sources, so that the usual forces of place-based concerns and community equity are not effective in motivating pollution abatement. The causes and effects transcend multiple political boundaries, across nation states and subnational governments – and so does the need for common solutions.

For all of these reasons, beginning in the late 1980s and the 1990s, abatement of large-scale coastal eutrophication has been addressed by a variety of ecosystem management efforts, including regional seas programs and national or watershed-scale assessments, management and regulation. Although eutrophication is often a major concern and a driving issue in these efforts, multiple issues are usually dealt with, including pollution by toxins, oil spills, fisheries management, habitat protection and restoration, and even social and livability issues.

The oldest of these multijurisdictional management endeavors focuses on the Baltic and North Seas. The North Sea Ministerial Conferences operating under the Paris Convention has set goals for a 50% reduction of contaminants discharged into the sea. The Baltic Marine Environmental Protection Commission, or Helsinki Commission (HELCOM) has nine nations plus the European Union as members. Although much of its effort has focused on activities in the Baltic itself and around its coasts, increasing attention is being paid to watersheds draining into the Baltic. HELCOM has issued recommendations for agricultural nutrient management that the member nations are expected to adopt. The Black Sea Environmental Programme has six littoral member nations, but the catchment area for the Black Sea includes some 17 nations and more than 160 million inhabitants. The Danube River Programme reaches upstream to involve 11 nations in improving the environmental quality of the river, its delta, and its effects on the nearshore Black Sea. In Japan, the five prefectures bordering the Seto Inland Sea have a similar multiparty compact for common commitments and shared resources.

In the United States, the National Estuary Program has engaged state and local governments and the public in the development of comprehensive conservation and management plans for 28 bays and estuaries around the coasts. Eutrophication is a major issue being addressed in most of them. The oldest and largest regional coastal ecosystem restoration is that of the Chesapeake Bay Program, which engages three states, the District of Columbia, and the federal government. Reversing eutrophication is the highest priority, with a commitment made in 1987 to reduce controllable sources of inputs of both nitrogen and phosphorus by 40% by 2000 (Boesch et al. 1999). A substantial element of the program seeks to provide technical assistance to farmers to reduce nutrient losses to the Bay. As the 2000 milestone approaches, models are being run and monitoring data analyzed to determine whether the goal would be met. It now appears that the phosphorus target will be met or exceeded and that – although N loadings have been reduced – progress will fall short of the goal for N (Boesch et al. 1999). Most documented gains have been made through reduction of point sources.

Even more aggressive commitments for nutrient reduction have been made in Denmark, where in 1987 the Parliament enacted the Action Plan on the Aquatic Environment, the main objectives of which were to reduce total nitrogen discharge to the aquatic environment by 50% and the total discharge of phosphorus by 80% by 1994 (Kronvang et al. 1993). While the goal for reductions of phosphorus, most of which were from point sources, has been met, it is estimated that the nitrogen load has been reduced by only about 15% (Conley and Josefson, 1999). These nitrogen reductions were mainly due to improved sewage treatment; little progress has been made on reductions of agricultural sources, which dominate the nitrogen loads. The second phase of the Action Plan was begun in 1998 and will include requirements for maximum livestock concentrations, manure storage facilities, mandatory fertilizer plans and budgets, cover crops, and set asides of agricultural lands.

Another large-scale ecosystem restoration focuses on the lower Florida peninsula and involves efforts to restore freshwater flows and improve water quality in the Everglades. Protection and restoration of Florida Bay, a shallow tropical ecosystem that has experienced discoloring algal blooms in recent years (Boesch 1996), is also an objective. Nutrient inputs from sugarcane production and other agricultural activities to the north are receiving particular attention and may be related to the problems seen in Florida Bay. Undoubtedly, the geographically largest challenge for the control of agricultural nutrient inputs concerns the eutrophication of the northern Gulf of Mexico. Ongoing assessments are evaluating the causes, consequences, sources, control options, and economics of agricultural nutrient reductions. It is clear that the majority of the nutrients discharged by the Mississippi and Atchafalaya rivers emanate from intensely farmed regions in the upper basin, over 1000 km upstream (Goolsby et al. 1999).

Some of these programs have demonstrated progress in arresting, if not reversing, coastal eutrophication. Most are just starting, and many face daunting challenges as a result of scale of the problems, fragmented responsibilities, and modest economic ability to address the needed controls. Most of the reductions in nutrient loading, even in the more advanced efforts such as the Chesapeake Bay Program, have come as a result of point source reductions. Pressures will intensify to redouble our efforts to control agricultural sources of nutrients – a major issue for agriculture during the first half of the 21st century.

References

Anderson DM (1995) Toxic red tides and harmful algal blooms: a practical challenge in coastal oceanography. Rev Geophysics 33:1189-1200

Angle JS (1990) Nitrate leaching losses from soybeans (*Glycine max* L. Merr.). Agric Ecosys Environ 31:91-97

Asman WAH, Larsen SE (1996) Atmospheric processes. In Jørgensen BB and Richardson K (eds.) Eutrophication in coastal marine ecosystems. American Geophysical Union, Washington, DC, p. 21-50

Baker JL, Johnson HP (1981) Nitrate-nitrogen in tile drainage as affected by fertilization. J Environ Qual 10:519-522

Balkas T, multiple authors (1990) State of the marine environment in the Black Sea region. UNEP Regional Seas Rep Studies 124:41 p

Boesch DF (1996) Science and management in four U.S. coastal ecosystems dominated by land-ocean interactions. J Coastal Conserv 2:103-114

Boesch DF, Brinsfield RB, Magnien RE (1999) Chesapeake Bay eutrophication: scientific understanding, ecosystem restoration, and challenges for agriculture. J Environ Qual, in press

Borum J (1996) Shallow waters and land/sea boundaries. In Jørgensen BB, Richardson K (eds) Eutrophication in coastal marine ecosystems. American Geophysical Union, Washington, DC, pp 179-203

Breeuwsma A, Reyerink JGA, Brus DJ, van heet Loo H, Schoumans OF (1989) Phosphate loading of soils, groundwater and surface waters in the catchment area of the Schuitenbeek (in Dutch). DLO Winand Staring Centre, Report 10, Wageningen, The Netherlands

Bricker SB, Clement CG, Pirhalla DE, Orlando SP, Farrow DRG (1999) National estuarine eutrophication assessment: a summary of condition, historical trend and future outlook. National Oceanic and Atmospheric Administration, Silver Spring, Maryland.

Burkholder JM, Glasgow HB, Jr (1997) *Pfiesteria piscicida* and other *Pfiesteria*-like dinoflagellates: behavior, impacts, and environmental controls. Limnol Oceanogr 42:1052-1075

Caddy JF (1993) Toward a comparative evaluation of human impacts on fishery ecosystems of enclosed and semi-enclosed seas. Rev Fish Sci 1:57-95

Cambardella CA, Moorman TB, Jaynes DB, Hatfield JL, Parkin TB, Simpkins WW, Karlen DL (1999) Water quality in Walnut Creek watershed: nitrate-nitrogen in soils, subsurface drainage water and shallow groundwater. J Environ Qual 28:25-34

Comin RA, Romero JA, Astorga V, Garcia C (1997) Nitrogen removal and cycling in restored wetlands used as filters of nutrients for agricultural runoff. Wat Sci Technol 35:255-261

Conley DJ, Josefson AB (1999) Hypoxia, nutrient management and restoration in Danish waters. In Rabalais NN, Turner RE (eds) Effects of hypoxia on living resources, with emphasis on the northern Gulf of Mexico. American Geophysical Union, Washington, DC, in press

Cooper SR (1995) Chesapeake Bay watershed historical land use: impact on water quality and diatom communities. Ecol Appli 5:703-723

Cooper SR, Brush GS (1991) Long-term history of Chesapeake Bay anoxia. Science 254:992-996

Costanza R (1992) Toward an operational definition of ecosystem health. In Costanza R, Norton BG, Haskell B (eds). Ecosystem health – New goals for environmental management. Island Press, Washington, DC, p 239-256

Council of the European Communities (1991) Council Directive 91/676/EEC concerning the protection of waters against pollution caused by nitrates from agricultural sources. Official Journal No. L 375; Finnish Special Edition (15) 10:192

Diaz RJ, Rosenberg R (1995) Marine benthic hypoxia: a review of its ecological effects and the behavioural responses of benthic macrofauna. Oceanogra Mar Biol Ann Rev 33:245-303

Doering OC, Diaz-Hermelo F, Howard C, Heimlich R, Hitzhusen F, Kamierczak R, Lee J, Libby L, Milon W, Prato T, Ribaudo M (1999) Evaluation of economic costs and benefits of methods for reducing nutrient loads to the Gulf of Mexico. In Hypoxia in the Gulf of Mexico. NOAA Coastal Ocean Program Decision Analysis Series No., National Oceanic and Atmospheric Administration, Silver Spring, Maryland. (www.nos.noaa.gov/Products/pubs_hypoxia.html)

Duarte CM (1995) Submerged aquatic vegetation in relation to different nutrient regimes. Ophelia 41:87-112

Fail JL, Jr, Haines BL, Todd RL (1987) Riparian forest communities and their role in nutrient conservation in an agricultural watershed. Am J Alternative Agric 2:114-121

GESAMP (1990) The state of the marine environment. UNEP Regional Seas Reports and Studies No. 115. United Nations Environmental Program, Paris

Goda T (1991) Management and status of Japanese public waters. Water Sci Technol 23:1-10

Goolsby DA, Battaglin WA, Lawrence GB, Artz RS, Aulenbach BT, Hooper RP (1999) Flux and sources of nutrients in the Mississippi-Atchafalaya River basin. In Hypoxia in the Gulf of Mexico. NOAA Coastal Ocean Program Decision Analysis Series No., National Oceanic and Atmospheric Administration, Silver Spring, Maryland. (www.nos.noaa.gov/Products/pubs_hypoxia.html)

Grattan LM, Oldach D, Perl TM, Lowitt MH, Matuszak DL, Dickson C, Parrott C, Shoemaker RC, Kaufman CL, Wasserman MP, Hebel JR, Charache P, Morris Jr JG (1998) Learning and memory difficulties after environmental exposure to waterways containing toxin-producing *Pfiesteria* or *Pfiesteria*-like dinoflagellates. Lancet 352:532-539

Groffman PM (1994) Denitrification in freshwater wetlands. Curr Top Wetland Biogeochem 1:15-35

Heip C (1995) Eutrophication and zoobenthos dynamics. Ophelia 41:113-136

Hill AR (1996) Nitrate removal in stream riparian zones. Journal of Environmental Quality 25:743-755

Houde ED, Rutherford ES (1993) Recent trends in estuarine fisheries: predictions of fish production and yield. Estuaries 16:161-176

Howarth RW, Billen G, Swaney D, Townsend A, Jaworski N, Lajtha K, Downing JA, Elmgren R, Caraco N, Jordan T, Berendse F, Freney J, Kudeyarov V, Murdoch P, Zhu Z-L (1996) Regional nitrogen budgets and riverine N & P fluxes for the draina-ges to the North Atlantic Ocean: natural and human influences. Biogeochemistry 35:75-139

Hunt PG, Stone KC, Humenik FJ, Matheny TA, Johnson MH (1999) In-stream wetland mitigation of nitrogen contamination in a USA coastal plain stream. J Environ Qual 28:249-256

Iserman K (1990) Share of agriculture in nitrogen and phosphorus emissions into surface waters of Western Europe against the background of their eutrophication. Fert Res 26: 253-269

Jansson B-O (1997) The Baltic Sea: current and future status and impact of agriculture. Ambio 26:424-431

Jansson AM, Andersson R, Berggren H, Leonardson L (1994) Wetlands and lakes as nitrogen traps. Ambio 23:20-325

Jansson B-O, Dahlberg K (1999) The environmental status of yhe Baltic Sea in the 1940's, today and in the future. Ambio 28:312-319

Jordan TE, Weller DE (1996) Human contributions to terrestrial nitrogen flux. BioScience 46:655-664

Jørgensen BB, Richardson K (eds) (1996) Eutrophication in coastal marine ecosystems. American Geophysical Union, Washington, DC

Jørgensen SE (1994) A general model for nitrogen removal by wetlands. In: Mitsch WJ (ed) Global wetlands: Old World and New. Elsevier, Amsterdam, p. 575-583

Justić D (1991) Hypoxic conditions in the northern Adriatic Sea: historical development and ecological significance. In Tyson RV, Pearson TH (eds) Modern and ancient continental shelf anoxia. Geological Society Special Publication 58, The Geological Society, London, pp 95-105

Justić D, Legovic T, Rottini-Sandrini L (1987) Trends in oxygen content 1911-1984 and occurrence of benthic mortality in the northern Adriatic Sea. Estuarine Coastal Shelf Sci 24:435-445

Justić D, Rabalais NN, Turner RE (1994) Riverborne nutrients, hypoxia and coastal ecosystem evolution: biological responses to long-term changes in nutrient loads carried by the Po and the Mississippi Rivers. In Dyer KR, Orth RH (eds) Changes in fluxes in estuaries: implications from science to management. Olsen & Olsen, Fredensborg, Denmark, pp 161-167

Keeney DR, DeLuca TH (1993) Des Moines River nitrate in relation to watershed agricultural practices: 1945 versus 1980s. J Environ Qual 22:267-272

Kronvang B, Ærtebjerg G, Grant R, Kristensen P, Hovmand M, Kirkegaard J (1993) Nationwide monitoring of nutrients and their ecological effects: state of the Danish aquatic environment. Ambio 22: 176-187

Larsson UR, Elmgren R, Wulff F (1985) Eutrophication and the Baltic Sea: causes and consequences. Ambio 14:9-14

Lowrance RL, Altier JD, Newbold R, Schnobel R, Groffman PM, Denver JM, Correll DL, Gillian JW, Robinson JM, Brinsfield RB, Staver KW, Lucas W, Todd AH (1997) Water quality functions of riparian forest buffers in Chesapeake Bay watersheds. Environ Management 21:687-712

Malone TC, Malej A, Harding Jr LW, Smodlaka N, Turner RE (1999) Ecosystems at the land-sea margin: drainage basin to coastal sea. American Geophysical Union, Washington, DC. 381 p

Mee LD (1992) The Black Sea in crisis: the need for concerted international action. Ambio 21:278-286

Mitsch WJ, Day JW, Gilliam Jr JW, Groffman PM, Hey DL, Randall GW, Wang N (1999) Reducing nutrient loads, especially nitrate-nitrogen, to surface water, groundwater, and the Gulf of Mexico. In Hypoxia in the Gulf of Mexico. NOAA Coastal Ocean Program Decision Analysis Series No., National Oceanic and Atmospheric Administration, Silver Spring, Maryland. (www.nos.noaa.gov/Products/pubs_hypoxia.html)

Nakata H (1995) Production enhancement and its implications for the restoration of marine biodiversity in the coastal waters of Japan. FAO Fisheries Circular No. 889. Food and Agriculture Organization, Rome

National Research Council (1994) Priorities for coastal ecosystem science. National Academy Press, Washington, DC

Nixon SW (1995) Coastal marine eutrophication: a definition, social causes, and future concerns. Ophelia 41:199-219

Nixon SW, Oviatt CA, Frithsen J, Sullivan B (1986) Nutrients and the productivity of estuarine and coastal marine systems. J Limnol Soc S Africa 12:43-71

Paerl HW (1995) Coastal eutrophication in relation to atmospheric deposition: current perspectives. Ophelia 41:237-259

Parker CA, O'Reilly JE (1991) Oxygen depletion in Long Island Sound: a historical perspective. Estuaries 12:248-264

Pearson TH, Rosenberg R (1978) Macrobenthic succession in relation to organic enrichment and pollution of the marine environment. Oceanogr Marí Biol Ann Rev 16:229-311

Pinckney JL, Paerl HW, Harrington MB, Howe KE (1998) Annual cycles of phytoplankton community-structure and bloom dynamics in the Neuse River Estuary, North Carolina. Marine Biology 131:371-381

Rabalais NN, Turner RE, Justic D, Dortch Q, Wiseman Jr WJ, Sen Gupta BK (1996) Nutrient changes in the Mississippi River and system responses on the adjacent continental shelf. Estuaries 19:386-407

Richardson K (1996) Carbon flow in the water column case study: the southern Kattegat. In Jørgensen BB, Richardson K (eds) Eutrophication in coastal marine ecosystems. American Geophysical Union, Washington, DC, pp 95-114

Richardson K, Heilmann JP (1995) Primary production in the Kattegat: past and present. Ophelia 41:317-328

Ryther JH, Dunstan WM (1971) Nitrogen, phosphorus, and eutrophication in the coastal marine environment. Science 171:1008-1013

Seagle SW, Pagnotta R, Cross FA (1999) The Chesapeake Bay and northern Adriatic Sea drainage basins: land cover and nutrient export. In Malone TC, Malej A, Harding LW, Smodlaka Jr N, Turner RE (eds) Ecosystems at the land-sea margin: drainage basin to coastal sea. American Geophysical Union, Washington, DC, p 7-28

Schepers J (1999) Precision farming. OECD Conference on Biological Resource Management, Paris. This volume

Sharpley A, Meisinger JJ, Breeuwsma A, Sims JT, Daniel TC, Schepers JS (1998) Impacts of animal manure management on ground and surface water quality. In Hatfield J, Steward BA (eds) Animal waste utilization: effective use of manure as a soil resource. Sleeping Bear Press

Shepared SA, McComb AJ, Bulthuis DA, Neverauskas V, Seffersen DA, West R (1989) Decline of seagrasses. In Larkum AWD, McComb AJ, Shepard SA (eds) Biology of seagrasses: a treatise on the biology of seagrasses with special reference to the Australian region. Elsevier, Amsterdam, p 346-393

Sims JT (1987) Agronomic evaluation of poultry manure as a nitrogen source for conventional and no-tillage corn. Agron J 79:563-570

Sims JT (1993) Environmental soil testing for phosphorus. J Prod Agric 6:501-507

Sims JT, Layahun MF, Malone GW, Saylor WW, Rabov V (1999) Effects of microbial phytase, high available phosphorus corn, and dietary phosphorus level on broiler production. 3. Effects on broiler litter composition. Southern Poultry Science Society

Smayda TJ, Shimizu Y (1993) Toxic phytoplankton blooms in the sea. Elsevier, Amsterdam

Smith VH (1998) Cultural eutrophication of inland, estuarine, and coastal waters. In Pace ML, Groffman PM (eds). Successes, limitations, and frontiers in ecosystem science. Springer Berlin, Heidelberg, New York

Staver KW, Brinsfield RB (1994) The effects of erosion control practices on phosphorus transport from Coastal Plain agricultural watersheds. Chesapeake Research Consortium (Publication No. 149), Solomons, Maryland

Staver KW, Brinsfield RB (1996) Seepage of groundwater nitrate from a riparian agroecosystem in the Wye River Estuary. Estuaries 19: 359-370

Staver KW, Brinsfield RB (1998) Use of cereal grain winter cover crops to reduce groundwater nitrate contamination in the Mid-Atlantic Coastal Plain. J Soil and Water Conserv 3:230-240

Svendsen LM, Kronvang B (1991) Phosphorus in the Nordic countries: methods of bioavailability, effects, and measures. National Environmental Research Institute, Roskilde, Denmark. 201 pp

Tolmazin (1985) Changing coastal oceanography of the Black Sea. I. Northwestern shelf. Prog Oceanogr 15:227-276

Turner RE, Rabalais NN (1991) Changes in Mississippi River water quality this century: implications for coastal food webs. BioScience 41:140-148

Ulanowicz RE (1997) Ecology: the ascendency perspective. Columbia University Press, New York

van Bennekom AJ, Salomons W (1981) Pathways of nutrients and organic matter from land to oceans through rivers. In Martin L J-M, Burton JD, Eisma D (eds.) River inputs to ocean systems. Food and Agriculture Organization, Rome, p 33-46

van der Ploeg RR, Ringe H, Machulla G, Hermsmeyer D (1997) Postwar nitrogen use efficiency in West German agriculture and groundwater quality. J Environ Qual 26:1203-1212

Vitousek PM, Aber JD, Howarth RW, Likens GE, Matson PA, Schindler DW, Schlesinger WH, Tilman DG (1997) Human alteration of the global nitrogen cycle: sources and consequences. Ecol Appl 7:737-750

Reducing Nonpoint Source Phosphorus Runoff from Poultry Manure with Aluminum Sulfate[1]

PA. Moore Jr[2], TC. Daniel[2], DR. Edwards[3]

Phosphorus (P) is generally considered to be the limiting nutrient for eutrophication in lakes and rivers. Phosphorus runoff from soils fertilized with animal manures, such as poultry manure, can be relatively high even when moderate application rates are used. Recent research has indicated that treating poultry manure with aluminum sulfate (alum) can reduce phosphorus runoff and decrease ammonia volatilization. The objectives of this study were to evaluate the effects of alum applications to poultry manure on (1) ammonia volatilization rates from manure, (2) atmospheric ammonia levels in poultry houses, (3) poultry performance (weight gains, feed conversion, etc.), (4) energy use, and (5) P runoff from small watersheds. Two farms in NW Arkansas, USA, were utilized for this study. Alum was applied at a rate of 1816 kg house^{-1} in half of the houses at each farm after each flock of birds and incorporated into the manure. The other houses were controls. Ammonia volatilization rates were reduced by 97% with alum applications for the first four weeks of each growout. Birds grown on alum-treated manure were significantly heavier and had better feed conversion that birds grown in control houses. Energy use was also lower in alum-treated houses due to reduced ventilation requirements to remove ammonia (NH3). An economic analysis indicated that this best management practice was very cost-effective, with a benefit/cost ratio of 1.96. Phosphorus runoff from normal and alum-treated poultry manure was evaluated from field-sized plots (1 acre each) for 3 years. Phosphorus concentrations in runoff water from alum-treated manure were 75% lower than normal manure. These results indicate that treating poultry manure with alum is a cost-effective best management practice that reduces nonpoint source P runoff.

1. Mention of a trade name, proprietary product, or specific equipment does not constitute a guarantee or warranty by the USDA and does not imply its approval to the exclusion of other products that may be suitable.

2. USDA-ARS, Plant Sciences 115, Agronomy Department, University of Arkansas, Fayetteville, Arkansas 72701.

3.USDA-ARS, 128 Agricultural Engineering Building, University of Kentucky, Lexington, KY 40546-0276

Introduction

Phosphorus (P) runoff from agricultural lands can be relatively high when animal manures, such as poultry manure, are used for fertilizer (Edwards and Daniel 1992a, 1992b, 1993). Phosphorus runoff is a problem since P is normally the limiting nutrient for eutrophication in freshwater systems (Schindler 1977). Accelerated eutrophication can result in degraded water quality, and can lead to fish kills and/or other problems, such as increased water treatment costs to remove odors and tastes associated with algal metabolites.

Most (80-90%) of the P in runoff from pastures fertilized with animal manures is in the soluble form, even though soluble P accounts for only about 20% of the total P in the manure being applied (Edwards and Daniel 1992a). Soluble P is the form that is most available for algal uptake (Sonzogni et al. 1982). Based on this, Moore and Miller (1994) hypothesized that the addition of chemical amendments containing Al, Ca, and/or Fe could be utilized to precipitate the soluble P in poultry manure. We showed soluble P concentrations in normal poultry manure were about 2000 mg P kg^{-1}. Additions of various compounds to the manure reduced soluble P contents down to 10 mg P kg^{-1} (Moore and Miller 1994). Later work by Shreve et al. (1995) showed that these reductions in soluble P would, in fact, result in reduced P runoff from land fertilized with poultry manure. These data indicate that P runoff from tall fescue (*Festuca arundinacea*) plots fertilized with poultry manure could be reduced by as much as 87%, indicating that reduction in soluble P could result in less P runoff.

Shreve et al. (1995) also found that tall fescue fertilized with alum-treated poultry manure had significantly higher yields than plots fertilized with manure treated with ferrous sulfate or with untreated manure. Tissue analyses of the tall fescue showed that the improved yields due to alum were probably due to increased N availability, since plots fertilized with alum-treated manure had significantly higher N contents and N uptake. Shreve et al. (1995) hypothesized that this increased N availability in alum-treated manure was probably due to a reduction in ammonia (NH_3) volatilization.

Evidence that alum applications to poultry manure can reduce NH_3 volatilization was provided in a laboratory study by Moore et al. (1995), who showed alum could reduce N loss via NH_3 volatilization by as much as 99% over a 6-week 6 period. Reductions in NH_3 emissions were accompanied by increases in the N content of the manure (Moore et al. 1995). Later work by Moore et al. (1996) showed that alum was more efficacious (and more cost-effective) than other compounds used for this purpose. The only compound that was as effective as alum in reducing NH_3 volatilization was phosphoric acid (H_3PO_4). Although phosphoric acid treatment of manure would be beneficial in areas where soil test P levels are deficient or low, it would not be recommended for areas that already have a buildup of soil test P to high levels. In this case, an amendment which reduces P availability (and P runoff) would be desirable. Recent research has also shown that treating poultry manure with alum will also reduce heavy metal runoff (Moore et al. 1998) and estrogen runoff (Nichols et al. 1997).

Reducing NH_3 volatilization from poultry manure is beneficial from both an environmental and an agricultural point of view. Ammonia volatilization from animal manures, like poultry manure, results in high levels of NH_3 gas in the atmosphere of animal rearing facilities, which is very detrimental to the health of both the animals and farm workers. Carlile (1984) stated that the critical atmospheric NH_3 level for poultry is 25 $\mu l\ l^{-1}$ (ppm). Above this concentration, levels of NH_3 can cause reduced growth rates, reduced egg production, poor feed efficiency, impairment of respiratory tract function, immunosuppression and retinal damage (Carlile 1984). High NH_3 concentrations in animal rearing facilities are more common in the cooler months (particularly winter) since high heating costs force growers to decrease ventilation rates during these periods (Anderson et al. 1964). Reducing NH3 emissions would also be desirable from an environmental point of view, since atmospheric NH_3 contamination results in acid rain and atmospheric N deposition (ApSimon et al. 1987; van Breemen et al. 1982; Schroder, 1985).

The objectives of this study were to evaluate the effects of alum applications to poultry manure on (1) ammonia volatilization rates from manure, (2) atmospheric ammonia levels in poultry houses, (3) poultry performance (weight gains, feed conversion, etc.), (4) energy use, and (5) P runoff from small watersheds.

Materials and Methods

Two poultry (*Gallus gallus domesticus*) farms were chosen for this study; one had six poultry houses, the other four poultry houses. The manure (referred to as poultry litter in the US) in the houses was removed at the beginning of the study and fresh bedding materials (wood shavings) were placed in each house. After each flock of birds, the manure was caked out (a process where the upper layer of manure is removed), using a commercial decaking machine. Afterwards, alum was applied in half of the houses on each farm, with the other houses serving as controls. The rate of alum application was 1362 kg $house^{-1}$ after the first flock (first year only) and 1816 kg $house^{-1}$ on all subsequent flocks. Alum was not applied after the last flock of birds prior to total house cleanout. For a detailed description of this study, see Moore et al. (1997a).

Manure pH and atmospheric NH_3 concentrations were determined weekly. Atmospheric NH_3 levels were measured using dragger tubes. Manure samples were taken with a soil corer (2.54 cm ID). Upon return to the laboratory, a 20-g sample of the manure was shaken with 200 ml of deionized water (1: 10 manure: water) for 2 h. The samples were centrifuged for 20 min at 7000 rpm and unfiltered aliquots were used for pH measurements.

In order to measure NH_3 fluxes from the manure surface, 18 plastic NH_3 flux chambers were constructed (Moore et al. 1997b). Three of the flux chambers were placed in each of six poultry houses at farm A during the 5th flock of birds and the concentration inside the chambers was measured immediately after placement and 1 h later using dragger tubes (Sensidyne ammonia detection tubes). These tubes contain a chemical compound which reacts quantitatively with NH_3, changing color in the process.

Although these flux measurements provide a good indication of the *relative* NH_3 emission rate, they underestimate the true NH_3 flux when calculations are simply based on the ideal gas law as done by Moore et al. (1997b). The technique was compared to that of Brewer (1998) in another study on the effects of alum on NH_3 volatilization. Although the fluxes obtained using the two methods were highly correlated to each other, the values obtained using Brewer's improved flux chamber were approximately 30 times higher. This was expected, since his method measures flux for much shorter time periods (less than 10 min), which reduces the chances of distorting the concentration gradient of NH_3 over the manure (Brewer's method also utilizes an infra-red NH_3 analyzer, which was more accurate than our method).

Our method used a 1 h flux measurement period, which resulted in a buildup of ammonia in the flux chamber, which reduces the concentration gradient of NH_3 at the manure/air interface, slowing diffusion. Therefore, the flux data were multiplied by a factor of 30, so the fluxes shown in this paper are 30 times higher than those reported by Moore et al. (1997b).

Bird performance (weight gains, feed conversion, condemnation, and mortality) data were obtained from the growers and poultry companies (integrators). Bird weights were determined by the poultry company for four growouts on farm A and three growouts on farm B. Feed conversion was determined three times at farm B and was not determined at farm A. Energy use measurements were also taken weekly (each house was equipped with an electric meter and propane tank).

Two watersheds (0.405 ha) were constructed at both farms (side by side) by building small (0.25-m) earthen berms using topsoil brought in from offsite. After the berm construction was completed, the watersheds were equipped with flumes and automatic watersamplers (American Sigma Corp., Medina, NY). The automatic water samplers were programmed to sample at 1, 3, and 7 min after runoff began, then the sampler switched to a volumetric mode and took a water sample after every 379 l (100 gallons). The samplers were checked after each heavy rain to determine if runoff had occurred and, if it had occurred, the information from the runoff event was downloaded from the water samplers to a portable computer and the water samples were returned to the lab for SRP (soluble reactive phosphorus) and TP (total phosphorus) analyses. Soluble reactive P was determined by the Murphy-Riley method on an auto-analyzer. Total P was determined by ICP following a nitric acid digestion (US EPA method 3030E).

Runoff samples were collected from each of the watersheds for 6 months prior to manure application to determine the background P concentrations in runoff water. Poultry manure applications to the watersheds were made each year in the spring (April-May). Manure application rates (on a fresh weight basis) were 4460 kg ha^{-1} in 1995 and 7136 kg ha^{-1} in 1996 and 1997.

Results and Discussion

Aluminum sulfate (alum) applications lowered the pH of the poultry manure significantly, with the greatest effects during the first 3-4 wks after the beginning of each flock (Fig. 1A). When alum was first applied to the manure, the pH would drop to around 5.5 or 6, then it would slowly increase to around pH 7.5, at which point it would plateau. The pH of the control manure remained fairly constant (near 8) throughout the study. The decrease in the pH of the manure when alum is applied is due to the following reaction:

$$Al_2(SO_4)_3.14H_2O + 6H_2O \rightarrow 2Al(OH)_3 + 3SO_4^{2-} + 6H^+ + 14H_2O. \quad (1)$$

Hence, 6 moles of protons are released for every mole of alum that is added to the manure. These protons reduce the manure pH and shift the NH_3/NH_4^+ towards $NH4^+$ (which is not volatile) as follows:

$$NH_3 + H^+ \rightarrow NH_4^+. \quad (2)$$

The effect of lower pH on ammonia volatilization (both atmospheric NH_3 and NH_3 flux from manure) was greatest during the first 4 weeks after application. Atmospheric NH_3 levels in untreated control houses were almost always above the critical NH_3 level for poultry (25 $\mu l\, l^{-1}$), whereas the NH_3 levels in alum-treated houses stayed

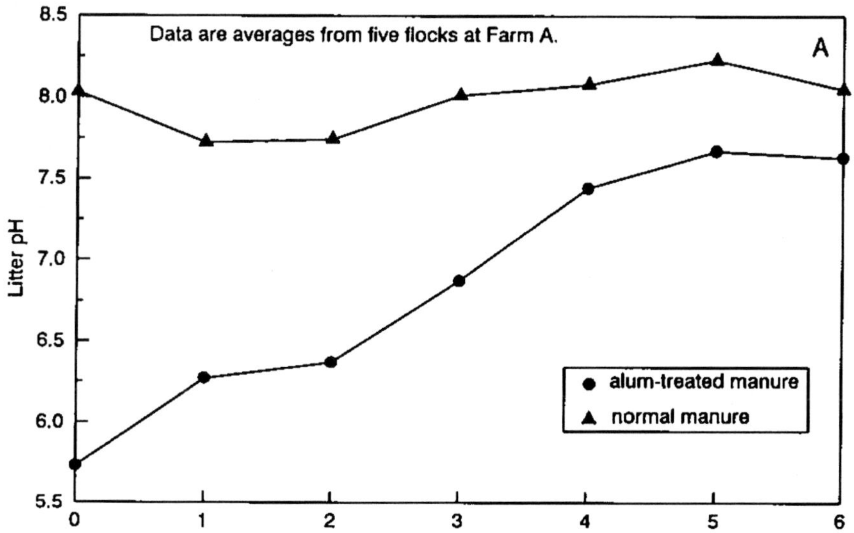

Fig. 1A: Manure pH as a function of time in alum-treated and control manure

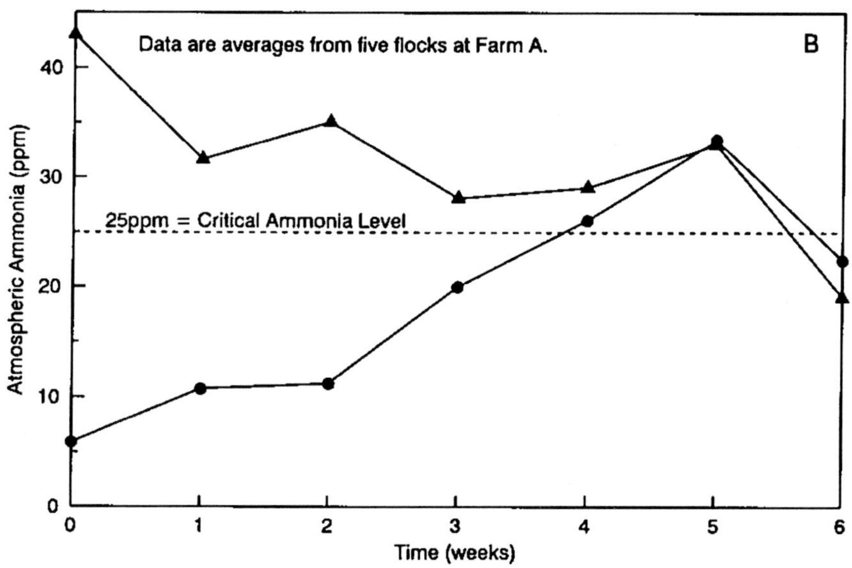

Fig. 1B: Atmospheric NH_3 as a function of time in alum-treated and control houses

relatively low for the first several weeks of each flock (Fig. 1B). This period is believed to be the time when chickens are most susceptible to high NH_3 levels. Reduced poultry performance (low weight gains and poor feed conversion) have been documented when atmospheric NH_3 is at or above 25 $\mu l\ l^{-1}$ (Carlisle 1984; Reece et al. 1981). Farm workers, who often spend over 8 h per day in poultry houses, are also vulnerable to the negative effects of atmospheric NH_3. In the European Community, the COSSH (Control of Substances Hazardous to Health) has set the limit of NH_3-N human exposure to ammonia at 25 µl l^{-1} for an 8 h day and 35 µl l^{-1} for a 10-min exposure (Williams 1992). Although the differences were large in atmospheric NH_3 levels in the houses, they would have been much greater had the ventilation in each house been the same. The farmers were in charge of the ventilation and almost always had the fans set so that the untreated houses were ventilated more (due to the high levels of NH_3 in these houses). Hence, atmospheric NH_3 concentrations do not paint an accurate picture of the rate of volatilization. Hence, NH_3 flux chambers were utilized to measure the relative differences in volatilization rates.

The NH_3 data indicated that fluxes from alum-treated manure were near zero for the first 28 days, whereas in the control manure the cumulative flux was over 200 g N m^{-2} (Fig. 2). By the end of the flock (day 43), the cumulative flux was 85 mg N m^{-2} for the alum-treated manure and 408 mg N m^{-2} for the untreated manure. Hence, the addition of alum to poultry manure reduced the NH_3 emission by 79% for the entire period.

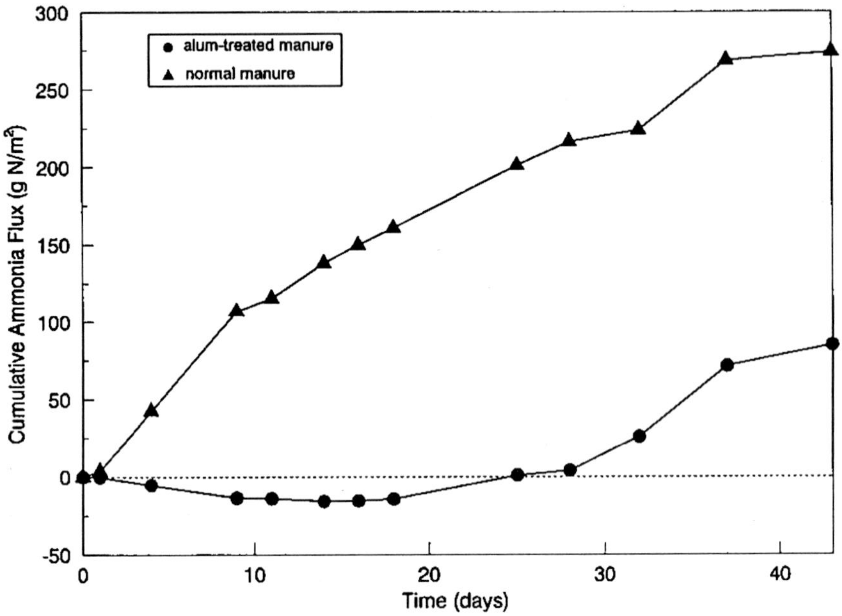

Fig. 2: Cumulative NH_3 fluxes from alum-treated and untreated manure

Brewer (1998) found similar results; alum applications applied at the recommended rate (0.091 kg $bird^{-1}$) resulted in a net flux of NH_3 of zero for the first 3 weeks of a growout, whereas alum applied at the 0.5x rate reduced NH_3 emissions by 52% compared to controls during this time. During a 6-week period (the time to grow one flock of broilers), Brewer (1998) estimated the total N release from a poultry house with 20 000 broilers would be 131 kg for an alum-treated house (at the recommended rate) and 296 kg for an untreated house. This is approximately the same order of magnitude of flux that would be predicted based on our data on total N after five flocks, which was 3.45% N for control and 3.85% N for alum-treated (Table 1).

Table 1. Chemical characteristics of alum-treated and normal poultry manure after five growouts. (After Moore et al. 1997a)

	Alum-treated manure		Normal manure	
Parameter	Average	Std. dev.	Average	Std. dev.
pH	7.59	0.77	8.04	0.18
EC	10 833	471	6611	311
		$g\ kg^{-1}$		
N	38.5	1.1	34.5	2.7
S	33.9	9.8	6.8	0.4
Ca	29.4	3.6	34.1	4.2
K	27.4	2.7	26.4	1.6
P	18.9	1.8	22.4	1.7
Al	18.7	6.0	1.18	0.2
Na	7.54	0.6	7.85	0.6
Mg	5.79	0.7	6.57	0.4
		$mg\ kg^{-1}$		
Fe	1717	312	1095	155
Mn	893	216	956	134
Cu	679	93	748	102
Zn	598	51	718	69
B	46	4	51	4
Ti	31	11	44	19
As	20	8	43	4
Ni	21	5	15	2
Pb	8	2	11	2
Co	6	2	6	1
Mo	5	0.5	6	0.5
Cd	3	0.4	3	0.2

It should be noted that the negative NH_3 fluxes shown in Fig. 2 are, in fact, real. When the pH of poultry manure is low (particularly below pH 6), the manure becomes a sink, rather than a source, for atmospheric NH_3. This phenomenon was also observed by Brewer (1998).

Production data, such as weight gains, indicated that broiler performance was significantly improved when the birds were grown on manure treated with alum.

The average weight of birds at 42 days was 1.66 kg for control birds and 1.73 kg for birds grown on alum-treated manure (Fig. 3). At this time, the reason for improved performance is unclear. There are several factors that could be causing increases in body weights due to alum, including; (1) a reduction in atmospheric NH_3; and (2) a change in the microbiology of the manure.

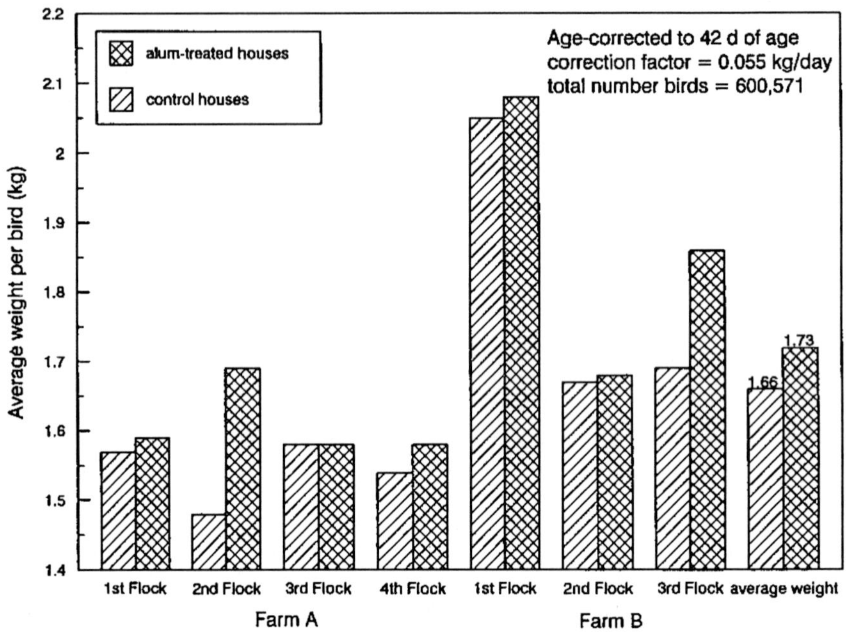

Fig. 3: Broiler weights at 42 days of age grown on normal manure and alum-treated manure

Earlier work by Scantling et al. (1995) showed *E. coli* and total coliform counts were significantly reduced when poultry manure was treated with alum. Likewise, Line (1998) found that aluminum sulfate applications to broiler manure significantly reduced both *Salmonella* and *Campylobacter* populations in the manure and completely eliminated the incidence of *Campylobacter* on poultry carcasses. However, the effects of high NH_3 levels have been well documented. Reece et al. (1981) showed that broilers exposed to concentrations of NH_3 as low as 25 $\mu l\ l^{-1}$ had body weights that were 4% lower than controls. The difference in weight gains between control birds and birds grown on alum-treated manure observed in this study was 4%.

Another important production parameter affected by alum use was feed conversion, which is the amount of feed consumed (kg) to produce a given body weight (kg); hence a lower number is desired. Feed conversion was 1.98 for birds grown on alum-treated manure and 2.04 for controls. Since the greatest cost associated with broiler production is the cost of the feed, lower feed conversions can greatly reduce production costs.

Electricity use was 13% lower and propane use was 11% lower for alum-treated houses than controls. Propane gas is used for heating animal rearing facilities in the US. Lower energy use in the houses treated with alum was a result of lower ventilation rates during cooler periods, such as winter, needed to reduce NH_3 levels.

The economic benefits of this practice are high. Moore et al. (1997a) found that the benefits obtained from the use of alum were \$940 per house per flock, while the cost to treat a house was \$480; resulting in a benefit/cost ratio of 1.96.

Although the economic benefits of alum treatment of poultry manure are encouraging, the biggest benefit is to the environment. The Al associated with alum binds soluble P in the manure, which reduces P runoff. The SRP concentrations in runoff were 1.05 and 3.23 mg P l^{-1} for the alum-treated and normal (control) manure, respectively, during the first year, indicating that alum reduced SRP runoff by 67% (Fig. 4). These differences were somewhat greater during year 2, with average SRP concentrations of 2.04 and 7.94 mg P l^{-1} for the alum-treated and normal manure, respectively; a 74% reduction. Phosphorus runoff during the third year followed the same trends, with 1.70 and 7.69 mg P l^{-1} for the alum-treated and control manure (a 78% reduction). Total P concentrations in runoff followed the same trends (Moore et al. 1997a).

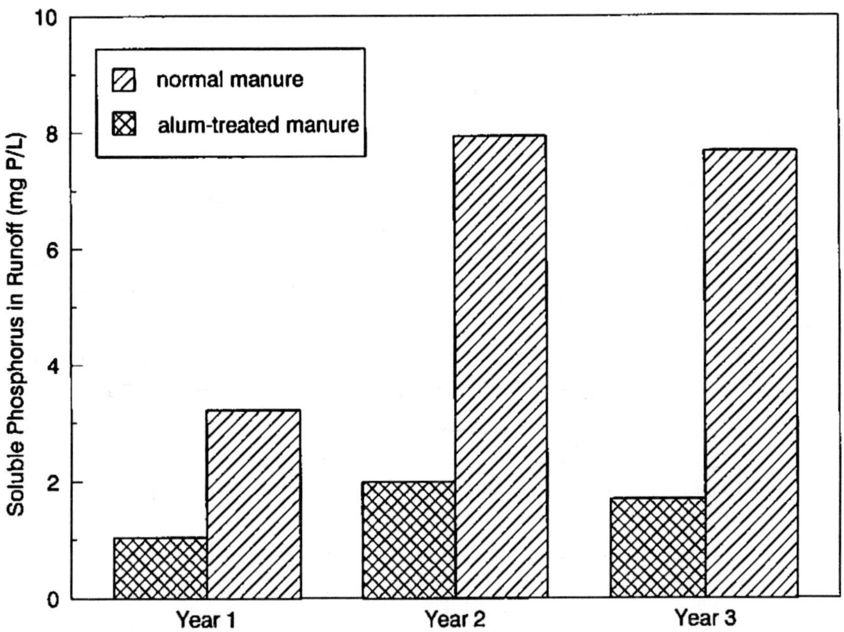

Fig. 4: Soluble reactive P concentrations in runoff water from fields fertilized with normal poultry manure or alum-treated manure

Conclusions

This study showed that the use of aluminum sulfate (alum) as a manure amendment in commercial broiler houses resulted in several important benefits. Among these were a reduction in NH_3 emissions, improved broiler performance (improved weight gains and feed conversion), lower energy use, and a reduction in P runoff. Other studies have shown that alum use reduces estrogen and heavy metal runoff as well. Hence, this management practice appears to be beneficial from both economic and environmental points of view.

References

Anderson DP, Beard CW, Hanson RP (1964) The adverse effects of ammonia on chickens including resistance to infection with Newcastle disease virus. Avian Dis 8:369-379

ApSimon HM, Kruse M, Bell JNB (1987) Ammonia emissions and their role in acid deposition. Atmosph Environ 21:1939-1946

Brewer SK (1998) Ammonia flux from fresh, re-used, and alum-treated broiler litter. MS Thesis, University of Arkansas, Fayetteville

Carlile FS (1984) Ammonia in poultry houses: a literature review. World's Poultry Sci J 40:99-113

Edwards DR, Daniel TC (1992a) Potential runoff quality effects of poultry manure slurry applied to fescue plots. Am Soc Agric Eng 35:1827-1832

Edwards DR, Daniel TC (1992b) Environmental impacts of on-farm poultry waste disposal – a review. Bioresour Technol 41:9-33

Edwards DR, Daniel TC (1993) Effects of poultry litter application rate and rainfall intensity on quality of runoff from fescue plots. J Environ Qual 22:361-365

Line JE (1998) Aluminum sulfate treatment of poultry litter to reduce *Salmonella* and *Campylobacter* populations. Poultry Sci Abstr S364

Moore PA Jr, Miller DM (1994) Decreasing phosphorus solubility in poultry litter with aluminum, calcium and iron amendments. J Environ Qual 23:325-330

Moore PA Jr, Daniel TC, Edwards DR, Miller DM (1995) Effect of chemical amendments on ammonia volatilization from poultry litter. J Environ Qual 24:293-300

Moore PA Jr, Daniel TC, Edwards DR, Miller DM (1996) Evaluation of chemical amendments to reduce ammonia volatilization from poultry litter. Poultry Sci 75:315-320

Moore PA Jr, Daniel TC, Gilmour JT, Shreve BR, Edwards DR, Wood BH (1998) Decreasing metal runoff from poultry litter with aluminum sulfate. J Environ Qual 27:92-99

Moore PA Jr, Haggard BE, Daniel TC, Edwards DR, Shreve BR, Sauer TJ (1997a) Demonstration of nutrient management for poultry litter using alum precipitation of soluble phosphorus. Final Report to US EPA for Federal Assistance Project No 9006749920. pp. 114

Moore PA Jr, Huff WE, Daniel TC, Edwards DR, Sauer TC (1997b) Effect of aluminum sulfate on ammonia fluxes from poultry litter in commercial houses. Proc 5th Int Symp Livestock Environ, II: 883-891, Bloomington, Minnesota

Nichols DJ, Daniel TC, Moore PA Jr, Edwards DR, Pote DH (1997) Runoff of estrogen hormone 17ß-estradiol from poultry litter applied to pasture. J Environ Qual 26:1002-1006

Reece FN, Lott BD, Deaton JW (1981) Low concentrations of ammonia during brooding decrease broiler weight. Poultry Sci 60:937-940

Scantling M, Waldroup A, Marcy J, Moore P (1995) Microbiological effects of treating poultry litter with aluminum sulfate. Poultry Sci Abstracts pp 216

Schindler DW (1977) The evolution of phosphorus limitation in lakes. Science 195:260-262

Schroder H (1985) Nitrogen losses from Danish agriculture – Trends and consequences. Agric Ecosyst Environ 14:279-289

Shreve BR, Moore PA Jr, Daniel TC, Edwards DR (1995) Reduction of phosphorus in runoff from field-applied poultry litter using chemical amendments. J Environ Qual 24:106-111

Sonzogni WC, Chapra SC, Armstrong DE, Logan TJ (1982) Bioavailability of phosphorus inputs to lakes. J Environ Qual 11:555-563

US Environmental Protection Agency (1979) Methods for chemical analysis of water and wastes (EPA-600/4-79-020). US EPA, EMSL, Cincinnati, OH

van Breemen N, Burrough PA, Velthorst EJ, van Dobben HF, de Wit T, Ridder TB, Reijinders HFR (1982) Soil acidification from atmospheric ammonium sulphate in forest canopy throughfall. Nature 299:548-550

Williams PEV (1992) Socio constraints on poultry production – addressing environmental and consumer concerns. In: 1992 Proc Arkansas Nutrition Conference, Fayetteville, pp 14-29

Precision Agriculture As a Tool for Sustainability

JS. Schepers[1], JF. Shanahan[1], A. Luchiari Jr[2]

Precision agriculture is the name given to an innovative approach for crop production that represents a unique blend of old thoughts and new tools. Established principles and processes support a treasure of knowledge and wisdom that provides a foundation for applying new technologies to proven concepts and practices. These new technologies include innovative computer software, a new generation of sensors, and ingenious control devices, as well as instrumentation and communication technologies from the defense industry. Had it not been for the development of global positioning systems (GPS) by the military and geographical information systems (GIS) by the mining and exploration companies, precision agriculture would not be happening. Incorporating the various monitors, communication devices, and computers into agricultural equipment has resulted in a variety of sophisticated implements intended to increase profitability and protect the environment through better management decisions. These new implements not only keep track of the geographical position in fields, but are also able to monitor what is happening and control production inputs and processes on-the-go. In essence, precision agriculture tools and devices are able to collect huge amounts of site-specific data that can readily be assimilated into useful information from which management decisions can be made. In the past, producers familiar with the land, crops, and climate subjectively integrated the various factors and made decisions as appropriate. The goal of precision agriculture is to systematically make intelligent site-specific management decisions based on objective information (data) gathered from various sources and collected at different times. The success of precision agriculture depends on being able to identify meaningful relationships between factors that are economically and environmentally important and parameters that can be easily measured with a reasonable degree of spatial resolution. The relative importance of these relationships is known to change with time (within growing seasons and between years) and the process of integrating the information into reliable and

1. Soil Scientist and Agronomist, USDA-Agricultural Research Service, Lincoln, Nebraska 58686-0915.
2. Visiting Scientist, LABEX-EMBRAPA/ARS, Brazil.

consistent decisions is complicated because the interactions between climate, soil, and the crop are complex.

Introduction

Precision agriculture is a concept that integrates technologies, knowledge, and wisdom to collect and analyze spatially and temporally variable data so as to make improved management decisions that are more profitable and environmentally sound. Precision agriculture is frequently called site-specific management, which essentially amounts to making management decisions at a higher level of spatial and perhaps temporal resolution than in the past. It goes without saying that information used to make such decisions must also be collected or estimated at a higher resolution. In some cases, the information may also be collected more frequently, especially when there is an opportunity to make spatial or more timely management decisions.

Precision agriculture is a relatively new discipline and, as such, many of the reports in the literature are in the form of progress reports or nonrefereed proceedings, but nonetheless they contain useful information (Robert et al. 1996; Stafford 1997). Over the past decade, several books that deal with specific aspects of precision agriculture have been published. Examples include spatial variability in soil and landforms (Mausbach and Wilding 1991), applications of GIS to modeling pollutants in the vadose zone (Corwin and Loague 1996), geospatial and information technologies for crop production (NRC 1997), soil-specific crop management (Robert et al. 1993), site-specific management for agricultural systems (Robert et al. 1995), and the state of site-specific management in agriculture (Pierce and Sadler 1997).

Global positioning system (GPS) technologies adapted from military applications make it possible to instantaneously determine the latitude and longitude of any position on the earth's surface. The accuracy of this positional data as used in agriculture is usually within 2 m and frequently < 1 m.

Merging positional data with other physical, chemical, and biological data from a given point or site provides the makings of a field map. In the case of soil chemical data, nutrient availability maps can be generated from tens to hundreds of sampling points (usually 1 to 5 samples ha^{-1}) per field. Yield maps generated by a yield monitoring combine will typically contain 600 to 1000 data points ha^{-1} (one yield measurement s^{-1}). Aerial photographs of agricultural fields typically have a resolution of $\sim$ 100 000 pixels (term referring to picture elements) ha^{-1}. Resolution of digital aircraft and satellite imagery currently ranges from 1 to 20 m (25 to 10 000 pixels ha^{-1}). The reliability of soil survey maps in the United States is such that areas of less than 1 ha are not delineated. Therefore, digitized soil survey maps have an inherent minimum area limitation as well as having unknown accuracy in defining the transition from one soil type to another. Some maps such as soil electrical conductivity generated with EM-38 or VARIS technologies, can be at virtually any scale. From these examples, it is clear that considerable disparity exists between the spatial resolution of the various sources of data used in site-specific management.

Even though the various sources of spatial data have different resolutions, geographic information system (GIS) technologies adapted from the mining and oil industry make it possible to interpolate between data points to generate maps. Many GIS software packages incorporate statistical programs that consider nearest-neighbor information to estimate values for intermediate points. Some of these GIS packages allow users to "resample" maps generated at various levels of resolution to establish a data set with a common resolution. This process is analogous to stacking a series of images or maps having the same outer boundary and then drilling through all layers to extract data for a specified point. In practice, the result is a data set providing the latitude, longitude, and a series of crop and soil data, from which one can calculate crop nutrient needs, make economic assessments, compare management practices, conduct computer simulations, or make environmental assessments.

One can visualize how GIS software works by thinking about graph paper with many square segments. Each square has a unique X (horizontal) and Y (vertical) coordinate relative to a designated reference point. This type of GIS is based on a raster approach much as is used with video systems. The other type of GIS software is based on a vector or polar coordinate system (angle and distance from a reference point). Raster-based systems result in blocky transitions between segments, but with high-resolution sampling will result in realistic maps with smooth transitions. Most experts agree that raster-based systems are more efficient for data collection and analysis, but vector-based systems result in the most pleasing maps. Many software packages are now available to capitalize on the strengths of each system by switching from raster to vector approaches.

Reasons to Consider Precision Agriculture

Observation of spatial variability in crop growth in agricultural fields has been the primary incentive for producers to consider site-specific management techniques. Producers are inclined to associate spatial variability with lost yield potential and reduced profitability. Agriculture is no different than any other business in that those involved would like to feel that they are in control to the greatest possible extent. Initial efforts to understand the causes of spatial variability in crop growth focused on nutrient availability. Assessing the nutrient status of soils requires extensive sampling, which is time-consuming, and chemical analyses are costly. Producers who follow this approach are able to generate maps of nutrient availability from which nutrient recommendation maps can be generated. The expectation is that variable rate nutrient application should result in uniform crop growth. Increased profitability is anticipated through increased yields in areas that were formerly depressed. The goal is that reduced input costs for parts of the field will offset increased input costs in other areas. Intuitively, environmental risks should be reduced with site-specific management because fertilizer and herbicides will only be applied where it is needed.

Approaches to Precision Agriculture

Early adapters of variable rate fertilizer application technology have been generally disappointed because spatial variability in crop yield has not disappeared. Reasons why crop yields remain variable include factors such as rooting depth limitations, unanticipated leaching or volatilization losses, drought, weeds, disease, or insect damage. Many of these factors cannot be detected even with intensive soil sampling.

Problems frequently encountered with soil sampling include reaching a compromise between what is practical and affordable. Producers and consultants frequently use a systematic grid pattern to represent all parts of the field when sampling. A grid size of 1 ha per sample (100 x 100 m) is typical for many applications. Deciding where within the cell to collect the sample may seem trivial, but sampling strategies that sample from the center vs. the corner of each cell can result in significantly different nutrient availability maps. This is because any standardized sampling strategy is bound to include samples from non-representative areas or samples from transition zones. The result can be maps with skewed and distorted patterns.

It stands to reason that more intensive soil sampling should result in more accurate maps. Gotway et al. (1996) demonstrated that high-intensity grid sampling (34 samples ha^{-1} in a 12 x 24-m alternate grid) resulted in an organic matter map that was considerably more accurate than when 8 samples ha^{-1} were used. The method of statistical interpolation (i.e., kriging or inverse distance) did not make as much difference in the patterns displayed as the sampling intensity. A related study by Varvel et al. (1999) on the same field showed that an aerial photograph of the bare soil had the same patterns as displayed in the map generated by high-intensity grid sampling. An increasing number of producers and consultants are looking to reduce soil sampling costs and are considering remote sensing to better define management regions within fields. A combination of yield maps, soil survey information, and crop and soil images can usually result in a series of well-defined sampling zones.

Another aspect of precision agriculture has focused on the end of the growing season rather than the beginning. Introduction of yield monitoring combines and harvesting equipment has provided producers with their first real glimpse of spatial variability through the generation of crop yield maps. Producers almost immediately translate the patterns seen in yield maps to lost productivity and reduced profits. The first questions asked by producers usually relate to why the yields are variable and what they can do to fix the problem. Answers to these questions are not usually obvious because crop yields represent an integration of climate and growing conditions over an entire growing season. As such, reduced yields can be due to cultural practices or any number of factors that result in the crop being stressed.

The limitation of crop yield maps is that they represent a post mortem evaluation and, unless other information is gathered before or during the growing season, it is difficult to determine the cause of yield variability. This is why remote sensing is gaining popularity as a tool for gathering temporal data. The remote sensing option offers a fast and inexpensive way to gather information about crop growth during the growing season. Remote sensing also permits real-time evaluation of spatial variability so that producers and consultants can make more timely management decisions.

Moving Precision Agriculture Forward

Precision agriculture has invigorated many groups within the agri-business community. In most cases, the interest is driven by the prospect of new business and increased profitability. One scientist summed it up by stating the "Everybody is looking for the opportunity to sell something." Those that stand to make the most are pushing the hardest. This is why the implement companies that provide the yield monitoring devices and variable rate application equipment are so interested in promoting the technologies. Suppliers of crop nutrients are also promoting the concept because of the perception that much of the spatial variability is due to nutrient availability. Virtually every segment of agriculture and many affiliated groups have found a niche and reason to support precision agriculture (Table 1).

Table 1. Reasons why various groups are interested in precision agriculture and services they might offer

Group	Activity
Producers	Fields with spatial variability, yield maps
Primary suppliers	Services for fertilizer, chemicals, and lime application
Secondary suppliers	Agronomic information (i.e., seed industry)
Manufacturers	Applicators, monitors, sensors, software, hardware
Consultants	Feld scouting and crop stress diagnosis
Finance	Profitability analysis, land valuation
Agency	Environmental emphasis
Military contractors	Software, hardware, remote sensing, sensors
Scientists	Curiosity, problem-solving skills
Educators	Passive association with other disciplines
Food processors	Identity preserved and specialty products

Reality

Precision agriculture has been in the implementation stage since the mid-1980s in the United States. Producers tend to explore new ideas and concepts in ways that minimize their risks or that are the least intimidating. Support and expertise from scientists, agribusiness, and industry often influence the approach that any given producer takes. One of the greatest challenges to producers has been the need to interface technologies from different disciplines (i.e., GPS, GIS, software). Initially, the potential for intimidation by the instrumentation and computer software was great because the technologies had been developed for different applications. As a result, communication between the various components was frequently not totally compatible. This frequently resulted in a high level of frustration on the part of the users. Over time, software and hardware have evolved that minimize the intimidation factor and shorten the time required to become proficient with the technologies. In fact, producers and consultants are now finding that technologies used to make site-specific management a reality are ahead of the science needed to interpret the data and make management decisions.

A number of producers now have several years' experience in generating yield maps for corn and soybean. More often than not, producer comments pertain to why they observe different patterns in yield maps between years. For this reason, consultants recommend accumulating several years' data before making too many management decisions. In reality, producers should not expect yield maps to look the same from year to year unless climatic conditions were the same for the entire growing season. The likelihood of this happening in nature is rare, with perhaps the exception of irrigated environments.

Detective skills are an integral part of site-specific management because producers are constantly faced with trying to explain what happened. This task can be made considerably easier with a series of aerial images or photographs to document when and where crop stresses occurred. Sometimes, features will appear in images or yield maps that cannot be explained. This is when a visit with retired producers in the area can be informative because these folks may be able to describe past field boundaries, long-term cropping practices, or location of farmsteads and livestock operations.

It is rare that patterns in any one map or photograph can be interpreted without the help of several more. The limitation seems to be in properly weighting the information contained in the various maps or photographs. This requires understanding how crop growth interacts with soil and climate features and identifying when specific factors have the greatest potential influence on yield. In the past, producers were better able to integrate all of these factors because they were more familiar with the characteristics of each part of a field. For years, larger equipment, automation of many farm operations, and larger farms has continued to compromise producer's time and ability to make site-specific management decisions. The information needed to make such detailed decisions simply was not available in a convenient form or timely manner. Precision agriculture offers the potential to gather temporal and spatial data on many parameters. What is needed is the time and ability to integrate the information into meaningful packages and decision aids for producers. Scientists must now develop guidelines or software packages that assimilate masses of data into recommendations much as our ancestors instinctively did as they worked their fields.

Integrating Science and Policy

The science involved in precision agriculture is not necessarily anything new in that the basic principles and concepts have been around for years. What is new is that precision agriculture offers producers and society the opportunity to be more quantitative in the decision-making process. The intuitive and qualitative approaches to management in the past did not offer the same level of evaluation and assessment as is possible today using site-specific management technologies. Precision agriculture today and in the future offers the potential to move away from the "one-size-fits-all" mentality and make better-informed management decisions. Times are sure to arise when profitability and environmental stewardship issues will place producers and society at odds. Precision agricultural technologies offer the potential for discussions about environmental issues to become more objective and perhaps less subjective and emotional.

Precision agriculture is offering an example for the rest of society to consider in that expertise from a number of disciplines must be integrated to make the concept work. This may not be too different from what is required to extend scientific information into the policy arena. Both require two-way communications and ample discussion to fully integrate the information and ideas into something that is meaningful and in the best interest of society. Such activities facilitate discussion of both short and long-term objectives and permit better allocation of resources. The approach also reduces the likelihood that the solution to one problem will cause two or three more problems that will need to be addressed in the future.

There could well come a time when society will decide that environmental consequences of whole-field management practices are not acceptable and insist that producers adapt site-specific management practices (Matson et al, 1998). If this happens, it may become necessary for society to provide the necessary incentives to entice producers to comply with forthcoming guidelines and regulations. Technologies are currently available to variably apply almost any input used to produce crops. The science to make these management decisions is moving ahead rapidly. It will soon be up to society to decide how far they are willing to go to protect the environment or provide a safe and high quality of food. Individuals from various disciplines and segments of society will need to be involved if we are to translate "Good Science" into "Good Policy".

References

Corwin DL, Loague K (1996) Applications of GIS to the modeling of non-point source pollutants in the vadose zone. Soil Science Society of America Spec Publ 48. Madison, Wisconsin

Gotway CA, Ferguson RB, Hergert GW, Peterson TA (1996) Comparison of kriging and inverse-distance methods for mapping soil parameters. J Soil Sci Soc Am 60:1237-1247

Matson PA, Naylor R, Ortiz-Monasterio I (1998) Integration of environmental, agronomic, and economic aspects of fertilizer management. Science 280:112-115

Mausbach MJ, Wilding LP (1991) Spatial variabilities of soils and landforms. Soil Science Society of America Spec Publ 28. Madison, Wisconsin

National Research Council (1997) Precision agriculture in the 21st century. National Academy Press, Washington, DC

Pierce FJ, Sadler EJ (1997) The state of site-specific management for agriculture. American Society of Agronomy, Madison, Wisconsin

Robert PC, Rust RH, Larson WE (1993) Soil-specific crop management. American Society of Agronomy Misc Publ, Madison, Wisconsin

Robert PC, Rust RH, Larson WE (1995) Site-specific management for agricultural systems. American Society of Agronomy Misc Publ, Madison, Wisconsin

Robert PC, Rust RH, Larson WE (1996) Precision agriculture. Proc 3rd Int Conf on Precision Agriculture, June 23-26, Minneapolis, Minnesota

Stafford JV (1997) Precision Agriculture '97. Proc 1st Eur Conf on Precision Agriculture, September 7-10, Warwick University, Warwick UK

Varvel GE, Schlemmer MR, Schepers JS (1999) Relationship between spectral data from an aerial image and soil organic matter and phosphorus levels. Precision Agric J (in press)

Part III
Managing Quality of Production

Jean-Pierre Toutant

D. Demeyer
A. Aguzzi
J. Jazzo et al.
R. Powell

Meat Quality and the Quality of Animal Production

D. DEMEYER[1]

The need for an integrated approach to both meat production and meat quality is discussed. The recent proportional increase in world meat production surpasses that of cereal production. Poultry and pig production are preferred over ruminant production. The effects of such intensification on environment and food quality are discussed and developed into fifteen critical points within an enlarged concept of meat quality.

Introduction

Apart from his task as coordinator of the animal biotechnology theme between 1985 and 1995, the author is responsible for the teaching of nutrition, ruminant nutrition, advanced nutrition, general and tropical animal production and meat technology at two Flemish universities. This environment has forced him from early on to adapt an "integrated" approach to both meat production and meat consumption, recognising the needs of the producer as well as of the technologist and consumer.

The difficulty and complexity of such an approach is apparent from:

- The variable definitions and concepts of "meat". Indeed, the (legal) definition of meat covers muscle tissue as well as, e.g. blood and all edible parts of a slaughter animal. On the other hand, dictionaries and consumers see meat primarily as red muscle tissue and evaluate it in concepts covering: an end product of animal production, as well as a food, a raw material for industry, and environmental risk and a merchandise. Concepts do not always involve consumption!
- The great number of conflicting interests associated with the meat production column, involve, e.g. animal breeder and environmental officer, feed manufacturer, butcher and veterinary surgeon, as well as meat technologist, food inspector, agricultural economist and ultimate consumer. Interests are defended in pressure groups and, in the present situation, pressure groups pay for the greater part of scientific research. Such a situation may shift emphasis from research and publication to development and promotion, respectively. It is therefore necessary to ensure that university

1. Department of Animal Production, Department of Food Technology and Nutrition, University of Ghent Institute of Biotechnology, Proef hoe vestraat 10, 9090 Melle, Belgium.

education provides objective information, conscious of, but not dependent on, conflicting interests. The word quality is crucial for this approach to the subject, as already stated 10 years ago by a Belgian government official: "We produce too many animal products to sell in a correct manner" and: "The problem of animal waste must be dealt with once and for all" (De Baerdemaeker 1988). Also, on the same occasion, carcass quality was confronted with meat quality (Demeyer et al. 1988) and there was less concern about the role of meat in nutrition. The recent evolution of concern seems to be mainly focused on public health and the sensory quality of products.

In a further non-specialist (!) approach to the subject, some aspects of the evolution in meat production and consumption will first be illustrated, using statistical data. Next, critical points in quality criteria will be identified and listed, based on earlier reports for OECD (Demeyer 1994, 1997).

A Summary of the Situation

The Economic Welfare and the Increase in Meat Consumption and Production

Table 1 shows that between 1980 and 1995 the percentage increase of meat production surpasses that of cereals. This is the case for developing countries in Africa as well as for the rich US, where the total meat production increased with 43 and 29%, respectively, and the production of cereals with 41 and 3%, respectively.

Table 1. Some indicators for the evolution of animal production (FAO, 1996)

Production (1000 ton)		Africa	US	France	Belgium
Cereals	1980	70 838	269 979	47 830	1975
	1995	100 093	276 936	53 607	2366
		(+ 41%)	(+ 3%)	(+ 12%)	(+ 20%)
Meat[a]					
Beef	1980	2909	10 002	1825	300
	1995	3391	11 612	1640	357
Pork	1980	360	7535	1860	690
	1995	751	8097	2144	1043
Poultry	1980	1186	8435	1186	123
	1995	2211	13 825	2081	274
Total	1980	4555	25 972	4871	1113
	1995	6353	33 534	5865	1674
		(+ 43%)	(+ 29%)	(+ 20%)	(+ 50%)

[a] Carcass weigh.

This evolution confirms the increase already found in consumption of animal products accompanied with the increase of economic growth (richness). Table 2 shows that the availability of food animal proteins was increased between 1961 and 1990 with 25% on average, with peaks of 300% (Korea) and 10% (Africa).

Table 2. Availability of animal and vegetable food: energy (kcal/p) and protein (g/p) (FAO, 1992)

		1961 – 1963 (1)		1988 – 1990 [% of (1)]	
		Vegetable	Animal	Vegetable	Animal
World	Energy	1927	359	118	118
	Protein	42.7	19.9	108	125
Africa	Energy	1989	166	109	107
	Protein	44.3	11.4	102	110
Korea	Energy	1882	74	130	505
	Protein	44.2	6.4	117	400
US	Energy	1919	1149	132	96
	Protein	32.8	65.0	118	109
Belgium	Energy	2104	1112	113	140
	Protein	40.4	50.1	101	131

The Preference for Pork and Poultry – The Promise of Aquaculture

Between 1980 and 1995, as between 1990 and 1996, the increase in production was more important for pork and poultry than for beef. The world production of poultry and pork increased by 42 and 23% respectively, compared with a 3% increase in beef production (Table 3).

Table 3. Recent evolution of animal production (FAO, 1996, 1997)

	World		Africa		Asia	
	1990[a]	1996[b]	1990[a]	1996[b]	1990[a]	1996[b]
Beef	52.6	102.6	3.3	102.3	5.1	180.2
Pork	69.5	123.3	0.5	126.6	29.4	159.8
Poultry	40.8	142.4	1.8	109.1	10.1	196.5
Total meat	178.5	120.7	8.5	98.3	51.1	166.5
Fisheries	98.0	115.2	1.9	146.4	10.4	158.5
Aquaculture[c]	12.4	168.7	0.06		10.1	181.4
% of total[d]	79.0	95.4				
% of mariculture	34.8	98.3				

[a] 10^6 ton year^{-1}
[b] as % of the 1990 values
[c] Fin fish and crustacea only
[d] The total production of aquaculture includes also the culture of alga.

This preference is strikingly associated with a higher efficiency of pork and poultry production, determined by a faster production rate and a higher reproduction capacity. This results in a lower fraction of the feed used for maintenance, for the individual animal as well as for the entire population, resulting in better food conversions (Table 4).

Fish culture, as part of aquaculture, deserves special attention. Fishes differ from mammals by their low maintenance requirement: ca 50 kJ kg^{-1} $G^{0.75}$ compared to

400 kJ kg^{-1} $G^{0.75}$. This is due to the lower energy requirements of cold-blooded animals, nitrogen excretion as ammonia instead of urea and less need of resistance against gravitation as reflected in the low feed conversions (Table 4). A second positive factor is the great number of descendants in comparison with mammals. These favourable production factors may partly explain the impressive development of aquaculture, clearly the greatest in animal production, with a 10% increase on a yearly basis.

Table 4. Production rates and feed conversions for some meat and fish production systems

	Final weight (kg)	Production (kg $year^{-1}$)		Descendants per year, per mother	Feed conversion[b]
Cattle	500	430	(0.2)[a]	0.9	7 – 8
Pork	100	200	(0.6)	8 – 24	3
Poultry	2	16	(2.4)	180	1
Carp	0.5	1.5	(0.8)	> 10^4	1
Salmon	1	1	(0.3)	> 10^4	1.5

[a] ( ) = % of final weight/day.
[b] kg feed used per kg of final weight

The Ruminant As Improver

The shift from beef to pork and poultry has led to an increased yield of meat production per animal unit present, in Africa as well as in the industrialised countries. Remarkable are the much lower values in Africa in comparison with the economically developed countries: the values are two to four times lower for pork and poultry and five to six times for beef (see also Table 6). This is partly explained by the general lower quality of production conditions in Africa but also, and especially for ruminants, by the lower feed quality. The unfavourable feed conversion with ruminants is mainly caused by the digestion in the rumen accompanied by large energy and protein losses.

The microbial predigestion allows the ruminant to transform the vegetable cell walls (lignocellulose) which cannot be used as food by humans, into milk and meat protein or wool, high quality products for humans. This digestion is slow, however, limiting feed intake. Replacement of lignocellulose (pasture) in the extensive land-requiring production by cereals such as corn (feedlot, stall feeding) in the intensive less land-requiring production of the industrialised countries accelerates and increases production. In the latter system, however, the ruminant loses his natural advantage as improver of lignocellulose and becomes, with pork and poultry, a competitor with man for the use of vegetable protein and cereals, raw materials for the food of men as well as for animals.

The conversion yields of these raw materials in cattle feed to milk, meat and eggs never exceed 50% and become only favourable when ruminants, mainly dairy cows, produce edible high-quality protein for man from feed that is not directly available to him (Table 5).

Table 5. Efficiencies of feed conversion in animal production (Rérat and Kaushik 1995)[a]

	Gross energy	Crude protein	Edible protein[b]
Dairy cow	12	17-25	> 170
Beef	4.5	6	> 90
Pork	17	12	30
Broiler	10	20	30
Layer	11	18	40
Salmon	> 30	> 30	

[a] % of feed in product.
[b] edible protein for man (Van Es, 1981).

It is obvious that such considerations are of secondary importance compared to cultural differences and a more favourable feed conversion and reproduction for pork and poultry in the development of animal production. It seems, however, indisputable that in the long run the ruminant must gain an important and perhaps the most important role as improver of lignocellulose, representing more than 90% of the plant biomass produced by photosynthesis.

Limits to Intensification

It is clear that in 1995, intensity of meat production, expressed as production ha^{-1} agricultural land, was 10 to 20 times higher for industrialised countries like the US and France in comparison with Africa (Table 6).

There was a 25 to 30% increase in intensity or intensification between 1980 and 1995 for Africa as well as for the industrialised countries shown. The Belgian situation is remarkable: in 1980 meat production ha^{-1} agricultural land amounted to ca 180 times that of Africa and it increased between 1980 and 1995 by 56%! When taking into account that the quasi total pig production is situated in Flanders, it is obvious that a limit may be reached or nearly reached.

Indeed, Table 6 shows that such a production intensity is accompanied by an average total N fertilisation of 334 kg ha^{-1} agricultural land, equivalent to maximal total nitrogen fertilisation levels recently imposed in Flanders: 325 à 450 kg ha^{-1}, dependent on the cultivation (Anonymous 1997). In relation to the actual concern with the nitrate pollution of surface waters, it can be noted that ca. 66% of total nitrogen excretion is of bovine origin. Milk and beef are, however, produced in a more land-bound manner, using more roughage produced on site.

Meat, Food and Public Health

In spite of important individual variation, it is general accepted that *on average*, better health means more food for the poor and less food for the rich countries. Limited, but rather convincing, data obtained from epidemiological research and animal experiments have clarified the relation between food and chronic, so-called civilisation diseases like obesity, cardio-vascular diseases (arteriosclerosis, stroke, coronary heart disease, etc.), cancers (mainly breast and colon cancer) and others. Awaiting the results of a more extensive epidemiological research such as undertaken in different countries, it can yet be stated with a probability, bordering on certainity, that a decrease in total food energy intake in the industrialised countries can significantly improve both health and lifespan (Weindruch 1996).

Table 6. Intensity and efficiency of animal production[a]

	Africa		US		France		Belgium	
	1980	1995	1980	1995	1980	1995	1980	1995
kg ha^{-1} [b]								
Meat	4	6	60	80	153	195	724	1129
Cow's milk	10	15	136	167	1059	849	2682	2459
Eggs	1	2	10	10	26	35	135	148
kg/au[c]								
Meat[d]	28	36	188	242	226	232	228	257
Cow's milk[e]	491	454	5386	7454	3365	5362	3905	4958
N-dose (kg ha^{-1})[b]								
Manure[f]	10	12	19	18	63	54	194	221
Fertiliser	2	2	25	26	68	80	126	113
Total	12	14	44	44	131	134	320	334

[a] FAO (1996).
[b] kg ha^{-1} agricultural land (permanent pastures and horticultural ground included).
[c] au = animal units : numbers present x 1 for cattle, x 0.5 for pork and x 0.01 for poultry.
[d] lactating cows are not included in the calculation.
[e] only lactating cows are taken into account.
[f] The N-excretion in manure was estimated based on the following values:

kg animal $year^{-1}$	Cattle			
	Dairy cow	Fattening cattle	Fattening pig	Poultry
Nitrogen[a]	97	56	13	5

[a] Average values of De Wilde (1997) and Tamminga and Verstegen (1996).

Especially fat as most energy-containing nutrient is important in this respect where intake of saturated fatty acids with 12 or 14 carbon atoms increases the risk of vascular diseases. The intake of *cis* poly and mono unsaturated fatty acids, on the other hand, is mainly associated with positive effects (Willett 1994; Nederlandse Voedingsraad 1995). These findings are reflected in the official food directives of most countries, including Belgium (Nationale Raad voor de Voeding 1996) stating that the maximal energy intake as fat must be reduced from ca. 40 to 30% of total energy intake. Intake of saturated fatty acid is limited to a maximum of 10% of total energy intake. It is obvious that especially pork is important in this respect. It was estimated for 1992 that fat and saturated fatty acids represent 43 and 11%, respectively, of the total energy intake in Belgium (Demeyer et al. 1995). Animal products were responsible for 63 and 76% of the total fat and saturated fat intake, respectively, as estimated in Table 7. It is clear that pork and dairy products each account for about a quarter of the fat in food. Dairy products only, however, account for an even higher proportion of saturated fatty acid intake; this is due to the hydrogenation activity of the microbiota in the forestomach of the cow. The recent shift to the greater consumption of poultry and fish will decrease the intake of saturated fatty acids. It is, however, obvious that, if the level of consumption of animal products is to be maintained, a decrease in the amount of saturated fatty acids in dairy products is the first aim for improvement. Indeed, selection has already decreased fatness of Belgium slaughter animals to the lowest levels in the world.

Table 7. Contribution of animal products to fat and fatty acid intake (% of total) in Belgium (1992)

		Fatty acids	
	Fat	Saturated	Other[a]
Dairy products	23	34	13
Beef	8	9	7
Pork	23	25	22
Poultry	6	6	6
Eggs	2	2	2
Fish	1	0.5	1.5
	63	76.5	51.5

[a] mono- and poly-unsaturated fatty acids.

Convincing evidence has been collected recently, linking the incidence of colon cancer with the consumption of (red) meat (Pernaud and Corpet 1997). Apart from the fat content (Slattery et al. 1997), the heating of meat seems to be determining this relationship (Kim and Mason 1996, Gaard et al. 1996). It is, however, nearly impossible to produce indisputable evidence for a causal relation in humans, because of the multifactorial and indivisible character of the factors determining illness.

Critical Points in Meat Quality

More than 10 years ago (Demeyer et al. 1988), an extended quality concept was proposed, distinguishing between indirect or production-related criteria, and direct criteria measurable on the product. Critical points in relation to these criteria were discussed in earlier work (Demeyer 1992), also related to the OECD project (Demeyer 1994, 1997). Further specification and problem identification is given in the following, with special attention to some newer insights in relation to the food value of beef.

Indirect Criteria

For the indirect criteria related to production systems, as defined earlier (Demeyer 1992, 1997), it is clear that in the present circumstances special attention should be given to environmental effects and animal welfare. Both these criteria, as well as the "meat image" and the economic justification of production should be expressed in concrete measurable values.

Environmental Effect

The acute problems in relation to the global animal waste production and methanogenesis by ruminants are paramount here. It is clear that these environmental pressure factors can be reduced by:

- Optimisation of the feed for improved N and P utilisation. Special attention should be given to:
- Better feed formulation in terms of ileal digestible essential amino acids (critical point no. 1).
- Minimisation of the N losses in grazing cattle (critical point no. 2).
- Reduction of the methanogenesis by adaptation of the feed and by manipulation of the rumen microbiota as discussed at an OECD workshop (Demeyer and Van Cleemput 1996) (critical point n° 3).

In this context it seems necessary to evaluate the N fertilisation of pasture in terms of the utilisation of grass protein by cattle. Efficiency of N utilisation is maximal at about 14% crude protein in the grass OM, whereas the present values amount to about 20%. Synchronisation of energy release and N utilisation in the rumen is difficult to establish on the pasture.

Animal Welfare

It becomes clear that this concept covers more than physiological and psychological definable and measurable criteria. Besides objective measurements in relation to, e.g. behaviour studies, philosophical-ethical considerations in relation to animal rights become more important. The increasing media interest for organisations like GAIA and the Animal Liberation Front is relevant in this respect. In an earlier approach at an OECD workshop (Wiepkema and Demeyer 1993), it was established that the concern about animal welfare is primarily a problem of sociology. The consumer needs to be clearly informed about the nature of production systems. Only then can sociological (and finally, political!) considerations decide whether such systems are socially acceptable. It would seem that providing such information, even at a primary school level, could be a well-defined task for producer as well as governmental organisations.

The Image of Meat Production

The meat sector is often seen as a closed, inaccessible sector, associated with white collar criminality. Two aspects can be distinguished:

- The slaughtering of animals and the processing of meat is considered dirty work. It would seem probable that in the near future working conditions in, e.g. slaughterhouses, may receive critical attention. The improvement of this image may be partly inhibited by the lack of communicative skills and of social integration of the people working in the meat sector. This situation may be caused by the hardness and the intensity of manual labour involved.
- Another aspect relates to the criminal image of the meat sector in connection with the illegal use of growth-promoting substances, and the great number of financial fraud cases. This situation is related to the commercial interest of fast lean muscle production and to the complex structure of subsidising organs within the EU. Widespread criminal, economical and political interests outside the sector are involved here.

Improvement of the working conditions in slaughterhouses (critical point no. 5), education of the staff (critical point no. 6) and transparency of the transactions (critical point no. 7) seem indicated for recommendation.

Justification of Meat Production

Animal production in the industrialised countries involves three to four times more use of cereal equivalents per head of the population than in developing countries. As an example, in 1991 the use of cereals in compound feeds in Belgium was equivalent with ca. 87% of local cereal production (Mottoulle 1992). These data reflect that, on a world basis about 60 to 70% of the cereal production is used in animal feeding. A large proportion is corn, which can also be used, however, as raw material for food technology. Besides, the transport of vegetable raw materials out of developing countries with minimum animal production to industrialised countries with developed intensive animal production leads to environmental pollution in the latter and soil degeneration in the former (Rérat and Kaushik 1995). Moreover, processing of the raw materials in compound feeds involves a (fossil) energy cost (e.g. ca. 11 MJ/100 VEM[1]) exceeding that of roughage (3.5 à 5.5 MJ/100 VEM[1]). Specification of the feeds used is already required in the production of label meat, and could include additional specification of the local origin of feed and roughage (critical point no. 8). The stimulation of extensification in Europe can also be justified in such requirement.

According to Rérat and Kaushik (1995), the increase in the world population to 7 billion in 2020 (Lutz et al. 1997) will demand an increase in food production of about 36% in the same period according with the associated need for distribution (!), if all men are to be fed to the standard of rich countries. This problem seems to be accompanied with the further development of intensive animal production systems with increasing yield in developing countries and the partial reduction of intensive systems in industrialised countries. The transfer to developing countries of bioindustrial know-how with associated costs, together with the necessary increase of vegetable production yields, is a tremendous challenge.

Direct Criteria

Direct quality characteristics are measurable on the product itself and determine the extent of satisfaction of the consumer, including the meat technologist. Again, earlier identified criteria (Demeyer et al. 1988; Demeyer 1997) are only mentioned briefly, with some specification in relation to recent insights concerning nutritive value.

- Usefulness, including tenability, manageability (dimensions) and the convenience (packaging) of meat gains more interest as a result of varying consumption habits and maximal choice possibilities in catering. Packaging gains considerable interest, also in connection with environmental effects whereas standardisation of the dimensions of meat cuts (critical point no. 9) can be seen as a task for both the animal breeder and the cutter.
- The importance of the technological and sensorial quality, especially for pork and beef, respectively, cannot be overestimated. Colour and water binding properties are concerned for pork and tenderness for beef. Although these characteristics will be discussed later, some critical points can be mentioned here.

1. VEM : Voedereenheid Melk (Dutch and Belgium Feed Unit).

- Tenderness between animals of the same breed, age, sex, fattening and slaughter conditions is subject to large variability. Optimisation of both slaughter and cooling processes is necessary (critical point no. 10).
- Sensorial and technological quality characteristics change with better conformation, i.e. more muscularity of carcasses. It concerns a paler colour and higher fluid losses in pork and, less obvious, a paler colour and lower tenderness in the more expensive cuts of beef. In addition, pigs and cattle with better conformation produce such lean meat (≤ 1% of intramuscular fat), that flavour may be impaired.
- The control of variability and optimisation of the contrast between carcass quality and meat quality requires, e.g. precise, rapid and cheap measurement of both quality characteristics (critical point no. 11).
- Technological as well as sensorial quality criteria are mostly related to the metabolic nature of the muscle tissue produced. It seems, therefore, indicated to use measurements concerning metabolic characterisation like LDH, protease and lipase activity, rather than the less accurate, direct or predicting and empirical quality measurements which are used at the present time (critical point no. 12). The pH_1 value used in pork carcasses is an example of metabolic characterisation. Within the empirical measurements, the optical methods (e.g. Near Infrared Reflectance Spectrometry) are attractive in relation to the speed of measuring.
- The safety of the product, i.e. the absence of harmful additives (consciously added, legal or not) and contaminants (accidentally added). Respective examples are sulphite, residues of growth promoters used in production (antibiotics and growth stimulating agents with hormonal action) and contamination with heavy metals, pesticides and health-threatening microorganisms. The latter involve bacteria inducing infections (*Salmonella*) as well as intoxication (*Staphylococcus*), as well as the terrifying prions which cause mad cow disease or BSE, and are also, as far as known, the probable responsible agents for the fatal softening of the brain in humans. The enormous media attention for hormones, BSE and Creutzfeldt Jacob transformed this direct quality characteristic into an image problem. The specialists should be asked to focus their attention on the *relative* health risks of these dangerous substances. It seems to me, e.g. that the illegality of hormonal growth stimulators is more related to politics than to public health, whereas risks related to food pathogens seem to be much greater than the risk of infection with BSE, in spite of the catastrophic results of the latter. Moreover, in general, it would seem that in residue problems more attention should be given to comparative risk analysis, more a mathematical than a biological science.
- The nutritive value of the product, i.e. the amounts and the ratio of nutrients in relation to health as discussed above. Recent results offer some possibilities for optimisation of fat quality in beef, possibly connected with extensification. It has indeed been shown that the *n-3*/*n-6* ratio in intramuscular lipids increases from 0.12 to 0.77 when feeding grass instead of concentrate (Wood and Enser 1997), while it is also known that γ-linolenic acid, an n-3 acid, is subjected to further chain elongation in muscle tissue. This means that analogous results are obtained with grass feeding as with the feeding of fish oils, a method involving problems with digestion and oxidation stability of both feed and product.

An important result is the demonstration of conjugated linoleic acids (CLA) in milk fat as well as in intramuscular fat of beef. In this group of isomers, mainly originating from the rumen, the anticarcinogenic fatty acid 18 : 2 *cis*-9, *trans*-11 is especially important. The data in Table 8 show that this bovinic or rumenic acid (? !) (Kramer et al. 1998) can amount to 2% of the fatty acids in the intramuscular fat of beef.

Table 8. ***Trans*** **18 :1 and *cis*-9, *trans*-11 18 :2 (CLA) in the fatty acids of duodenum contents, milk and beef (% w/w). (Demeyer and Doreau 1999)**

Fatty acids in	*Trans* 18 :1	CLA
Duodenum contents	4.7 ± 1.1	
Milk fat	2.0 ± 0.9	1.5 ± 0.3
Beef	2.5 ± 0.9	> Milk fat

An optimisation of the fat digestion with grass feeding may increase the *n-3/n-6* ratio as well as the CLA amount in meat. This means an increased nutritive and functional value of meat (critical point no. 14) (Demeyer and Doreau 1998). Such interventions at the level of feeding give an additional value to the benefit of the farmer, as opposed to an intervention at the food industry level (Doreau et al. 1997).

The Need for Integration and the Possibilities of the Newer Biotechnology

In the foregoing, 14 more or less precisely formulated critical points were identified with suggestions for possible improvement (Table 9). Eight critical points concern the indirect quality, i.e. the quality of the production systems, rather than the quality of the product. The relative importance of these bottlenecks at the level of the consumer must be further investigated. Consumer behaviour may not be the only determining factor for further research and for development in animal production. Every one will agree, for example, that animal production should contribute to the resolution of the (acute) food problem in the world. This may mean that priority should be given in research to the intensification of animal production in developing countries on the basis of lignocellulose feeding. Manipulation of rumen microbial digestion, the reduction of the infection pressure for ruminants and also an optimisation and diversification of fish culture could be targets (Demeyer et al. 1998).

Table 9. Critical points in meat quality : a summary

1. Feed formulation using ileal amino acid digestibilities
2. Minimise nitrogen losses by grazing cattle
3. Limit methanogenesis in cattle
4. Information in (primary) schools about slaughter conditions and meat production systems
5. Improvement of the working conditions in slaughterhouses
6. Communication training of the staff in the 'meat sector'
7. Transparancy of transactions in the 'meat sector'
8. Specification of the feed used
9. Information on the evaluation of the *relative* risks for public health
10. Standardisation of the dimensions of meat cuts by selection and cutting
11. Optimisation of slaughtering and cooling processes
12. Developing of acurate, rapid and cheap methods for determining carcass and meat quality, usable in the slaughter chain
13. Metabolic characterisation of muscle tissue produced
14. Optimisation of the fatty acid composition in beef
15. Sociological support for the applications of biotechnology in animal production

Furthermore, it must be stressed that any attempt to improve critical quality characteristics should be evaluated within an integrated approach of the whole production system and in a broad social context. The following examples illustrate this view:

- N-utilisation at pasture should be balanced with grass biomass production and, ultimately, with the farmer's income and environmental effects.
- The need for developing animal production in Africa should be evaluated against the associated increase of methanogenesis.
- An improvement of feed conversion through, e.g. immunomodulation, should be accompanied by a characterisation of the muscle tissue produced. Immunomodulation could, indeed, induce changes in protein turnover, possibly affecting meat texture.
- Possibilities for processing of animal waste should be investigated not only in relation to financial criteria, but also in consideration of an increased utilisation of fossil energy. The latter indeed increases CO_2 production, contributing to the now generally recognised greenhouse effect.

Finally, reference should also be made to the dual task of the farmer both as environmental manager and food producer, recognised by governments.

The translation of the different optimised quality characteristics in objective measurable and controllable criteria, is not always self-evident for indirect criteria (e.g. specification of feed). Such criteria are, however, needed with the further development of label meat and chain management. The modern technologies of molecular genetic tracability (Portetelle et al. 1998) can also play an important role in this respect.

Special attention in relation to critical points for direct as well as indirect meat quality deserve the rapidly developing possibilities of the newer biotechnologies. Briefly, the following can be mentioned (Demeyer et al. 1998) :

- The possibilities of reproduction technology ranging from the classic in vitro insemination, to embryo transplantation, sexing of embryos, in vitro fertilisation and embryo splitting to cloning, recently involving also non-embryonic tissue cells. Cloning is barely available for practical application and has a (very) low success frequency. This technology seems however perfect for controlling the animal variability in the feed utilisation as well as in direct meat quality, with specification of the muscle tissue produced.
- Vaccin technology with production and increasingly accurate specification of monoclonal antibodies offers, e.g. also possibilities for regulation of the fat/meat ratio in carcasses through passive immunisation of animals against adipose tissue (Wood et al. 1994).
- Genetic manipulation theoretically offers excellent possibilities, including gene identification with applications in selection (MAS or marked assisted selection) as well as the more problematical development and application of transgenic animals. At this moment, and as far as is known, directed insertion of foreign genetic material in a fertilised ovum is still impossible. This means that transgenic animals are unique. Also, gene transfer is at the present time limited to monogenic characteristics, coding for the synthesis of one protein only. Transgenic animals with, e.g. expression of increased growth hormone synthesis, show too many harmful side effects and their development is strongly questioned socially as well as legally (Solomon et al. 1997).

- The use of production improvers to manipulate food, digestion or tissue metabolism probably involves the most numerous practical applications at this moment. The effects of these products, however, often lack specificity and their use is submitted to legal limitations or is subject to increasing suspicion. Classic production improvers as feed antibiotics as well as newer substances such as probiotics, prebiotics and enzymes are concerned. Simultaneous control of direct meat quality when using such products seems necessary.

Beside the need for specification of the muscle tissue produced, it seems necessary for introduction of newer biotechnologies to establish a centre for sociological support of such introduction (critical point no. 15 in Table 9).

References

Anonymous (1997) Gids bij het nieuwe mestdecreet. Vlaamse Landmaatschappij, Brussel

De Baerdemaeker A (1988) Toekomstige productiesystemen in de veeteelt. Verh Fac Landbouwet Gent 27:21-37

Demeyer DI (1992) Biotechnology and the quality of animal production in sustainable Agriculture. J Appl Anim Res 1:65-80

Demeyer DI (1994) Sustainable agriculture: biotechnology and the quality of animal production. In: Towards sustainable agricultural production:cleaner technologies. OECD document, OECD, Paris, pp 57-58

Demeyer D (1997) An introduction to the OECD programme: meat quality and the quality of animal production. Food Chem 59, 4:491-497

Demeyer D, Van Cleemput O (1996) Methane emission through animals and from the ground. Environ Monit Assess Spec Issue 42:1-210

Demeyer D, Doreau M (1999) Targets and means for altering ruminant meat and milk lipids. Invited lecture, Proc British-French Nutrition Symp, Nancy, Sept 30- Oct 3, 1998, Proceedings Nutrion Society (in press)

Demeyer D, Fiems L, Vandendriessche F (1988) Produktontwikkeling in de vleesproduktie en Vleestechnologie. Verh Fac Landbouwet Gent 27:67-95

Demeyer D, Van Nevel C, Fiems L (1988) Nutritional engineering of beef fat composition: motive, target and approach. In: Lundström K, Hansson I, Wiklund E (eds) Composition of meat in relation to processing, nutritional and sensory quality. ECCEAMST Utrecht, pp 15-36,1995

Demeyer D, Burny A, Renaville R (1998) Dierlijke Productie in België. Intern werkdocument

De Wilde R (1997) Recente evaluatie van de dierlijke mestuitscheidingsnormen. Verslag van de technische werkgroep uitscheidingsnormen. Studiedag Mestproblematiek Beveren 30 oktober 1997, Technologisch Instituut KVIV, Antwerpen

Doreau M, Demeyer DI, Van Nevel CJ (1997) Transformation and effects of unsaturated fatty acids in the rumen. Consequences on milk fat secretion. In: Welch RAS (ed) Milk Composition, Production and Biotechnology. CAB International, New York, pp 73-92

FAO (1992) Production Yearbook, vol 46, FAO Statistics ser 112. FAO, Rome

FAO (1996) Production Yearbook, vol 50, FAO Statistics ser 135, FAO, Rome

FAO (1997) Fishery statistics:catches and landings, 1995 vol 80. FAO, Rome

Gaard M, Tretli S, Loken EB (1996) Dietary factors and risk of colon cancer: a prospective study of 50,535 young Norwegian men and women. Eur J Cancer Prevent 5:445-454

Kim Y.-I, Mason JB (1996) Nutrition chemoprevention of gastrointestinal cancers. Nutr Rev 54:259-279

Kramer J, Parodi PW, Jensen RG, Mossoba MM, Yurawecz MP, Adloff RA (1998) Rumenic acid: a proposed common name for the major conjugated linoleic acid isomer found in natural products. Lipids 33:835

Lutz W, Sanderson W, Scherbov S (1997) Doubling of world population unlikely. Nature 387:803-805

Mottoulle A (1993) Landbouwstatistisch jaarboek 1992. Landbouweconomisch Instituut (LEI), Brussels, 98 pp

Nationale Raad voor de Voeding (1996) Voedingsaanbevelingen voor België, Hoge Gezondheidsraad, Brussel

Nederlandse Voedingsraad (1995) Enkele aspecten van de voedingswaarde-aanduiding van vet(zuren), Den Haag

Pernaud G, Corpet DE (1997) Cancer colorectal: le rôle controversé de la consommation de viande. Bull Cancer 84:899-911

Portetelle D, Renaville R, Mortiaux F, Van Zeveren A, Bouquet Y, Peelman L (1998) Genetische identificatie en Traceerbaarheid bij rundvee d.m.v. Fingerprinting. Mededeling Ministerie van Middenstand en Landbouw, Brussel, pp 3-51

Rérat A, Kaushik SJ (1995) Nutrition, animal production and the environment. Wat Sci Technol 31:1-19

Slattery ML, Caan BJ, Potter JD, Berry TD, Coates A, Duncan D, Edwards SL (1997) Dietary energy sources and colon cancer risk. Am J Epidimiol 145:199-210

Solomon MB, Pursel VG, Campbell RG, Steele NC (1997) Biotechnology for porcine products and its effects on meat products. Food Chem 59:499-504

Tamminga S, Verstegen MWA (1996) Implication of nutrition of animals on environmental pollution. In: Garnsworthy PC, Cole DJA (eds) Recent developments in animal nutrition 3. Nottingham University Press, Nottingham, p 213-228

Van Es AJH (1981) De voeding van mens en dier. Tijdschr. Diergeneeskunde 106:159-165

Weindruch R (1996) Caloric restriction and aging. Sci Am 1996: 32-38

Wiepkema PR, Demeyer D (1993) Biotechnology, animal welfare and ethics: remarks on a conference. Livestock Prod Sci 36:117-119

Willett WC (1994) Diet and health: what should we eat? Science 264:532-537

Wood JD, Enser M (1997) Factors influencing fatty acids in meat and the role of antioxidants in improving meat quality. Br J Nutr 78, Suppl 1:S49-S60

Wood PR, Willadsen P, Vercoe JE, Hoskinson RM, Demeyer D (1994) Vaccines in agriculture: immunological applications to animal health and production. CSIRO Australia, 219 pp

Prion Diseases: an Update

A. AGUZZI[1]

Prion diseases are fatal neurodegenerative disorders affecting a variety of vertebrates, including humans. Recently, a new variant of Creutzfeldt-Jakob disease (CJD) in unusually young people appeared in Great Britain. Considerable evidence argues that this novel disease phenotype is caused by dietary exposure to BSE-contaminated food products. In addition to this, the past year's literature holds tantalizing clues to the mechanisms of prion spread in the organism after peripheral infection. This should eventually lead to the development of diagnostic and therapeutic strategies for prion diseases.

Introduction

Prion diseases or transmissible spongiform encephalopathies (TSE) are neurologic disorders caused by novel transmissible pathogens termed prions. While the controversy about the exact physical nature of the transmissible agent is still ongoing, a wealth of experimental data now supports the protein only hypothesis, which postulates that the agent is devoid of nucleic acid and consists solely of an abnormal conformer (PrPSc) of the cellular prion protein, PrPC. This view was certainly reinforced by the award of the 1997 Nobel Prize in Physiology or Medicine to Stanley Prusiner.

While the prototype of all prion diseases, scrapie in sheep and goats, has been known for more than two centuries, a new form of animal prion disease, designated bovine spongiform encephalopathy (BSE), has since its first recognition in 1986, developed into an epizootic. The emergence of a new variant form of Creutzfeldt-Jakob disease (vCJD) in young people in the UK has raised the possibility that BSE has spread to humans by dietary exposure. This fearful scenario has recently been supported by experimental evidence claiming that the agent causing BSE is indistinguishable from the vCJD agent. Currently little is known about the unusual features of BSE-vCJD prions that allow them, in striking contrast to all other known prions, to cross species barriers with ease.

This chapter focuses first on new aspects of the molecular biology of prions and then highlights some of the most recent evidence demonstrating a relationship between BSE and vCJD. Furthermore, it summarizes important implications of stud-

1. Institute of Neuropathology, University Hospital of Zürich, 8091 Zürich, Switzerland

ies on the pathogenesis and on the spread of prions from peripheral sites to the central nervous system (CNS).

Prion Protein (PrP) and the Molecular Biology of Prions

The hallmark of prion diseases is the accumulation in brain of an abnormal isoform (PrPSc or PrP-res) of the host encoded prion protein (PrPC or PrP-sen). The protein-only hypothesis predicates that the partially protease-resistant and detergent-insoluble PrPSc is congruent with the infectious agent. Furthermore, agent propagation is thought to occur by a PrPSc-mediated conformational conversion of PrPC into new PrPSc molecules (for reviews, see Aguzzi 1997; Aguzzi and Weissmann 1997; Prusiner 1997).

A wealth of experimental evidence supports linkage of PrP and prion diseases. The primary aminoacid sequence of PrP is a main determinant of the species barrier of prions (Prusiner et al. 1990; Scott et al. 1997). Linkage to mutations in the PrP gene was demonstrated for all familial forms of human prion diseases (for a review, see Aguzzi and Weissmann 1996). Mice devoid of PrP (*Prnp0/0*) are resistant to infection by prions (Büeler et al. 1993) and PrPC is required for prion spread in the CNS (Brandner et al. 1996). Recently, it has been demonstrated that prion strain characteristics change depending on the sequence of PrP encoded by the host during multiple serial transmissions. These studies further imply that prion strain diversity is limited to a finite and highly restricted number of conformations of PrPSc that can be adopted by the sequence of PrP encoded by the host (Scott et al. 1997).

The ultimate proof for a protein being the only or major component of the infectious particle would be to produce prions *in vitro* from recombinant PrP. Kocisko et al. (1994) showed for the first time that PrPC can be converted in a cell-free system into a protease-resistant isoform by incubation with PrPSc. Even more strikingly, some of the biological properties of the transmissible agent such as species barrier and strain specificity could be reproduced to some extent in this *in vitro* cell-free conversion system (Bessen et al. 1995; Kocisko et al. 1995).

The physiological function of the prion protein has so far resisted elucidation. The NMR structure determination for the full-length, unglycosylated PrPC revealed the presence of a highly flexible N-terminal polypeptide of 98 and 96 amino acids for murine (Riek et al. 1997) and hamster PrP (Donne et al. 1997), respectively. This polypeptide segment comprises five glycine-rich octapeptide repeats which are among the best-conserved regions of mammalian PrP. There is recent evidence *in vitro* and *in vivo* that the N-terminal domain of PrPC exhibits binding sites for copper (Hornshaw et al. 1995; Brown et al. 1997). In addition, copper was shown to modulate synaptic transmission in cerebellar Purkinje cells of *Prnp0/0* mice but not of wild-type mice (Brown et al. 1997). Altogether, this raises the possibility that PrPC may play a role in copper homeostasis in the brain. Whether perturbations in copper physiology or other downstream effectors such as Cu/Zn superoxide dismutase are involved in prion-elicited pathogenesis remains to be seen (Brown et al. 1997).

By virtue of its location at the outer surface of cells, anchored by phosphatidylinositol glycolipid, PrP is a candidate for a signaling, cell adhesion or (less likely) transport function. Recently, several candidate proteins that bind PrPC have been reported. Among them are the amyloid precursor-like protein 1 (Aplp1) (Yehiely et al. 1997), the human laminin receptor precursor (Rieger et al. 1997) and an uncharacterized 66-kDa membrane protein (Martins et al. 1997). Evidence that some of these molecules interact with PrPC *in vivo* is, however, still missing.

Human Prion Diseases and BSE: a Complex Case

The human prion diseases comprise Creutzfeldt-Jakob disease (CJD), Gerstmann-Straeussler-Scheinker disease (GSS), fatal familial insomnia (FFI) and kuru. These diseases illustrate the three manifestations of prion diseases in general, namely the sporadic forms of the disease (90% of all CJD cases), the inherited forms linked to mutations in the human PrP gene (familial GSS, familial CJD and FFI) and the infectious forms which are acquired by transplantation, injection or ingestion of prion-contaminated tissue-derived products (iatrogenic CJD, vCJD and kuru). A new hereditary form of prion disease presenting with psychiatric symptoms that is linked to a new mutation resulting in a Asn>Ser change at codon 171 of the PrP gene extends the range of disease phenotypes associated with the prion protein gene (Samaia et al. 1997)

Among the human prion diseases, most attention currently focuses on vCJD and its relationship to BSE (Aguzzi and Brandner 1999). New variant CJD was first reported in March 1996 affecting unusually young people in the UK and was shown to be different from sporadic CJD (sCJD) in regard to its clinical and neuropathological features (Will et al. 1996). So far, 21 cases of vCJD in the UK and one case in France have been reported (Zeidler et al. 1997). All these cases share a unique histopathological phenotype consisting of numerous florid amyloid plaques, which are composed of PrP.

Several lines of evidence argue for a causal relationship between BSE and vCJD: Firstly, all but one case of vCJD have been identified in the UK, with the initial reporting taking place 10 years after the BSE outbreak. With no obvious risk factors identified in these patients, a causal link to the exposure of BSE-contaminated food products seems very likely. Secondly, cynomolgus macaques inoculated intracerebrally with brain homogenate from a BSE-afflicted cow developed a CJD-like disease with multiple florid amyloid plaques (Lasmezas et al. 1996). Thirdly, classical strain-typing experiments have convincingly demonstrated that transmission of vCJD or BSE prions into a panel of inbred mouse strains produced incubation times and brain lesion profiles which were indistinguishable, thereby establishing a single ancestral prion strain causing vCJD and BSE (Bruce et al. 1997). Even more strikingly, comparison of incubation times and lesion patterns produced by prions derived from sCJD, vCJD or BSE revealed that vCJD prions are more closely related to BSE prions than to sCJD prions. Although this study is not yet complete – because incubation periods in some of the inbred mice strains used are extremely long – the final results are likely to confirm this picture. Finally, supportive evidence for a link between BSE and vCJD was obtained by comparing patterns of protease-resistant PrPSc on Western blots. Protease-treated PrPSc displays three distinct bands corresponding to different glycoforms of the same protein. Heritable, stable differences in the heterogeneity of the PrPSc glycoforms have been observed in various CJD isolates (Parchi et al. 1996) and also in different strains of mink encephalopathy (Bessen and Marsh 1994). Based on the fragment size and the relative abundance of the individual bands, three distinct patterns (PrPSc types 1-3) were defined for sporadic and iatrogenic CJD cases. In contrast, all cases of vCJD exhibited a novel pattern, designated type-4 pattern. Moreover, extracts from the brains of BSE-infected cattle, cats or kudu that were thought to have acquired BSE, and macaques that were infected experimentally with BSE, all showed type-4 pattern (Collinge et al. 1996). Even more intriguingly, transmission of BSE or vCJD to mice produced mouse

PrPSc with a type-4 pattern indistinguishable from the original inoculum (Hill et al. 1997a). These findings, besides providing a compelling case for vCJD being the human counterpart of BSE, lend further support to the concept that prion strains are linked to the tertiary structure of PrP (Collinge et al. 1996; Telling et al. 1996).

It has been shown that molecular strain typing by Western blot analysis can be used in the differential diagnosis of vCJD (Hill et al. 1997). However, some concern towards the general applicability of glycoform ratio analysis for strain typing has been expressed (Somerville et al. 1997a, b). More powerful methods for resolution of various glycoforms, perhaps in conjunction with a PrPSc-specific reagent such as the recently described antibody 15B3 (Korth et al. 1997), may eventually serve as diagnostic tool for prion strains.

Is there evidence arguing against vCJD being the human counterpart of BSE? Using the *in vitro* conversion assay, Raymond and colleagues (Raymond et al. 1997) have found similar low efficiencies for BSE or scrapie-mediated conversion of human PrPC to PrPSc. This finding suggests that BSE would be no more transmissible to humans than is sheep scrapie, for which there is no evidence of spreading to humans. However, these studies have to be treated with caution for several reasons. Firstly, the *in vitro* conversion assay generates protease-resistant PrP, which is not necessarily identical with infectivity. Secondly, the influence of other factors such as dose, route of inoculation and the effect of protein X (Telling et al. 1995) on cross-species transmission are not addressed in these experiments.

Prion Spread and Neuroinvasion

Prion diseases are disorders of the CNS and experimental transmission is most efficiently achieved by direct inoculation of the agent into the host brain. The primary target cell in the CNS appears to be the neuron (Race et al. 1995; Brandner et al. 1998), yet neuronal loss is not always seen in prion diseases. Instead, astrocytic activation occurs very early, leading to physiological effects such as impairment of the blood-brain barrier. Evidence has now been provided that not only neurons but also astrocytes can sustain prion propagation (Raeber et al. 1997). How this cell type is involved in the pathogenesis of TSE and whether indirect effects, perhaps mediated by cytokines, are playing a role in the disease process remains to be elucidated.

Epidemiologically more relevant than the intracerebral transmission is the oral uptake of prions which is thought to be responsible for the BSE epidemic and for transmission of BSE to a variety of species including humans. Prions can find their way through the body to the brain of their host, yet histopathological changes have not been identified in organs other than the CNS. However, the prions may multiply silently in reservoirs during the incubation phase of the disease. One such reservoir may be the immune system, and many studies point to the importance of prion replication in lymphoid organs (Eklund et al. 1967). Recently, it was shown that vCJD prions accumulate in the lymphoid tissue of tonsils in such large amounts that PrPSc can easily be detected with antibodies on histological sections (Hill et al. 1997b).

Although a wealth of early studies points to the importance of prion replication in lymphoid organs, little is known about the cells supporting prion propagation in the lymphoreticular system. Splenectomy experiments after intraperitoneal infections have suggested that the critical cells are long-lived. The follicular dendritic cell (FDC) would be a prime candidate, and indeed PrPSc accumulates in such cells of wild-type and nude mice (which have a selective T-cell defect) (Kitamoto et al. 1991). However, adoptive bone marrow transfer (which does not efficiently replace follicular dendritic cells) of wild-type cells to prion-resistant *Prnp0/0* mice restores accumulation of the infectious agent in the spleen (Blättler et al. 1997). This finding argues that a cell type other than FDC is responsible for prion propagation in the LRS. Using a panel of immune-deficient mice inoculated intraperitoneally with prions, Klein et al. (1997) found that defects affecting T cells had no apparent effect, but that all mutations that disrupted the differentiation of B cells prevented the development of clinical scrapie. Furthermore, immune-deficient mice with no germinal centers in lymphatic organs and virtually no FDCs but functional B and T cells succumbed to scrapie. These results argue for a crucial role of B cells in the development of scrapie after peripheral infection and disprove a prime role for FDCs in peripheral pathogenesis.

Conclusions

A little more than a decade ago, prion diseases – except for scrapie which has been endemic in sheep for more than 250 years – were regarded as rare neurodegenerative disorders with no serious impact on public health issues and no immediate need for the development of diagnostic or therapeutic measures. This has radically changed with the emergence of BSE and its human counterpart, vCJD. Because of the long incubation times and other unknown factors such as genetic predisposition and exposure criteria, it is difficult to predict whether the incidence of vCJD will increase, and to what extent (Cousens et al. 1997).

In the past years, there has been a considerable increase in our understanding of prion-host interactions, prion spread and mechanisms of disease pathogenesis. We have learned that the agent first interacts with components of the immune system, thus providing us with a possibility of interference with the very early stage of the disease. Since B cells seem to be involved in the transport of prions from the periphery to the CNS, a true opportunity may arise for development of diagnostic tools and interference with prion spread aiming at B cell depletion strategies (Aguzzi and Collinge 1997).

Acknowledgments. This chapter is dedicated to the memory of the late Prof. Volker Henn. As President of the Committee of Experts of the Swiss National Research Program *Diseases of the Nervous System* (NRP38), we remember him for his continuous support and expert leadership devoted to the advancement of neuroscience.

The author is supported by the Kanton of Zürich and by grants from the Swiss National Research Program. (NFP38 and NFP38+).

References and Recommended Reading

Aguzzi A (1997) Prions and antiprions. Biol Chem 378:1393-1395

Aguzzi A, Brandner S (1999) Shrinking prions: new folds to old questions. Nat Med 5:486-487

Aguzzi A, Collinge J (1997) Post-exposure prophylaxis after accidental prion inoculation. Lancet 350:1519-1520

Aguzzi A, Weissmann C (1996) Sleepless in Bologna: transmission of fatal familial insomnia. Trends Microbiol 4:129-131

Aguzzi A, Weissmann C (1997) Prion research: the next frontiers [news] [see comments]. Nature 389:795-798

Bessen RA, Marsh RF (1994) Distinct PrP properties suggest the molecular basis of strain variation in transmissible mink encephalopathy. J Virol 68:7859-7868

Bessen RA, Kocisko DA, Raymond GJ, Nandan S, Lansbury PT, Caughey B (1995) Non-genetic propagation of strain-specific properties of scrapie prion protein. Nature 375:698-700

Blättler T, Brandner S, Raeber AJ, Klein MA, Voigtländer T, Weissmann C, Aguzzi A (1997) PrP-expressing tissue required for transfer of scrapie infectivity from spleen to brain. Nature 389:69-73

Brandner S, Raeber A, Sailer A, Blattler T, Fischer M, Weissmann C, Aguzzi A (1996) Normal host prion protein (PrPC) is required for scrapie spread within the central nervous system. Proc Natl Acad Sci USA 93:13148-13151

Brandner S, Isenmann S, Kuhne G, Aguzzi A (1998) Identification of the end stage of scrapie using infected neural grafts. Brain Pathol 8:19-27

Brown DR, Qin K, Herms JW, Madlung A, Manson J, Strome R, Fraser PE, Kruck T, von Bohlen A, Schulz-Schaeffer W, Giese A, Westaway D, Kretzschmar H (1997) The cellular prion protein binds copper *in vivo*. Nature 390:684-687

Bruce ME, Will RG, Ironside JW, McConnell I, Drummond D, Suttie A, McCardle L, Chree A, Hope J, Birkett C, Cousens S, Fraser H, Bostock CJ (1997) Transmissions to mice indicate that 'new variant' CJD is caused by the BSE agent (see comments). Nature 389:498-501

Büeler HR, Aguzzi A, Sailer A, Greiner RA, Autenried P, Aguet M, Weissmann C (1993) Mice devoid of PrP are resistant to scrapie. Cell 73:1339-1347

Collinge J, Sidle KC, Meads J, Ironside J, Hill AF (1996) Molecular analysis of prion strain variation and the aetiology of 'new variant' CJD (see comments). Nature 383:685-690

Cousens SN, Vynnycky E, Zeidler M, Will RG, Smith PG (1997) Predicting the CJD epidemic in humans (see comments). Nature 385:197-198

Donne DG, Viles JH, Groth D, Mehlhorn I, James TL, Cohen FE, Prusiner SB, Wright PE, Dyson HJ (1997) Structure of the recombinant full-length hamster prion protein PrP(29- 231): the N terminus is highly flexible. Proc Natl Acad Sci USA 94:13452-13457

Eklund CM, Kennedy RC, Hadlow WJ (1967) Pathogenesis of scrapie virus infection in the mouse. J Infect Dis 117:15-22

Hill AF, Desbruslais M, Joiner S, Sidle KC, Gowland I, Collinge J, Doey LJ, Lantos P (1997a) The same prion strain causes vCJD and BSE (letter) (see comments). Nature 389:448-450

Hill AF, Zeidler M, Ironside J, Collinge J (1997b) Diagnosis of new variant Creutzfeldt-Jakob disease by tonsil biopsy. Lancet 349:99

Hornshaw MP, McDermott JR, Candy JM (1995) Copper binding to the N-terminal tandem repeat regions of mammalian and avian prion protein. Biochem Biophys Res Commun 207:621-629

Kitamoto T, Muramoto T, Mohri S, Doh ura K, Tateishi J (1991) Abnormal isoform of prion protein accumulates in follicular dendritic cells in mice with Creutzfeldt-Jakob disease. J Virol 658:6292-6295

Klein MA, Frigg R, Flechsig E, Raeber AJ, Kalinke U, Bluethmann H, Bootz F, Suter M, Zinkernagel RM, Aguzzi A (1997) A crucial role for B cells in neuroinvasive scrapie. Nature 390:687-690

Kocisko DA, Come JH, Priola SA, Chesebro B, Raymond GJ, Lansbury PT, Caughey B (1994) Cell-free formation of protease-resistant prion protein. Nature 370:471-474

Kocisko DA, Priola SA, Raymond GJ, Chesebro B, Lansbury PT Jr, Caughey B (1995) Species specificity in the cell-free conversion of prion protein to protease-resistant forms: a model for the scrapie species barrier. Proc Natl Acad Sci USA 92:3923-3927

Korth C, Stierli B, Streit P, Moser M, Schaller O, Fischer R, Schulz-Schaeffer W, Kretzschmar H, Raeber A, Braun U, Ehrensperger F, Hornemann S, Glockshuber R, Riek R, Billeter M, Wuthrich K, Oesch B (1997) Prion (PrPSc)-specific epitope defined by a monoclonal antibody. Nature 390:74-77

Lasmezas CI, Deslys JP, Demaimay R, Adjou KT, Lamoury F, Dormont D, Robain O, Ironside J, Hauw JJ (1996) BSE Transmission to Macaques. Nature 381:743-744

Martins VR, Graner E, Garcia-Abreu J, de Souza SJ, Mercadante AF, Veiga SS, Zanata SM, Neto VM, Brentani RR (1997) Complementary hydropathy identifies a cellular prion protein receptor [see comments]. Nat Med 3:1376-1382

Parchi P, Castellani R, Capellari S, Ghetti B, Young K, Chen SG, Farlow M, Dickson DW, Sima AAF, Trojanowski JQ, Petersen RB, Gambetti P (1996) Molecular basis of phenotypic variability in sporadic Creutzfeldt-Jakob disease. Ann Neurol 39:767-778

Prusiner SB (1997) Prion diseases and the BSE crisis. Science 278:245-251

Prusiner SB, Scott M, Foster D, Pan KM, Groth D, Mirenda C, Torchia M, Yang SL, Serban D, Carlson GA, et al. (1990). Transgenetic studies implicate interactions between homologous PrP isoforms in scrapie prion replication. Cell 63:673-686

Race RE, Priola SA, Bessen RA, Ernst D, Dockter J, Rall GF, Mucke L, Chesebro B, Oldstone MB (1995) Neuron-specific expression of a hamster prion protein minigene in transgenic mice induces susceptibility to hamster scrapie agent. Neuron 15:1183-1191

Raeber AJ, Race RE, Brandner S, Priola SA, Sailer A, Bessen RA, Mucke L, Manson J, Aguzzi A, Oldstone MB, Weissmann C, Chesebro B (1997) Astrocyte-specific expression of hamster prion protein (PrP) renders PrP knockout mice susceptible to hamster scrapie. EMBO J 16:6057-6065

Raymond GJ, Hope J, Kocisko DA, Priola SA, Raymond LD, Bossers A, Ironside J, Will RG, Chen SG, Petersen RB, Gambetti P, Rubenstein R, Smits MA, Lansbury PT Jr, Caughey B (1997) Molecular assessment of the potential transmissibilities of BSE and scrapie to humans [see comments]. Nature 388:285-288

Rieger R, Edenhofer F, Lasmezas CI, Weiss S (1997) The human 37-kDa laminin receptor precursor interacts with the prion protein in eukaryotic cells [see comments]. Nat Med 3:1383-1388

Riek R, Hornemann S, Wider G, Glockshuber R, Wüthrich K (1997) NMR characterization of the full-length recombinant murine prion protein, mPrP(23-231). FEBS Lett 413:282-288

Samaia HB, Mari JJ, Vallada HP, Moura RP, Simpson AJ, Brentani RR (1997) A prion-linked psychiatric disorder (letter). Nature 390:241.

Scott MR, Safar J, Telling G, Nguyen O, Groth D, Torchia M, Koehler R, Tremblay P, Walther D, Cohen FE, DeArmond SJ, Prusiner SB (1997) Identification of a prion protein epitope modulating transmission of bovine spongiform encephalopathy prions to transgenic mice. Proc Natl Acad Sci USA 94:14279-14284

Somerville RA, Birkett CR, Farquhar CF, Hunter N, Goldmann W, Dornan J, Grover D, Hennion RM, Percy C, Foster J, Jeffrey M (1997a) Immunodetection of PrPSc in spleens of some scrapie-infected sheep but not BSE-infected cows. J Gen Virol 78:2389-2396

Somerville RA, Chong A, Mulqueen OU, Birkett CR, Wood SC, Hope J (1997b) Biochemical typing of scrapie strains (letter; comment). Nature 386:564

Telling GC, Scott M, Mastrianni J, Gabizon R, Torchia M, Cohen FE, DeArmond SJ, Prusiner SB (1995) Prion propagation in mice expressing human and chimeric PrP transgenes implicates the interaction of cellular PrP with another protein. Cell 83:79-90

Telling GC, Parchi P, DeArmond SJ, Cortelli P, Montagna P, Gabizon R, Mastrianni J, Lugaresi E, Gambetti P, Prusiner SB (1996) Evidence for the conformation of the pathologic isoform of the prion protein enciphering and propagating prion diversity (see comments). Science 274:2079-2082

Will R, Ironside JW, Zeidler M, Cousens SN, Estibeiro K, Alperovitch A, Poser S, Pocchiari M, Hofman A, Smith (1996) A new variant of Creutzfeldt-Jakob disease in the UK. Lancet 347:921-925

Yehiely F, Bamborough P, Da Costa M, Perry BJ, Thinakaran G, Cohen FE, Carlson GA, Prusiner SB (1997) Identification of candidate proteins binding to prion protein. Neurobiol Dis 3:339-355

Zeidler M, Stewart GE, Barraclough CR, Bateman DE, Bates D, Burn DJ, Colchester AC, Durward W, Fletcher NA, Hawkins SA, Mackenzie JM, Will RG (1997) New variant Creutzfeldt-Jakob disease: neurological features and diagnostic tests. Lancet 350:903-907

Animal Welfare and Product Quality

J. Jago[1], A. Fisher[1], P. Le Neindre[2]

As we enter the 21st century, the sustainability of modern agricultural farming systems is being questioned. Historically, sustainability has been defined primarily in terms of profit, with less emphasis placed on ethical and environmental issues. It is increasingly apparent that for modern farming systems to be sustainable into the next century, they must not only be profitable, but also both ethically and environmentally sound. As a consequence of this shift in emphasis, the quality of products arising from the farming of animals is being more closely scrutinised by consumers.

In recent years, the definition of quality has widened beyond purely physico-chemical/organoleptic factors to now include such factors as hygiene/safety, nutrition, traceability and quality of the production environment. The welfare of the production animal is an increasingly important factor in the quality equation. We are beginning to question more openly what is fair and reasonable in the way we derive benefit from animals. Suboptimal animal welfare can detrimentally affect the quality of products either directly, or indirectly through consumer's perception of animal well-being. Animals living in harmony with their environment may give rise to products of higher quality, on health, nutritional and organoleptic criteria, than those under "stressful" conditions. Such products will probably also have a better "image", enhancing marketability. For example, in red meat-producing animals, high levels of stress and exertion are known to produce meat which is tough, dark in colour, and more prone to microbial spoilage, whereas calm animals are much more likely to provide high-quality meat.

Constraints that we may impose on the management of production animals due to ethical considerations can also have consequences for product characteristics. This relationship may often be positive, but can also have adverse implications. The provision of greater space and freedom for interaction for egg-producing birds, may, for example, necessitate greater care in some systems in avoiding the spread of potential zoonotic agents such as salmonellosis.

This chapter examines the relationship between animal welfare and product quality within the context of the need for sustainable agricultural systems. It discusses the impacts of animal welfare on the quality of animal products, including

---

1. Animal Behaviour and Welfare, AgResearch, Private Bag 3123, Hamilton, New Zealand
2. Unité de Recherche sur les Herbivores, INRA Theix, 63122 Saint-Genès-Champanelle, France

meat, milk and fibre quality, and outlines some of the challenges facing producers, consumers, researchers and policy makers as we move into the 21st century.

Introduction

Historically, the sustainability of agricultural farming industries has been defined primarily in terms of profit, with less emphasis on ethical and environmental issues. Given that consumer's attitudes and beliefs are changing, it is increasingly apparent that for modern farming systems to be sustainable into the next century, they must not only be economically viable, but also both ethically and environmentally sound.

The profitability of farming enterprises is determined in part by the quality of the products produced. In recent years, the definition of quality has extended beyond purely organoleptic factors. Consumers now consider a broad range of criteria when assessing product quality, including the quality of life or welfare status of the animal during its lifetime (Matthews 1996). This shift in consumer focus is reflected in the increasing emphasis that some major food producers are placing on animal welfare when sourcing and marketing products (Brooks 1997). The increase in demand for products produced using organic farming techniques, of which sound animal welfare practices are a focal point (Dunn 1997), is also indicative of the change in consumer beliefs and demands.

The welfare status of animals can have a major influence on product quality, either directly or indirectly via consumer perceptions. As a consequence, animal welfare is influencing change in modern agricultural production, marketing and consumption. We are beginning to question more openly what is fair and reasonable in the way we derive benefit from animals. As a result, in the past two decades there has been considerable scientific research examining the issues surrounding the welfare of animals within modern farming systems. Animal welfare research has been part of the Quality of Animal Production Theme within the 1995-1999 OECD Cooperative Research Programme. This inclusion recognises the relationship between animal welfare and product quality, its importance to modern, sustainable agricultural production, and the fact that there are gaps in our scientific knowledge.

In this chapter we examine the relationship between animal welfare and product quality within the context of the need for sustainable agricultural systems. The impacts of animal welfare on the quality of animal products are considered and the challenges facing producers, consumers, researchers and policy-makers are outlined.

Concepts of Welfare

Most concepts of animal welfare include avoidance of undue suffering, optimising animal health and vigour and are aimed at achieving practices and environmental conditions which are fair and reasonable for the animal. Although the concept of animal welfare is widely regarded as being important, currently there is no single definition of animal welfare that has met with universal approval. People's beliefs and understanding of what is meant by "welfare" and what is optimal or suboptimal welfare will vary, depending on such factors as their cultural, scientific, religious and political backgrounds (Swanson 1995). According to Kellert (1988) the attitudes people have towards animals can be classified into nine categories including naturalistic, ecologistic, humanistic, moralistic, scientific, aesthetic, utilitarian, dominionistic and negative; and that differences exist between countries in the predominant attitude. Despite these differing attitudes towards animals, there is a biological basis for evaluating animal health and welfare, and widespread acceptance that decisions about animal welfare should be based on good scientific evidence (Duncan 1981; Simonsen 1996).

It is helpful to have basic guidelines or rules to refer to when making decisions that may impact on an animal's welfare. Probably the most widely utilised set of guidelines is the *Five Freedoms* (Farm Animal Welfare Council 1993). These state that for an animal's welfare not to be compromised it must have: freedom from thirst, hunger and malnutrition; freedom from discomfort; freedom from pain, injury and disease; freedom to express normal behaviour; and, finally, freedom from fear and distress. Sometimes slight modifications are made to these basic freedoms (e.g. fear is sometimes omitted from the final freedom); however, they generally serve as a set of goals towards which animal owners and handlers should strive. The Five Freedoms have been used by many legislators and frequently appear as the basis upon which animal welfare codes and practices have been established.

Concepts of Quality

Products that arise from the farming of animals include milk, fibres, meat and pelts, as well as a large variety of by-products and speciality items. With such a broad range of products, a simple concept of quality can be difficult to define. One approach is to view quality as a measure of fitness for purpose. Most products generally have to address a number of purposes or criteria during various stages of the production process in order to be successful. Thus, the quality of a product may be assessed as the difference between its expected performance and its true performance across a range of criteria.

The different purposes or quality criteria which animal products need to address may be classified as physicochemical/organoleptic, hygiene/safety, nutritional, quality of production environment, and traceability/assurance. Traditionally, consumers have purchased products based on perceptions of quality related to physicochemical/organoleptic criteria, in conjunction with price, and with an expectation that nutritional and hygiene/safety aspects were fulfilled. More recently, in many markets, fitness for purpose in terms of physicochemical and organoleptic properties in combination with an appropriate price, have not always been sufficient to address the quality concerns of the consumer. Highly publicised food safety issues such as *E. coli* O157 and BSE, nutritional concerns related to diet and health, and an increasing public interest in animal welfare and environmental issues associated with farming, have all influenced a shift in how we measure and present the quality of products.

As a result of the greater complexity of the fitness for purpose criteria used to assess product quality, a need has arisen for the additional criterion of quality assurance and traceability. These criteria allow the consumer and intermediaries to judge the quality of the information provided on other aspects of product quality. Put simply, consumers need assurance that information they receive regarding the organoleptic hygiene/safety, nutritional and animal welfare aspects of product quality are accurate and reliable. Quality assurance schemes have evolved to fulfil this role. In some cases, large supermarket chains have developed specific labels or brands that indicate to the consumer that a particular product has satisfied a number of quality criteria during production. A specific example of this is the Freedom Foods label, common in the United Kingdom, and under which food is produced in accordance with standards set by the Royal Society for the Protection and Care of Animals (RSPCA). The standards draw heavily on the concept of the Five Freedoms. The label informs the consumer that the product has been produced under conditions that achieved a specified standard of animal welfare.

The various quality criteria contribute to our overall perception of product quality in different ways. Where quality assurance procedures for animal products have addressed some of the fitness-for-purpose criteria demanded by the market, but have failed to provide the basic requirements of satisfactory organoleptic and usability properties, they have invariably failed (Brooks 1997). By comparison, where the physicochemical/organoleptic properties of a product are highly regarded, it can be very successful, but may be at risk from problems or controversies associated with other fitness-for-purpose criteria. While product quality problems associated with dietary and safety/hygiene issues clearly may have a direct impact upon the end user, concerns over the quality of the production environment in terms of animal (and human) welfare and environmental impact may also have a strong influence on perceived fitness for purpose within the marketplace. Thus, in addition to animal welfare interacting with and influencing the other product quality criteria, animal welfare issues themselves may contribute directly to the quality equation (Hughes 1995).

The Impact of Animal Welfare on Product Quality

It has been suggested by Gregory (1998) that there are three major reasons for being concerned about animal welfare:

- Respect for animals and a sense of fair play.
- Risk of loss of market share for products that acquire a poor welfare image.
- Poor welfare can lead to poor product quality.

Suboptimal animal welfare can detrimentally affect the quality of products either directly or indirectly via the consumer's perception of animal well-being. Animals living in harmony with their environment are more likely to give rise to products of higher quality, on health, nutritional and organoleptic criteria, than those that are maintained under stressful conditions. Such products will probably also have a better image, enhancing marketability. There are numerous examples of poor welfare directly affecting the quality of products in a large number of farmed animals, some of which are detailed below.

Animal Welfare and Meat Quality

A large number of animals are reared ultimately for meat products. In beef and sheep meat, the colour and tenderness of the uncooked product are greatly influenced by the final pH of the muscle after slaughter. If muscle glycogen stores are inadequate or depleted prior to death, final meat pH is increased and the meat is often dark in colour and markedly less tender in texture before cooking. In addition to possessing less desirable organoleptic properties, high pH meat is also more susceptible to bacterial proliferation and microbial spoilage (Newton and Gill 1981).

Poor welfare, as a result of excessive exertion and stress before slaughter, is likely to deplete muscle glycogen, leading to impaired meat tenderness and unattractive colour.

A similar relationship between ante mortem animal stress and meat quality exists in pigs. Increased stress associated with transport conditions, mixing of unfamiliar animals and poor handling can contribute to the development of PSE (pale, soft, exudative) or DFD (dark, firm, dry) pork. Whereas DFD pork is largely influenced by preslaughter conditions, PSE meat is also induced by the presence of the porcine stress gene, which, where present, renders animals more susceptible to the physiological effects of stressors (Geers et al. 1994). The development of the ability to identify pigs with this genotype has allowed selection to greatly reduce the prevalence of the deleterious gene in the population.

In aquaculture, although issues associated with animal welfare are not as prominent as in the traditional livestock industries, there are defined relationships that exist between stress levels in the living organism and the quality of the flesh produced. This is particularly the case for farmed fish (Pickering 1998). Studies with farmed Atlantic salmon and rainbow trout have shown that stress and exertion induced by lack of water and being chased around the tank before capture can reduce flesh quality (Thomas et al. 1999).

In all classes of livestock, rough handling and transport can lead to bruising, which requires trimming and which may also cause carcass downgrading. This problem is more prevalent for cattle than for sheep or pigs, and is also a concern in deer meat production. Deer are generally more reactive than other livestock, and may require particular care during handling and transport to minimise physical trauma and associated bruising (Jago et al. 1996).

In some circumstances, allowing animals complete freedom to express their normal behaviour may lead to adverse effects on product quality. For example, bulls that are kept in groups in lairage before slaughter often display a high level of mounting or riding behaviour. Excessive mounting during lairage or yarding can compromise the welfare of the bulls being ridden. The physical exertion associated with mounting is known to be a key cause of meat that has poor quality. A management procedure designed to curb/reduce such behaviour, such as an electrified wire grid erected above the lairage pen (Kenny and Tarrant 1987), reduces the expression of such behaviour and consequently increases the quality of the meat product. In this example, freedom to express normal behaviour compromises both welfare and product quality. Another example of a management procedure suppressing the expression of normal behaviour leading to improved product quality is the use of immunosuppressing vaccines such as anti-GnRH vaccines. These have been shown to reduce typical bull behaviour and result in improved meat quality (Jago et al. 1999).

Animal Welfare and Milk Quality

In dairy animals, the composition of the milk can have a profound effect on processing qualities. There are strong relationships in dairy cows between the animal's health and welfare, and desirable processing and hygienic properties of their milk. The somatic cell content of the milk is strongly negatively correlated with its suitability for cheese making and subsequent cheese quality (Rogers and Mitchell 1994). Milk somatic cell content also acts as an indicator of milk hygiene and the quality of the production environment, and many countries have regulatory limits for acceptable somatic cell content. Milk somatic cell levels are elevated in cows suffering from clinical or subclinical mastitis or which are kept in unhygienic conditions. Stressors such as strenuous physical activity or walking excessive distances may also increase somatic cell counts (Coulon et al. 1998). A similar relationship exists between animal stress and milk quality in small ruminants, such as sheep and goats (Lerondelle et al. 1992).

Animal Welfare and Fibre Quality

In wool-producing sheep, periods of nutritional stress will cause the wool grown to be of reduced tensile strength (Thwaites 1972). Fluctuating nutritional levels over a period of time will also induce variation in the diameter of fibre in the wool clip, limiting the usefulness of the fleece for manufacturing.

Indirect Effects of Animal Welfare on Product Quality

Conditions which affect animal welfare may also, as an indirect consequence, have positive or negative effects on product quality. For example, constraints that we may impose on the management of production animals due to ethical considerations may have adverse implications for product characteristics. One example of this is the provision of greater space and freedom for interaction for egg-producing birds, which may result in an increased risk of the occurrence of diseases and parasitic infections including worms and coccidia (Branius 1989; Bosch and van Niekerk 1995). In some circumstances, the hygiene quality of eggs produced in "welfare-friendly" systems, such as free-range or deep litter barns, may be poorer than that of eggs produced in cage systems (Torges et al. 1976; Mattes 1985). As a consequence, under a barn or free-range management system, in which there is increased interaction among birds, greater care must be taken to avoid the spread of potential infectious agents.

Other indirect influences of animal welfare on product quality may result from consumer's perceptions of welfare status during an animal's life or of particular management practices to which the animal is subject. These influences can be either positive or negative. For example, there is a perception among the public that extensive systems in which animals are farmed outdoors offer a higher standard of welfare than intensive systems in which animals are primarily housed indoors (Rogers et al. 1989; Matthews et al. 1994). This is despite the fact that many indoor housing systems offer animals protection from extremes in environmental conditions and provide a high standard of welfare. Extensive systems are probably rated more highly because they are seen to offer the animals a greater freedom to express their normal behaviour and more closely resemble what the public believe to be "natural" conditions for the animals (Matthews 1996).

Animal Welfare, Product Quality and Sustainability of Modern Agricultural Farming Methods: Challenges for the Future

There are a number of challenges currently facing researchers, producers, consumers, policy drafters and legislators working in agricultural industries within the OECD.

- To include animal welfare as part of a balanced quality equation.

Animal welfare is clearly an integral part of the quality equation. Animal welfare can influence product quality either directly or indirectly, however the relationship between welfare and quality is complex. If welfare is not considered in the quality equation, animal stress may adversely impact on quality. In addition, product quality is at risk from adverse consumer perceptions of the quality of the production environment. However, if animal welfare becomes an overemphasised focus of the quality equation, there is the potential in some systems for organoleptic and hygienic quality attributes to be compromised.

- To gain the public's confidence that product quality satisfies their desires for animal welfare along with safe, wholesome and ecologically sustainable food
- To construct a policy and market framework in which these challenges can be met.

Having balanced the quality equation, it is important to validate it and communicate it to the public in a way that they trust. There has been a large increase in the number of quality assurance schemes operating in many of the OECD countries, both at an industry level and within major food producing companies. A major challenge to the scientific community is to provide scientifically based information to assist the policy drafters to formulate animal welfare codes that will operate within farming systems that are sustainable. It therefore becomes crucial to better define the relationship between animal welfare and product quality and understand the many factors that can influence product quality, including how consumer perceptions influence purchasing decisions. Agricultural industries must satisfy consumer's demands for competitively priced and high-quality food that has been produced under acceptable and humane conditions.

References

Bosch JGM, van Niekerk TGCM (1995) Health. In: Blokhuis HJ, Metz JHM (eds) Aviary housing for laying hens. ID-DLO Lelystad and IMAG-DLO Wageningen, p 59-71

Branius WW (1989) Coccidiosis in poultry housed in various systems. In: Juit AR, Ehlhardt DA, Blokhuis HJ (eds) Alternative improved housing systems for poultry. CEC Report, Luxembourg, pp 115-119

Brooks P (1997) Consumers and the food chain. Agric Prog 72:45-65

Coulon JB, Pradel P, Cochard T, Poutrel B (1998) Effect of extreme walking conditions for dairy cows on milk yield, chemical composition, and somatic cell count. J Dairy Sci 81:994-1003

Duncan IJH (1981) Animal right – animal welfare: a scientist's assessment. Poultry Sci 60:489-499

Dunn N (1997) Organic farming is growing in Europe. Hoard's Dairyman 142:21

Farm Animal Welfare Council, 1993. 2nd Report on Priorities for Research and Development in Farm Animal Welfare. MAFF, Tolworth, UK

Geers R, Bleus E, van Schie T, Ville H, Gerard H, Janssens S, Nackaerts G, Decuypere E, Jourquin J (1994) Transport of pigs different with respect to the halothane gene: stress assessment. J Anim Sci 72:2552-2558

Gregory NG (1998) Animal welfare and meat science. CAB International, Wallingford

Hughes D (1995) Animal welfare: the consumer and the food industry. Brit Food J 97:3-7

Jago JG, Hargreaves AL, Harcourt RG, Matthews LR (1996) Risk factors associated with bruising in red deer at a commercial slaughter plant. Meat Sci 44:181-191

Jago JG, Matthews LR, Trigg TE, Dobbie P, Bass JJ (1999) The effect of immunocastration 7 weeks before slaughter on the behaviour, growth and meat quality of post-pubertal bulls. Anim Sci, 68:163-171

Kellert X (1988) Human-animal interaction: a review of American attitudes to wild and domestic animals in the twentieth century. In: Rowan AM (ed) Animals and people sharing the world. University Press of New England, Hanover, New Hampshire, pp 137-175

Kenny FJ, Tarrant PV (1987) The behaviour of young Friesian bulls during social regrouping at an abbatoir. Influence of an overhead electrified wire grid. Appl Anim Behav Sci 18:233-246

Lerondelle C, Richard Y, Issartial J (1992) Factors affecting somatic cell counts in goat milk. Small Ruminant Res 8:129-139

Matthes S (1985) Hygiene and egg quality. Proc 29th Symp on Egg Quality. Harper Adams Agriculture College, Newport, UK

Matthews LR (1996) Animal welfare and sustainability of production under extensive conditions: a non-EU perspective. Appl Anim Behav Sci 49:41-46

Matthews LR, Loveridge AM, Guerin B (1994) Animal welfare issues and attitudes in New Zealand. Animal Behaviour and Welfare Research Centre, AgResearch Ruakura, Hamilton, New Zealand

Newton KG, Gill CO (1981) The microbiology of DFD fresh meats: a review. Meat Sci 5:223-232

Pickering AD (1998) Stress responses of farmed fish. In: Black KD, Pickering AD (eds) Biology of farmed fish, Sheffield Academic Press, Sheffield, U.K. pp 222-255

Rogers CS, Appleby MC, Keeling L, Roberton ES, Hughes BO (1989) Assessing public opinion on commercial methods of egg production: A pilot study. Res Dev Agric 6:19-24

Rogers SA, Mitchell GE (1994) The relationship between somatic cell count, composition and manufacturing properties of bulk milk. 6. Cheddar cheese and skim milk yoghurt. Aust J Dairy Technol 49:70-74

Simonsen HB (1996) Assessment of animal welfare by a holistic appraoch: behaviour, health and measured opinion. Acta Agric Scand Sect A Animal Sci Suppl 27:91-96

Swanson JC (1995) Farm animal well-being and intensive production systems. J Anim Sci 73:2744-2751

Thomas PM, Pankhurst NW, Bremner HA (1999) The effect of stress and exercise on post-mortem biochemistry of Atlantic salmon (*Salmo salar*) and rainbow trout (*Oncorhynchus mykiss*). J Fish Biol 54:1177-1196

Thwaites CJ (1972) The effects of short-term undernutrition and adrenocortical stimulation on wool growth. Anim Prod 15:39-46

Torges HG, Matthes S, Harnisch S (1976) Vergleichende Qualitätsuntersuchungen an Eiern aus kommerziellen Legehennenbeständen in Freiland-, Boden- und Käfighaltung. Arch Lebensmittelhyg 27:107-112

Biotechnology in Aquaculture

R. POWELL[1]

Aquaculture, or the farming of aquatic animal and plants species, represents one of the fastest-growing food production sectors with an annual average growth rate of almost 10%. Along with important issues including environmental impact concerns and the promotion of aquaculture as a successful long-term industry, producers are increasingly examining the role biotechnology can play in sustained growth. Appropriate research areas include disease control, and reproduction, growth, development and nutrition of the cultured species. Potential commercial applications include enhanced growth rates, appropriate stock maturity regimes, more efficient food conversion, improved disease resistance and control, and efficient selective breeding programmes. Although it is important to consider the diversity of species exploited in aquaculture and the worldwide range of these activities, including both developed and developing nations, biotechnology holds much promise for the production of both new and improved products to meet consumer demands.

Introduction

Although average human consumption of fish and seafood is reported to have remained static over the past decade, the source of these products is gradually but inexorably moving from harvesting of wild stocks to aquaculture. This trend is irreversible for the foreseeable future as the traditional fisheries sector is unlikely to increase substantially and is also faced with decreasing stocks of many fresh- and sea-water-harvested fish species (New 1991; Csavas 1995). Conversely, aquaculture has shown an annual growth rate of almost 10%, generating valuable employment and foreign exchange while responding to the challenge of improved food security (Rana 1997a). The contribution of cultured fish and shellfish production to total fish and shellfish production grew from 11 to 18% between 1989 and 1995, and, when cultured plant species are included, the total contribution of aquaculture to total aquatic production grew from 14 to 23%. Aquaculture is currently one of the fastest-growing food production sectors, with an estimated value of US$42 billion in 1995 due to the production of 27 million tonnes of cultured fish, shellfish and plants.

1. Department of Microbiology, National University of Ireland, Galway, Ireland.

Table 1 notes the estimated increases (high and low) in production of the different cultured groups expected by the year 2010. Production of cultured fish, crustacean and molluscan groups is expected to double by 2010, while cultured plant production is projected to grow threefold. Combined, production of all groups is expected to grow from approximately 19 million tonnes to approximately 50 million tonnes by 2010 (Pedini and Shehadeh 1997).

Table 1. Projected production values for the year 2010 in million tonnes (mt) for the various cultured groups. (After Pedini and Shehadeh 1997)

Group	1992	2010	(Low - high)
	(mt)	(mt)	(mt)
Fish	9.4	20.8	(17.9 - 25.5)
Crustacea	0.9	1.8	(1.5 - 2.1)
Molluscs	3.5	9.9	(7.0 - 10.0)
Plants	5.3	14.2	(10.9 - 16.3)
Total including plants	19	47	(37 - 55)
Total excluding plants	13	33	(27 - 39)

Aquaculture is a worldwide activity spanning both developed and developing nations. Asia, and in particular China, with a long tradition in aquaculture, is the dominant force in current aquaculture, with the production of 25 million tonnes in 1995 (Rana 1997a). This geographical range, allied to the variety of exploited species, explains the large variation in the estimated contribution of aquaculture to total aquatic production among the major aquaculture nations (Table 2). However, future expansion of aquaculture in developing nations may be limited due to constraints such as water quality and supply, technical and financial support, and low national priority.

Table 2. Contribution of aquaculture to the national total aquatic production in 1995 for the major aquaculture nations. (After Rana 1997a)

China	60%	Taiwan	22%
Italy	36%	Japan	18%
France	32%	Viet Nam	18%
India	32%	Indonesia	17%
Rep. Korea	30%	Thailand	13%
Philippines	29%	Norway	10%
Bangladesh	27%	USA	7%

In terms of environment usage, in 1995 freshwater aquaculture resulted in the production of 13.2 million tonnes (65% of total production) valued at approximately US$ 17 billion (63% of total value), while marine aquaculture produced 6.3 million tonnes (31%) valued at US$ 12 billion (27%). Coastal aquaculture in brackish waters resulted in the production of 1.5 million tonnes (4%) valued at US$ 7 billion (10%) (Rana 1997b). Financial costs associated with marine aquaculture and environmental and public concerns with coastal and freshwater aquaculture represent significant challenges for the aquaculture industry. Table 3 lists the percentage of total production tonnage and the percentage of total value for the various cultured fish groups (based on aquaculture environment) and molluscs, plants and crustacea. Finfish production is the dominant aquaculture component, accounting for 53% of production and 55% of value.

Table 3. Percentage tonnage and monitory values of the various cultured groups in 1995. (After Rana 1997a)

Cultured group	% Tonnage	% Value
Freshwater fish	45.8	33.4
Plants	24.5	13.8
Molluscs	18.3	13.2
Diadromous fish	5.0	13.2
Crustacea	4.1	17.3
Marine fish	2.1	8.0

While it is estimated that between 200 and 300 species are exploited in aquaculture, in 1995 the majority (i.e. 65%) of cultured species contributed less than 10 000 million tonnes in total, while 32 species contributed between 0.1 and 1 million tonnes. Three of the ten species with the largest production were plants, and the major species in descending order of production were: kelp (*Laminaria japonica*), 4.06 million tonnes; silver carp (*Hypophthalmichthys molitrix*), 2.56 million tonnes; grass carp (*Ctenopharyngodon idellus*), 2.10 million tonnes; common carp (*Cyprinus carpio*), 1.78 million tonnes; bighead carp (*Aristichthys nobilis*), 1.26 million tonnes; yesso scallop (*Pecten yessoensis*), 1.14 million tonnes; and Pacific oyster (*Crassostrea gigas*), 1 million tonnes. Penaeid species currently constitute the cultured group of highest monetary value. The giant tiger prawn (*Penaeus monodon*) was ranked 14th in the 1995 tonnage census along with white shrimp (*P. vannamei*) in Ecuador and *P. chinensis* in China at 39th and 42nd positions, respectively (Rana 1997b).

The major cultured finfish can be divided into three classes. Production of carp rose from 2 000 000 million tonnes (90% of total carp product) in 1984 to greater than 10 000 000 million tonnes (95%) by 1995. This was followed by the tilapias, with almost 200 000 million tonnes (38% of total tilapia product) produced in 1984 rising to over 600 000 million tonnes (57%) in 1995, and salmonids (50% of which is Atlantic salmon) with 34 000 million tonnes (5% of total salmonid product) in 1984 increasing to 542 000 million tonnes (34%) in 1995.

Aquaculture, with an annual growth rate between 1984 and 1995 of 10.9%, has seen food fish production grow from 6.7 million tonnes to approximately 21 million tonnes during this period. This compares favourably with the total food fish production from fisheries which grew at a rate of 1.5% over the same period to produce 61 million tonnes in 1995. Annual growth rates for the other major agricultural livestock sectors were estimated as: chicken meat production, 5.3%; pig meat production, 3.4%; mutton and lamb meat production, 1.4%; and beef and veal meat production, 0.9% (Tacon 1997). In 1995, it is estimated that aquaculture produced 13.7 million tonnes of aquatic animal meat for direct human consumption, or 6.2% of the total farmed animal meat production of 222.1 million tonnes (Laureti 1996).

Maintaining the current rates of increase in production presents aquaculture with a variety of challenges. In order to progress towards a successful and economically valuable entity, sustained growth in the longer term is required. However, aquaculture is currently faced with a variety of environmental impact concerns, including land and water usage (Pimental et al. 1996), potentially detrimental disease and genetic interactions between cultured and wild stocks, negative impacts on wild fisheries due to the collection of seed material for cultured stocks, requirements for fish-feed and fishoil (Chamberlain 1993; Chamberlain and Hopkins 1994), and concern over the misuse and effects of chemicals including antibiotics (Barg and Lavilla-Pitogo 1996; Subasinghe 1997). Therefore, the aquaculture industry must strive for increased public awareness and acceptance if the rewards of greater employment, larger foreign exchange earnings, and improved food security are to be gained. Biotechnology can play an important role in aiding the resolution of this situation. Some of the problems including disease control and chemical usage may be resolved by biotechnology-based research including the development of novel pathogen detection systems and more efficient vaccines. This can complement the current interest in the use of biotechnology to improve cultured stocks allowing sustained growth and exploitation. The areas of aquaculture where both biotechnology-based research is currently in progress, and where biotechnology is likely to provide the greatest benefits, are improved pathogen detection and disease diagnosis, the production of novel, more efficient vaccines, and genetic modification of cultured stocks.

Molecular Pathogen Detection Systems

Disease is a major constraint of aquaculture, with an estimated cost of several US$ billion annually. The large variety of cultured species means a large variety of parasitic, bacterial and viral pathogens to contend with. Chemotherapy, or the treatment of fish stocks with chemicals and antibiotics, has been the major weapon of farm management for disease control. However, there are several reasons to suggest that current chemotherapy regimes need to be replaced with alternative disease control management strategies. Disadvantages associated with chemotherapy include: (1) the poor efficiency of many drugs in the aquatic environment including the short-term nature of protection, (2) the high costs associated with large scale use, (3) the development of drug- or chemical-resistance by pathogenic organisms, (4) the limited number of chemicals licensed for use in the aquatic environment, (5) public concern with respect to the fate of chemical residues in the environment, (6) public and market-place concerns with respect to the fate of chemical residues in aquaculture produce, and (7) international regulations governing stock health and the movement of fish stocks and products. Faced with this trend to limit the use of chemotherapy in the culture of many species, one alternative approach is analagous to the use of public health measures for human disease control. Improved stock husbandry practises dealing with stock densities, production rates and site fallowing for defined periods have been used successfully to tackle disease. One crucial aid for these disease control practises is the availability of efficient detection systems for the numerous pathogens. Useful detection systems should allow the farm management to detect the presence of pathogen in the farm environment before signs of disease are apparent in the cultured stocks. Such early-warning systems would allow farm management sufficient time to react to the disease threat and also to monitor the subsequent efficiency of stock treatments, i.e. limited chemotherapy and/or stock husbandry practises.

Molecular-based diagnostic tests for disease-causing organisms such as the immunological enzyme-linked immunosorbant assays (ELISA) and immunofluorescence antibodies, and the genetical DNA probe and polymerase chain reaction (PCR) technology rose to prominence in the 1990s as rapid and sensitive alternatives for pathogen detection in comparison to traditional pathogen culture or tissue histochemical analysis. Table 4 lists recent reports of a variety of different immunological- and DNA-based detection systems developed to detect bacterial, viral and parasitic infections of different cultured fish, shellfish and crustacean species. In general, both ELISA- and PCR-based detection systems have the advantages of sensitivity and speed of use. ELISA systems demonstrate the ability to detect approximately 1000 pathogenic organisms, while PCR sensitivity can be sufficient to detect approximately 10 pathogenic organisms. Testing times are measured in hours to days.

ELISA-based technology has a successful track record as "off the shelf" commercially available pathogen detection kits, but obviously requires a viable market demand. PCR-based technology requires specialised and experienced personel. The current size of the aquaculture industry suggests that both these detection systems are more likely to be performed by central veterinary laboratories, individual research laboratories or commercial veterinary services for the foreseeable future. These molecular-based pathogen detection systems can provide new and useful information to both the aquaculture and fisheries sectors on pathogen ecology and disease epizootiology. They have been reported to detect pathogen in apparently healthy cultured fish stocks (O'Brien et al 1994) and also in wild fish stocks (Mooney et al. 1995). Such information can be used to support and monitor farm disease prevention and control strategies and also potential disease interactions between cultured and wild stocks. Finally, these diagnostic tests are likely to play a larger role in stock health and quarantine controls, including international regulations on the movement of aquaculture stocks and products.

Recombinant DNA Vaccines

Immunoprophylaxis, or vaccination, is another powerful tool for disease control that has the potential to modify or replace current chemotherapy regimes. Since the first scientific report of a salmonid antifurunculosis vaccine in 1942 by D.C.B. Duff, aquaculture has relied on chemotherapy as its primary disease defence strategy. However, by the late 1970s with increased interest in the culture of finfish and with the disadvantages of chemotherapy becoming readily apparent, commercially available vaccines for finfish culture reached the market-place. Effectiveness in terms of reduced chemotherapy is obvious. In 1987, 48.8 tonnes of antibiotic were used in Norwegian fish farming, while in 1994, only 1.9 tonnes were used although fish production had increased from approximately 55 000 tonnes to 249 000 tonnes in the same period. The current challenges for the development of effective fish vaccines include: (1) effective means of delivery, (2) economic costs for research and development, and (3) a relatively poor knowledge of fish immunity when compared to mammalian species. Fish, along with other vertebrates, exhibit the adaptive response or the ability to distinguish between self and non-self. In recent years, much reseach emphasis has being placed on the exploration of fish immunology. The evidence

Table 4. Reports of the development of molecular-based pathogen detection systems

Pathogen class	Host class	Immuno-based test	DNA-based test
Bacteria			
Aeromonas salmonicida	Salmonids	Hiney et al. (1994)	Hiney et al. (1992)
Renibacterium salmoninarum	Salmonids	Elliot and McKibben (1997)	Brown et al. (1994)
			Mauel et al. (1996)
Piscirickettsia salmonis	Salmonids	Lannan et al. (1991)	
Yersinia rickerii	Salmonids		Argenton et al. (1996)
Vibrio penaeicida	Prawns	de la Pena et al. (1995)	Genmoto et al. (1996)
Cytophaga psychrophila	Salmonids		Toyama et al. (1994)
Virus			
Infectious salmon anaemia	Salmonids	Falk et al. (1998)	
Taura syndrome virus	Shrimp		Mari et al. (1998)
Red sea bream irido virus	Sea bream		Oshima et al. (1998)
Penaeid baculovirus	Shrimp	Lewis (1986)	SY Wang et al. (1996)
Yellow head virus	Shrimp		Wongteerasupaya et al. (1998)
Grass carp hemorrhagic virus	Carp		Jun et al. (1997)
Infectious hematopoietic necrosis	Salmonids	Ristow et al. (1993)	WS Wang et al. (1996)
Infectious pancreatic necrosis	Salmonids	Dixon and Hill (1983)	Wang and Lee (1997)
Viral haemorrhagic septicaemia	Salmonids	Way and Dixon (1988)	Miller et al. (1998)
Parasites			
Enterocytozoon salmonis	Salmonids		Barlough et al. (1995)
Loma salmonae	Salmonids		Docker et al. (1997)
Myxobolus cerebralis	Salmonids	Wolf and Markiw (1975)	Andree et al. (1998)
Haplosporidium nelsoni	Oysters		Stockes et al. (1995)

suggests that in comparison to mammals, the fish immune response is much less aggressive, which could be due to several factors including (1) specific genes may be missing or modified, (2) important anatomical structures may be missing or modified, (3) fish immunity may be profoundly affected by environmental factors such as stress and nutrition, and (4) the physiology of fish may differ considerably from mammals (Warr 1996). Thus, the fish immune system remains somewhat of a mystery, and fish vaccine production remains dictated to by a trial-and-error system to monitor effectiveness. Research into the immune process of invertebrates is at an even earlier stage, but for example, among crustaceans, a blood cell-localised system for the recognition of foreign microbial polysaccharides, termed the prophenoloxidase activating system, has been identified (Soderhall 1996).

Curently, almost all fish vaccines are based on the production of dead pathogen cells or components. However, biotechnology is expanding the scope of vaccine production, including vaccines based on recombinant subunit antigens, live attenuated strains of pathogen, and naked DNA (Leong 1993; Munn 1994). Recombinant subunit vaccines are currently commercially available. A vaccine composed of *E. coli*-produced infectious pancreatic necrosis viral proteins was licensed in 1995 in Norway for salmon aquaculture, and was the first fish-virus vaccine to be available commercially (Christie 1996). Similar research on the production of bacteria-expressed antigens for salmonid viral haemorrhagic septicaemia and infectious haematopoietic necrosis is now concentrated on correct expression of the viral antigens using eukaryotic gene expression systems. Vaccines based on live attenuated viral and bacterial pathogens have also been developed and tested in trials. This has included the development of polyvalent vaccines carrying epitopes for both bacterial and viral pathogens. Initial data are optimistic but economic costs, regulations regarding the use of recombinant DNA products, and sustained virulence or fears of reversion to virulence remain to be overcome. Recently, nucleic acid immunization strategies have come to the attention of fisheries scientists. DNA, containing genes coding for antigens, can elicit host antibody production and protective immunity when delivered directly into the tissues of live animals. This approach has the potential to remove much of the costs associated with antigen production, the fears of virulence, and the design of adjuvants and boosters. Trial experiments using markers genes, as opposed to pathogen antigens, have been tested in fish by both direct injection and particle bombardment and gene expression has been observed (Hansen et al. 1991; Gomez-Chiarri et al. 1996).

Genetic Modification of Cultured Species

Generally, as aquaculture can be considered a young industry, genetic-based improvement of cultured species is relatively recent, and for the most part, cultured species are essentially similar to their wild counterparts. Finfish apart, most other cultured species are derived from wild seeds or eggs collected at the beginning of each production cycle. Therefore, there is great scope for increased productivity through genetic manipulation of cultured species. Potential genetic modification mechanisms include: (1) selective breeding, (2) species hybridization, (3) chromosome manipulation, (4) mono-sex production, (5) transgenic technology and (6) genomics. Biotechnology can play an important role in the elucidation, and subsequent modification, of host genetic factors governing traits of economic interest such as growth rate, stock maturity, efficient food conversion and nutrition, disease resistance, body shape, flesh quality and coloration, temperature tolerance and sex determination.

Selective breeding represents probably the most appropriate strategy for genetic enhancement of cultured stocks. Although long term by nature and with associated economic costs for broodstock maintenance, selective breeding-based genetic improvement programmes have successfully resulted in increased production of Atlantic salmon in Norway, tilapia in the Philippines, catfish in the USA and Thailand and oysters in the USA (Gjedrem 1997). It is likely that similar programmes have begun, or will begin soon, for sea-bream and sea-bass culture in Europe, and marine shrimp, tilapia, common carp and rohu culture in Asia and Africa (Bartley 1997).

Short-term strategies for increased production have also been applied. It is estimated that approximately 80% of Venezuelan culture of cachama (*Colossoma macropomum*) and morocoto (*Piaractus brachypoma*) is composed of hybrids of both species. Predominantly male tilapia, produced through hormone-induced sex reversal of broodstocks, result in better growth rates for breeders. Unwanted reproduction can be lessened by this approach, and also by the production of triploid individuals resulting from pressure treatment of eggs.

Direct genetic intervention can play a role in both these long- and short-term stock enhancement strategies. Transgenic technology has been used to produce transgenic common carp, catfish, coho salmon, Atlantic salmon and tilapia with increased growth characteristics due to the insertion of growth hormone genes into the fertilised egg. However, public and scientific concerns must be met before full commercialisation of these products occurs. These concerns include: (1) the impact of transgenic escapees on wild stocks, (2) potential health risks with respect to human consumption of genetically modified foods, (3) welfare of species involved in transgenic research, and (4) ethical or moral objections to transgenic research.

Recombinant DNA technology is also likely to play a less controversial but very positive role in long-term selective breeding programmes. One of the most rapidly developing sectors in genetic engineering is genomics, or large- or global-scale analysis of genomes. Genome mapping programmes have been underway for several years for commercially exploited species including cattle, sheep, goats, pigs and chickens using genetic techniques such as chromosome isolation, large-insert capacity cloning vectors (YAC & BAC), radiation hybrid mapping, sequence-tagged sites (STS), expressed sequence tags (EST), random amplification of polymorphic DNA (RAPD), arbitrary fragment length polymorphism (AFLP), fluorescence in situ chromosome hybridisation (FISH) and chromosome painting. The rapid pace and progress of recombinant DNA technology means that it is now possible to generate large amounts of genetic data on cultured species using these methods.

Current fish genomics-based projects include pufferfish (*Fugu rubripes*) (Fugu Genome WWW site 1999), zebra fish (*Danio rerio*) (Zebra fish Genome WWW site 1999), tilapia (Tilapia Genome WWW site 1999), salmonid species (SALMAP Project 1996), and Atlantic salmon (SALGENE Project 1998), while similar oyster and channel catfish projects are reported to be underway. Genetic data derived from genome mapping projects have the ability to markedly improve selective breeding programmes for the enhancement of traits of commercial interest. Large-scale gene expression data will complement the genome mapping results by providing new data on candidate genes associated with traits of interest, and also will allow the study of global cell responses using new DNA-chip technology to a variety of parameters including organism development, pathogen infection and disease progression, the immune system, sex and reproduction, and environmental stresses. Finally, these technical advances provide new research tools to monitor inbreeding of broodstocks (Reilly et al. 1999) and also to assess the diversity of natural stocks from which they are derived (Sanchez et al. 1996). They also have the ability to trace and monitor any interactions between cultured and wild species with respect to many eukaryotic parasites (Todd et al. 1997).

Concluding Remarks

As stated previously, aquaculture is currently one of the fastest-expanding sectors of food production. However, in the short to medium term, aquaculture must successfully resolve some critical challenges if the result is to be a sustainable and successful industry. Some of the more important considerations for aquaculture are public acceptance, environmental impact concerns, and safety of produce. Also, aquaculture must respond successfully to the developing regulatory frameworks, including those of international trade. If resolved, biotechnology can be expected to play an ever-increasing role in aquaculture. This promise, already tagged as the blue revolution (Entis 1997), is likely to result in increased production of aquaculture products through stock enhancement and disease control.

References

Andree KB, MacConnell E, Hedrick RP (1998) A nested polymerase chain reaction for the detection of genomic DNA of *Myxobolus cerebralis* in rainbow trout *Oncorhynchus mykiss*. Dis Aquat Org 34:145-154

Argenton F, De Mas S, Malocco C, Dalla Valle L, Giorgetti G, Colombo L (1996) Use of random DNA amplification to generate specific molecular probes for hybridisation tests and PCR-based diagnosis of *Yersinia ruckerii*. Dis Aquat Org 24:121-127

Barg U, Lavilla-Pitogo CR (1996) The use of chemicals in aquaculture: A brief summary of two international expert meetings. FAO Aquacult Newsl 14:12-13

Barlough JE, McDowell TS, Milani A, Bigornia L, Slemenda SB, Pieniazek NJ, Hedrick RP (1995) Nested polymerase chain reaction for detection of *Enterocytozoon salmonis* genomic DNA in chinook salmon *Oncorhynchus tshawytscha*. Dis Aquat Org 23:17-23

Bartley D (1997) Biodiversity and genetics. Review of the state of world aquaculture. FAO Fisheries Circular 886. FAO, Rome

Brown LL, Iwama GK, Evelyn TPT, Nelson WS, Levine RP (1994) Use of the polymerase chain reaction (PCR) to detect DNA from *Renibacterium salmoninarum* within individual infected salmonid eggs. Dis Aquat Org 18:165-171

Chamberlain GW (1993) Aquaculture trends and feed projections. World Aquacult 24 (1):19-29

Chamberlain GW, Hopkins JS (1994) Reducing water use and feed costs in intensive ponds. World Aquacult 25 (3):29-32

Christie KE (1996) Immunization with viral and parasite antigens: infectious pancreatic necrosis. Proc Int Symp on Fish Vaccinology, Norway, Oslo, 5-7 June 1996, 29 pp

Csavas I (1995) The status and outlook of world aquaculture. In: Nambiar KPP, Singh T (eds) Aquaculture towards the 21st century. Proc Infofish-Aquatech'94. 29-31 August 1994, Colombo, Sri Lanka

de la Pena LD, Nakai T, Muroga K (1995) Dynamics of *Vibrio* sp. PJ in organs of orally infected kuruma prawns, *Penaeus japonicus*. Fish Pathol 30:39-45

Dixon PF, Hill BJ (1983) Rapid detection of infectious pancreatic necrosis virus (IPNV) by the enzyme-linked immunosorbent assay (ELISA). J Gen Virol 64:321-330

Docker MF, Devlin RH, Richard J, Khattra J, Kent ML (1997) Sensitive and specific polymerase chain reaction assay for detection of *Loma salmonae* (Microsporea). Dis Aquat Org 29:41-48

Elliot DG, McKibben CL (1997) Comparison of two fluorescent antibody techniques (FATs) for detection and quantification of *Renibacterium salmoninarum* in coelomic fluid of spawning chinook salmon *Oncorhynchus tshawytscha*. Dis Aquat Org 30:37-43

Entis E (1997) Aquabiotech: a blue revolution? World Aquacult 28:12-15

Falk K, Namork E, Dannevig BH (1998) Characterisation and applications of a monoclonal antibody against infectious salmon anaemia virus. Dis Aquat Org 34:77-85

FUGU Genome WWW Site. 1999. http://fugu.hgmp.mrc.ac.uk/

Genmoto K, Nishizawa T, Nakai T, Muroga K (1996) 16S rRNA targeted RT-PCR for the detection of *Vibrio penaeicida*, the pathogen of cultured kuruma prawn *Penaeus japonicus*. Dis Aquat Org 24:185-189

Gjedrem, T. 1997. Selective breeding to improve aquaculture production. World Aquacult 28:33-45

Gomez-Chiarri M, Livingston SK, Muro-Cacho C, Sanders S, Levine RP (1996) Introduction of foreign genes into the tissue of live fish by direct injection and particle bombardment. Dis Aquat Org 27:5-12

Hansen E, Fernandes K, Goldspink G, Butterworth P, Umeda PK, Chang KC (1991) Strong expression of foreign genes following direct injection into fish muscle. FEBS Lett 290:73-76

Hiney M, Dawson MT, Heery DM, Smith PR, Gannon F, Powell R (1992) DNA probe for *Aeromonas salmonicida*. Appl Environ Microbiol 58:1039-1042

Hiney MP, Kilmartin JJ, Smith PR (1994) Detection of *Aeromonas salmonicida* in Atlantic salmon with asymptomatic furunculosis infections. Dis Aquat Orga 19:161-167

Jun L, Tiehui W, Yonglan Y, Hanqin L, Renhou L, Hongxi C (1997) A detection method for grass carp hemorrhagic virus (GCHV) based on a reverse transcriptase-polymerase chain reaction. Dis Aquat Org 29:7-12

Lannan CN, Ewing SA, Fryer JL (1991) A fluorescent antibody test for the detection of the rickettsia-causing disease in Chilean salmonids. J Aquat Anim Health 3:229-234

Laureti E (1996) Fish and fishery products. World apparent consumption statistics based on food balance sheets (1961-1993). FAO Fisheries Circular 821. FAO, Rome

Leong JAC (1993) Molecular and biotechnological approaches to fish vaccines. Curr Opin Biotechnol 4:286-293

Lewis DH (1986) An enzyme-linked immunosorbent assay (ELISA) for detecting penaeid baculovirus. J Fish Dis 9:519-522

Mari J, Bonami JR, Lightner DV (1998) Taura syndrome of penaeid shrimp: cloning of viral genome fragments and development of specific gene probes. Dis Aquat Org 33:11-17

Mauel MJ, Giovannoni SJ, Fryer JL (1996) Development of polymerase chain reaction assays for detection, identification and differentiation of *Piscirickettsia salmonis*. Dis Aquat Org 26:189-195

Miller TA, Rapp J, Wastlhuber U, Hoffmann RW, Enzmann PJ (1998) Rapid and sensitive reverse transcriptase-polymerase chain reaction based detection and differential diagnosis of fish pathogenic rhabdoviruses in organ samples and cultured cells. Dis Aquat Org 34:13-20

Mooney J, Powell E, Clabby C, Powell R (1995) Detection of *Aeromonas salmonicida* in wild Atlantic salmon using a specific DNA probe test. Dis Aquat Orga 21:131-135

Muir JF (1995) Aquaculture development trends: perspectives for food security. Proc Government of Japan and FAO Int Conf on Sustainable Contribution of Fisheries to Food Security. 4-9 December 1995, Kyoto, Japan

Munn CB (1994) The use of recombinant DNA technology in the development of fish vaccines. Fish Shellfish Immun 4:459-473

New MB (1991) Turn of the millenium aquaculture: navigating troubled waters or riding the crest of a wave? J World Aquacult Soc 22 (3):28-49

O'Brien D, Mooney J, Ryan D, Powell E, Hiney M, Smith PR, Powell R (1994) Detection of *Aeromonas salmonicida*, causal agent of furunculosis in salmonid fish, from the tank effluent of hatchery-reared Atlantic salmon smolts. Appl Environ Microbiol 60:3874-3877

Oshima S, Hata JI, Hirasawa N, Ohtaka T, Hirono I, Aoki T, Yamashita S (1998) Rapid diagnosis of red sea bream iridovirus infection using the polymerase chain reaction. Dis Aquat Org 32:87-90

Pedini M, Shehadeh ZH (1997) Global outlook. Review of the state of world aquaculture. FAO Fisheries Circular 886. FAO, Rome

Pimentel D, Shanks RE, Rylander RC (1996) Bioethics of fish production: energy and the environment. J Agricult Environ Ethics 9 (2):144-164

Rana KJ (1997a) Trends in Global Production, 1984-1995. Review of the state of world aquaculture. FAO Fisheries Circular 886. FAO, Rome

Rana KJ (1997b) Aquatic environments and use of species groups. Review of the state of world aquaculture. FAO Fisheries Circular 886. FAO, Rome

Reilly A, Elliot NG, Grewe PM, Clabby C, Powell R, Ward RD (1999) Genetic differentiation between Tasmanian cultured Atlantic salmon (*Salmo salar* L.) and their ancestral Canadian population: comparison of microsatellite DNA and allozyme and mitochondrial DNA variation. Aquaculture 173:457-467

Ristow SS, de Avila J, LaPatra S, Kauda K (1993) Detection and characterisation of rainbow trout antibody against infectious hematopoietic necrosis virus. Dis Aquat Org 15:109-114

SALGENE Project (1998) Generation of a genetic body map for Atlantic salmon. E.U. Contract FAIR6-CT98-4314. Project Coordinator: Richard Powell, Galway, Ireland. Email: richard.powell@nuigalway.ie

SALMAP project (1996) Generation of highly informative DNA markers and genetic maps of salmonid species. E.U. Contract FAIR3-CT96-1591. Project Coordinator: Bjorn Hoyheim, Oslo, Norway. Email: bjorn.hoyheim@veths.no

Sanchez JA, Clabby C, Ramos D, Blanco G, Flavin F, Vasquez E, Powell R (1996) Protein and microsatellite single locus variability in *Salmo salar* L. (Atlantic salmon). Heredity 77:423-432

Soderhall K (1996) Crustacean immunity. Proc Int Symp on Fish Vaccinology 5-7 June 1996, Oslo, Norway, p 14

Stokes NA, Siddall ME, Burreson EM (1995) Detection of *Haplosporidium nelsoni* (Haplosporidia: Haplosporidiidae) in oysters by PCR amplification. Dis Aquat Org 23:145-152

Subasinghe R (1997) Fish health and quarantine. Review of the state of world aquaculture. FAO Fisheries Circular 886. FAO, Rome

Tacon AGJ (1997) Contribution to food fish supplies. Review of the state of world aquaculture. FAO fisheries Circular 886. FAO, Rome

Tilapia Genome WWW Site. 1999. http://www.ri.bbsrc.ac.uk/tilapia/

Todd CD, Walker AM, Wolff K, Northcott SJ, Walker AF, Ritchie MG, Hoskins R, Abbott RJ, Hazon N (1997) Genetic differentiation of populations of the copepod sea louse *Lepeophtheirus salmonis* (Kroyer) ectoparasitic on wild and farmed salmonids around the coasts of Scotland: evidence from RAPD markers. J Exp Mar Biol Ecol 210:251-274

Toyama T, Kita-Tsukamoto K, Wakabayashi H (1994) Identification of *Cytophaga psychrophila* by PCR targeted 16S ribosomal RNA. Fish Pathol 29:271-275

Wang SY, Hong C, Lotz JM (1996) Development of a PCR procedure for the detection of *Baculovirus penaei* in shrimp. Dis Aquat Org 25:123-131

Wang WS, Lee JS (1997) Single tube, non-interrupted reverse transcription PCR for detection of infectious pancreatic necrosis virus. Dis Aquat Org 28:229-233

Wang WS, Lee JS, Shieh MT, Wi YL, Huang CJ, Chien MS (1996) Detection of infectious hematopoietic necrosis virus in rainbow trout *Oncorhynchus mykiss* from an outbreak in Taiwan by serological and polymerase chain reaction assays. Dis Aquat Org 26:237-239

Warr GW (1996) Adaptive immunity in fish. Proc Int Symp on Fish Vaccinology. 5-7 June 1996, Oslo, Norway, p 11

Way K, Dixon PF (1988) Rapid detection of VHS and IHN viruses by the enzyme-linked immunosorbent assay, ELISA. J Appl Ichthyol 4:182-189

Wolf K, Markiw ME (1975) *Myxosoma cerebralis*: serological identification by indirect fluorescent antibody test. Fish Health News 4:8

Wongteerasupaya C, Tongchuea W, Boonsaeng V, Panyim S, Tassanakajon A, Withyachumnarnkul B, Flegel TW (1997) Detection of yellow-head virus (YHV) of *Penaeus monodon* by RT-PCR amplification. Dis Aquat Orga 31:181-186

Zebra fish Genome WWW site. 1999. http://zfish.uoregon.edu/index.html

Part IV
Biotechnology

Ervin BALÁZS

A. TSAFTARIS et al.
L.M. HOUDEBINE
H. NAGASHIMA
C. KUBICEK

Transgenic Crops: Recent Developments and Prospects

AS. TSAFTARIS, AN. POLIDOROS, M. KARAVANGELI, I. NIANIOU-OBEIDAT, P. MADESIS, C. GOUDOULA[1]

It is now more than 15 years since the first transgenic plants were generated experimentally. In this period there have been dramatic advances in our understanding on both basic and applied aspects of plant biology.

Transgenic plant research depends on the availability of procedures for plant transformation. Two types of method for plant transformation exist, the use of *Agrobacterium* as a biological vector for foreign gene transfer, and direct gene transfer techniques, in which DNA is introduced into cells by the use of physical, electrical or chemical means. *Agrobacterium* can be used to transform a wide range of plants, but there are number of species which are of interest for basic or applied research in which of *Agrobacterium*-mediated transformation is not reproducible or efficient. Using this procedure, thousands of transgenic crops have been developed experimentally or field-tested, while a few of them are currently being cultivated worldwide, predominately on temperate-zone crops and under conditions prevailing in industrial countries with potential increasing and improving food production capacity while limiting the use of agrochemicals and protecting the environment. The first generation of transgenic crops was aimed at improving traits involving single genes. Now we are on the verge of a new step in crop modification, fueled by the rate at which new genes (important for plant growth and development metabolism and stress tolerance) are characterised. Transgenic technology has been pivotal in the full spectrum of these new developments, from gene identification to an improved understanding of their regulation, as well as genetic transformation involving more complex transfers of many genes simultaneously. This will further help in managing natural resources like water, soil, etc. in a better way.

Our view of the nature of crop products can also be expected to change in the short to medium term, as plants are exploited for the production of novel compounds such as biodegradable plastics and new pharmaceuticals. However, the extent to which the potential of transgenic research is realised will depend on public acceptance. To a significant extent this will require that the biology of transgenics be

1. Department of Genetics and Plant Breeding, University of Thessaloniki, Greece, 54006

fully understood, and that a maximum degree of predictability of transgene effect, both phenotypic and genotypic, can be ensured. There is a need to spread this technology to tropical plants and adapt it to benefit the small farmer in the developing world, where food demands will increase.

To achieve this requires: to find ways integrating biotechnology research into their national agricultural research activities on the one hand, and horizontal and vertical networking cooperation on the other.

Finally, the implication of advancement in this relatively new technology, especially in the area of biosafety, production patterns, biodiversity, property rights and other critical factors should be sufficiently discussed and understood.

Introduction

Genetic modification of crops has enabled plant breeders to modify plants in novel ways and has the potential to overcome important problems of modern agriculture. Introduction of genes into plants has been made possible using *Agrobacterium* as a biological vector, and direct gene transfer techniques. *Agrobacterium*-based methods are more efficient and simple but have the disadvantage that they are not applicable in every plant species (Christou 1995). Recent developments indicate that these host-range limitations can be overcome by developing specific plant cell culture procedures and defining inoculation and cocultivation conditions (Park et al. 1996). Some important non-host species such as maize and rice have now been stably transformed by *Agrobacterium*. Although plant transformation was initially experimental, the potential of commercialization of new improved varieties was early realized and a fast-growing international ag-biotech market has already been formed. New transgenic varieties have been produced that are resistant to pathogens, insects, and herbicides, or express novel characters that improve product quality and agronomic traits. The new opportunities to modify plants in novel ways with genetic modification present new responsibilities for safe use to avoid adverse effects on human health and the environment (Dale and Irwin 1998). Risk assessment studies are an integral part in producting and placing a transgenic variety on the market. Different countries have adopted different approaches in biosafety assessment. International harmonization of biosafety standards is an important challenge as we face the international trade of transgenic plant products. The role of international organizations such as the OECD and United Nations may be critical in archieving this goal.

In the following sections we will present the methods and techniques that are utilized in the production of transgenic plants, emphasizing recent developments for multiple gene transfers; we will consider the major achievements of the new technology and the future prospects and challenges, as well as the remaining technological gaps; we will enumerate and discuss some of the possible risks involved in the unrestrained use of transgenic plants and their products; and finally, we will present the current status of the regulatory framework pertaining to the field release of transgenic plants.

Methods of Gene Transfer in Plants

Transgenic plant research depends on the availability of procedures of plant transformation. There are two types of effective gene transfer to plants, the first based on the use of *Agrobacterium* as a biological vector and the second on the use of physical, electrical or chemical treatments to introduce isolated DNA into cells alleviating the need for vector use. The latter techniques are commonly termed direct gene transfer methods.

Indirect Gene Transfer Using Agrobacterium As Vector

The most widely used method for the introduction of new genes into plants is based on the natural DNA transfer capacity of *Agrobacterium tumefaciens*. In nature, this soil bacterium causes tumor formation (called crown gall) on a large number of dicotyledonous plant species. During this infection a part of the Ti-plasmid of *Agrobacterium*, called T-DNA, is transferred and integrated into the plant genome. This natural capacity made us use this bacterium as a natural vector of foreign genes (inserted into the Ti-plasmid) into plant chromosomes.

Agrobacterium-based and direct gene transfer techniques were developed in parallel, but the former is today the most widely used method because of its simplicity and efficiency in many plants, although it still suffers limitations in terms of the range of species which are amenable to transformation. These limitations are due to the natural host range of *Agrobacterium*, which generally infects herbaceous dicotyledonous species most efficiently and is less effective on monocotyledonous and woody species (De Cleene and De Ley 1976). In these plants, direct gene transfer techniques offer the means to establish transformation systems, but many of these techniques suffer from a relatively low efficiency of transformation. Attempts are therefore being made to exploit and adapt the relatively simple and convenient *Agrobacterium* system to transform recalcitrant plant species. Recent work has shown that these host-range limitations are not absolute and by developing specific plant cell culture procedures and defining inoculation and cocultivation conditions, some important non-host species have now been stably transformed by *Agrobacterium*, although there are still many plant species for which *Agrobacterium* transformation cannot be used.

The development of reliable transformation protocols for recalcitrant species depends on the establishment of an efficient regeneration procedure, a high transformation rate of the regenerable cells, and an effective selection for regenerating transformed cells (Gheysen et al. 1998). The plant genotype is an important factor, determining both the regeneration capacity and the efficiency of *Agrobacterium* transformation. Equally important is the choice of the bacterial strain, and the external conditions during the preculture and cocultivation of agrobacteria and plant material.

The *Agrobacterium* transformation methods use two different procedures. The first transformation is dependent on a regeneration procedure while the second is not. The purpose of the regeneration procedure is twofold: it allows the recovery of uniformly transformed shoots and the selection of such shoots. For many plant species, the lack of suitable regeneration method is one of the main bottlenecks in developing a transforming procedure. A particular regeneration method is usually only efficient with a limited number of genotypes even within a species. Somaclonal variation may also be problematic with some regeneration procedures. Therefore, many efforts have been devoted to the development of regeneration-independent transformation procedures, such as meristem transformation and in planta transformation techniques.

The shoot apex has been used in meristem transformation as an attractive target for transformation since it contains the meristematic cells from which all the aerial parts of the plant are derived. Because meristems are multicellular organs, primary transformants are expected to be chimeric, consisting of transformed and untransformed sectors. This has two important consequences. First, it does not always result in germline transformation and transmission of the transgenes to the offspring and second, a stringent selection procedure cannot be applied (Gheysen et al. 1998). These have as a result that this transformation method is labor-intensive and very inefficient. Several reports clearly give evidence for stable transformation that has been achieved through meristem transformation with *Agrobacterium* infection in important crops, such as *Musa acuminata* (May et al. 1995) and *Oryza sativa* (Park et al. 1996). *Oryza sativa is* the first cereal species that has been stably transformed via *Agrobacterium*. Targeting cells of meristem for transformation has, therefore, the advantage that transformed cell lineages can be obtained without the involvement of a regeneration pathway which involves dedifferentiation and reorganization of cells, so somaclonal variation is not a problem and the transformation is rather genotype-independent. These are the main advantages of such an approach. Nevertheless, additional manipulations (e.g., hormonal treatments) are necessary to obtain transformants with acceptable frequencies, reintroducing a factor of genotype dependence in the procedure (Gheysen et al. 1998).

Over the past decade, several in planta methods for *Agrobacterium*-mediated transformation of *Arabidopsis thaliana* have been developed that do not involve any tissue culture steps. In the first described procedure (Feldmann and Marks 1987), imbibed seeds are infected with *Agrobacterium*, allowed to grow into mature plants and finally transformants were identified among the seeds harvested from these plants. Bechtold *et al.* (1993) inoculated flowering *A. thaliana* plants by vacuum infiltration with an *Agrobacterium* suspension and managed to obtain transformants at even higher frequencies. Another technique which has been developed recently (Clough and Bent 1998) is floral dip. It is a simple dipping of developing floral tissues into an *Agrobacterium* suspension. The absence of any tissue culture step (so somaclonal variation does not occur), the simplicity and the relatively high efficiency of the transformation procedure would make such techniques attractive to adapt the technique to other plant species recalcitrant to regeneration procedure.

Direct Gene Transfer

The development of novel direct gene transfer methodology, bypassing limitations imposed by *Agrobacterium*-host specificity and cell culture constraints, has allowed the engineering of almost all major crops, including formerly recalcitrant cereals, legumes and woody species. Direct gene transfer transformation methods are species and genotype-independent in terms of DNA delivery, but their efficiency is influenced by the type of target cell, and their utility for the production of transgenic plants in most cases depends on the ease of regeneration from the targeted cells, as most methods operate on cells cultured in vitro. Direct gene transfer methods are particle bombardment, DNA uptake into protoplasts, treatment of protoplasts with DNA in the presence of polyvalent cations, fusion of protoplasts with bacterial spheroplasts, fusion of protoplasts with liposomes containing foreign DNA, electroporation-induced DNA uptake into intact cells and tissues, silicon carbide fiber-induced DNA uptake, ultrasound-induced DNA uptake, microinjection of tissues and cells, electrophoretic DNA transfer, exogenous DNA application and imbibition, and macroinjection of DNA (Barcelo and Lazzeri 1998; Walden and Schell 1990). The most significant direct gene transfer methods are presented in Table 1.

Table 1. The most significant methods of direct gene transfer.

Methods	Description
Particle bombardment	Delivery of DNA into cells using microscopic gold or tungsten particles coated with DNA as carriers accelerated into target cells by gunpowder, gas or air pressure or by electrical discharge (Christou 1995)
Protoplast transformation	DNA introduction into protoplasts using PEG-mediated DNA uptake and electroporation, liposome (containing plasmid DNA) fusion. (Krens et al. 1982; Caboche 1990)
Tissue electroporation	Transformation of plant organs or regenerable cell cultrures. (Li et al. 1991; D'Halluin et al. 1992)
Ultrasound-induced transformation	DNA uptake into protoplasts, suspension cells, and tissues induced by ultrasound waves. (Joersbo and Brunstedt 1990; Joersbo 1990; Zhang et al. 1991)
Silicon-carbide fiber or whisker transformation	The fibers perforate cell walls and allow DNA to penetrate the cells. (Frame et al. 1994)
Laser-mediated transformation	Laser beams are used to create openings in cell components and organelles allowing DNA insertion (Weber et al. 1989)
Microinjection	Direct delivery of DNA into plant cells using a microsyringe. (Schnorf et al. 1991)

Most workers in transgenic plant research are interested primarily in applying a transformation technique rather than in its mechanism of operation, so there is a general wish for technically simple methods which are easily transferred between laboratories and which ideally do not require expensive, specialized equipment. Of the above, direct gene transfer techniques, particle bombardment, and protoplast transformation are today the most widely used. The former most closely satisfies the criteria of technical simplicity and reproducibility, although it requires a specialized particle gun, the commercial version of which uses relatively expensive consumables. Protoplast transformation can be highly efficient, but demands more complicated cell culture techniques and is limited by the difficulty of regenerating plants. Tissue electroporation is relatively simple, applicable to regenerable tissues, and has produced stably transformed plants in several systems after only a relatively short period of development. These results suggest that the method should receive further attention to evaluate its potential for wider application. Ultrasound and silicon carbide fiber-mediated techniques are newer methods, which are again technically quite simple. They have been tested in few laboratories and need more research to determine their limitations. Microinjection and laser-mediated transformation are specialized techniques, which are at present inefficient. Electrophoretic transfer, to date, gives no evidence that the gene transfer actually occurs. Whole-plant direct gene transfer methods would be methods of choice for most users, but, despite several claims of high transformation efficiencies, most critical studies have not produced evidence for integrative transformation (Barcelo and Lazzeri1998).

Plant genetic engineering is now at a crucial cross roads. The gene transfer constraints appear to have been removed from a number of important crops. Technical problems still remain, but they are not insurmountable. The attention of the scientific community is gradually shifting to other areas such as identification and cloning of genes responsible for multigenic traits. The study of genomes (known as genomics) involves the mapping, sequencing, and analysis of genomes in order to determine the structure and function of every gene in an organism. This has already been accomplished in several microorganisms, and much effort has been devoted to the complete sequencing of the genome of higher eukaryotes including plants. Information derived from analysis of such data will be used to map entire biochemical pathways, which will then be easier to transfer and incorporate in transgenic organisms. Thus, genomic information can be used to improve important plant traits through genetic engineering, such as high and stable yield and product quality. Utilization of genomics in transgenic technology will also require establishment of routine techniques for simultaneous multiple gene transfer in a single transformation event. In most cases, one or a few genes are transferred to the plant genome along with a selectable marker that facilitates selection of transgenic tissues. Genetic transformation with a single target gene has been used for the production of transgenic crop plants that expressing herbicide tolerance, resistance to fungal, viral, and bacterial diseases and insect pests. In addition, improved agronomic characteristics have been achieved by manipulating metabolic pathways through overexpression of a specific gene or the use of antisense sequences. As most agronomic characteristics are polygenic in nature plant genetic engineering will require manipulation of complex metabolic or regulatory pathways involving multiple genes or gene complexes. Redirecting complex biosynthetic pathways and modifying polygenic agronomic traits requires the integration of multiple transgenes into the plant genome, while ensuring their stable inheritance and expression in succeeding generations. Transfer of multiple genes via *Agrobacterium*-mediated transformation, although possible, is technically demanding and becomes increasingly problematic as the number of genes and the size of the transferred DNA increases; but this transfer could be achieved through cobombardment in a simple process in which genes carried on separate plasmids are mixed prior to transfer by particle bombardment. In this manner numerous genes can be transferred simultaneously (Fig. 1) using a single selectable marker (Chen et al. 1998). This will certainly be one of the major future goals of transgenic technology in plants (see also below).

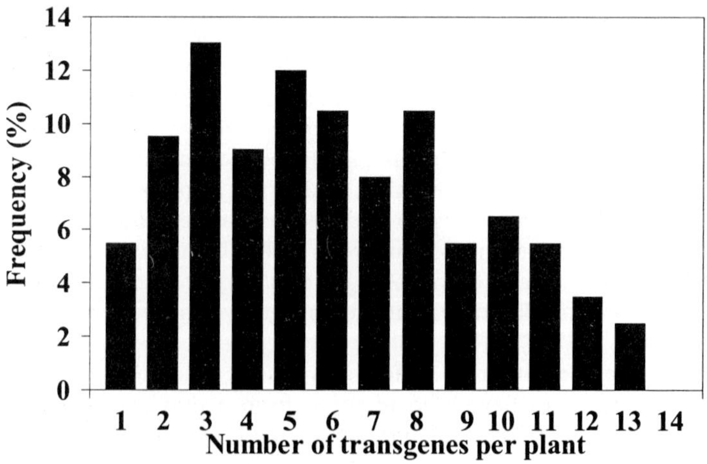

Fig. 1: Frequency of transgenic plants containing 1-14 genes

Technological Gaps

Although considerable progress in gene transfer technology has already been accomplished, and techniques are consistently improved for more efficient plant transformation, there are several obstacles preventing application of the new technology in certain circumstances. In such cases, transgenic technology needs radical improvement and perhaps reformation of the underlying strategies to enable breeding transgenic varieties in certain plant species and for certain uses. The barriers that new developments in transgenic technology will have to overcome in the near future include:

- The need for genotype-independent protocols for regeneration and transformation in a number of mainly tropical vegetables and trees. International organizations should promote networking of interested regional countries and labs to work together towards this goal. Also, companies advanced in transgenic technologies could participate in such networks, providing solutions and expert assistance to overcome difficulties that will arise during these efforts.
- For standing tree populations transgenic technology will face the considerable time that is needed for such trees to be productive. For example, breeding olive trees in Mediterranean countries with the new technology will require several years from a successful transformation to a new, fully productive transgenic plantation. In such cases, alternatives will have to be devised, that may include grafting of mature trees in the plantation, and in situ transformation of mature trees.
- Securing the stability of the transgenic phenotype, especially under stress, will be of immense importance. There are already examples where significant losses were observed in transgenic varieties grown under adverse environments. Stability of transgene expression can be dependent on the position the transgene has landed in the genome, the number of copies incorporated, the promoter used, the presence of repeated elements or cryptic remnants of transposable elements. Novel techniques allowing precise manipulation of transgene incorporation and advancement of our knowledge on genome function can help to secure stable expression in a wide range of environmental conditions.

Achievements and Future Prospects

Products Developed

In conventional plant-breeding, genes can be transmitted only by crossing in the same or closely related species. Transgenic techniques have allowed genetic material to be transferred between completely unrelated organisms, so that breeders can incorporate characteristics that are not normally available within a species. The modified organisms exhibit properties that would be impossible to obtain by conventional breeding techniques. Modern biotechnology makes plant-breeding programs more effective in two important ways. Firstly it allows transfer of specific genes, incorporating into the new variety only those traits that are wanted. This makes the process of trait transfer faster, more exact, cheaper and less likely to fail than traditional cross breeding methods. Secondly, it gives breeders the freedom to incorporate genes from unrelated species into the target plant, a possibility that is unprecedented in plant breeding.

Transgenic methods have been employed over the past 15 years in a number of important crop plants such as maize, cotton, soybean, oilseed rape, and a variety of vegetable crops like tomato, potato, cabbage, and lettuce. In the EU, 1255 field tests involving transgenic plant varieties have been approved. This number has surpassed 5000 field releases (permits and notifications) in the US. A summary of permits per European Country is shown in Table 2. The commercial production of transgenic crops shows a rapid increase the last few years. The global area (excluding China) of transgenic crops from zero in 1995 reached 27.8 million ha in 1998 (Fig. 2). During this period, the larger part of global transgenic crops was grown in industrial countries, with significantly less in developing countries (James 1998). The proportion of transgenic crops grown in industrial countries in 1998 was 84%, slightly less than 1997 (86%), and only 16% grown in the developing countries, with most of that area in Argentina, and the balance in Mexico and South Africa.

Table 2. Permits for field trials of genetically modified plants in European countries from October 1991 to June 1998. (*http://biotech.jrc.it*)

Country	Total
Austria	3
Belgium	90
Denmark	30
Finland	13
France	391
Germany	77
Greece	12
Ireland	4
Italy	206
Netherlands	103
Portugal	10
Spain	117
Sweden	35
United Kingdom	164
Total for the European Union	1255

The most frequently modified plants from a total of 50 or more plant species are shown in Fig. 3. Among these, almost 50% of the new varieties are modified maize and oilseed rape plants. The most frequent traits that have been modified using genetic engineering are shown in Fig. 4. The major achievements of transgenic plant technology up to now concern tolerance to insect or disease pests, herbicide tolerance, and improved product quality. A description of the major categories of modified traits with characteristic examples will follow.

Insect Tolerance

New varieties of maize, cotton and tobacco, for example, have been developed utilizing a gene from the bacterium *Bacillus thuringiensis* to produce a protein (the Bt protein) that is specifically toxic to certain insect pests including bollworm, but not to animals or humans (Carozzi et al. 1992; Liang et al. 1994). This protein has been

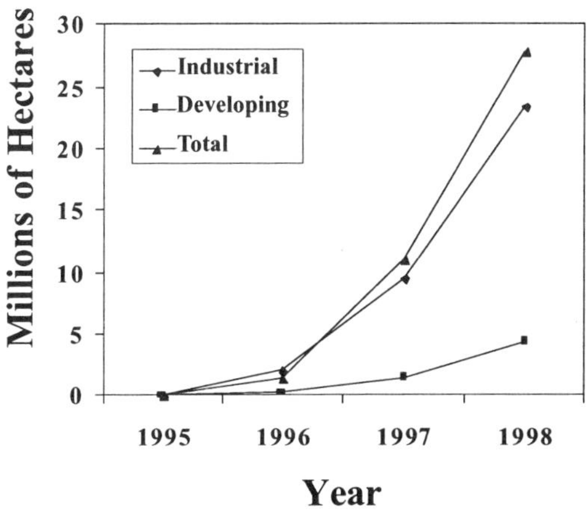

Fig. 2: Global area cultivated with commercial transgenic crops (excluding China) from 1995 to 1998. The distribution of the area between industrial and developing countries is also shown. (James 1998)

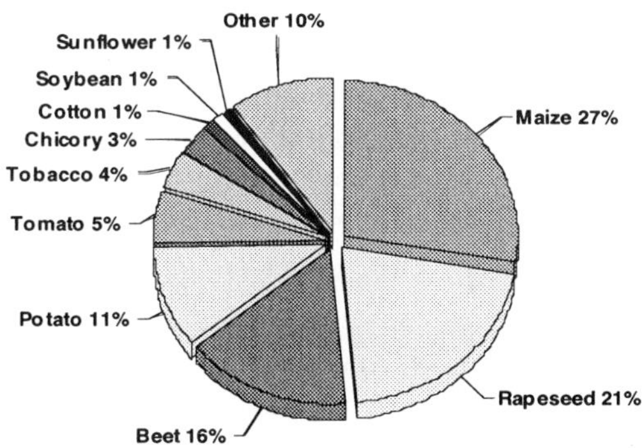

Fig. 3: The most frequently modified plants from a total of 50 or more plant species in European Union. (http://biotech.jrc.it)

used as a pesticide spray for many years. Cultivation of these transgenic plants should help reduce the use of chemical pesticides in cotton production, as well as in the production of many other crops, which could be engineered to contain the *Bacillus thuringiens* gene. For a more detailed discussion on insect tolerance and uses of the transgenic technology for production of pesticidal crops see the chapter by Dr. David Andow in this Volume.

Disease Resistance

Tobacco mosaic virus (TMV) causes the leaves of some important crop plants to wither and die. Incorporation into the plant of a gene that encodes the coat protein of the virus protects it from disease (Clark et al. 1995). This approach

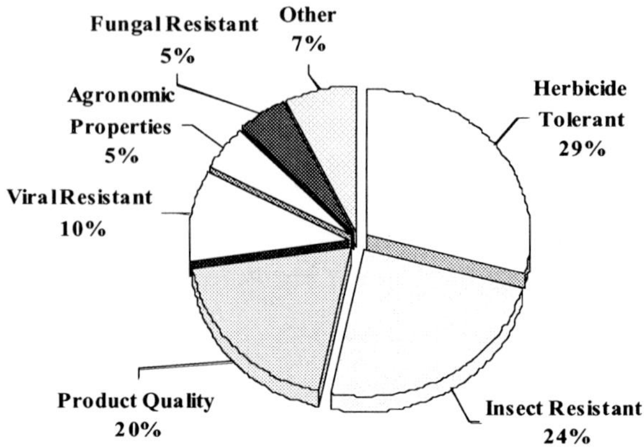

Fig. 4: The most frequent traits that have been modified using genetic modification in European Union. (http://biotech.jrc.it)

has also been applied to other viral diseases in crops. More progress in development of disease-resistant transgenic plants will be seen in the near future. Over the past decade, many efforts have been focused on understanding plant-pathogen interactions in molecular terms. This led to the identification of disease resistant plant genes that specify race-specific resistance to pathogens. The tomato disease-resistant gene *Pto*, for example, confers resistance to the bacterial pathogen *Pseudomonas syringae* pv *tomato* carrying the *avrPto* gene. Recently, Tang et al. (1999) reported that overexpression of the *Pto* gene in transgenic tomato plants activated defense responses and conferred broad resistance to several bacterial pathogens.

Herbicide Tolerance

Engineering herbicide tolerance in transgenic plants has been accomplished by exploiting at least three different mechanisms: overexpression of the target enzyme, modification of the target enzyme, and herbicide detoxification (Tsaftaris 1996). Examples of transgenic plants developed based on each mechanism are as follows.

Glyphosate is an environmentally more benign, widely used broad-spectrum herbicide. It is easily degraded in the agricultural environment and works by interfering with the EPSPS enzyme system that is present only in plants. Unfortunately, the herbicide kills crop plants as well as weeds. Transgenic plants including maize, soybean, and cotton have been developed, overexpressing an additional copy of the EPSPS gene from *Petunia hybrida* under the strong 35S promoter and exhibiting increased tolerance to glyphosate.

Alternatively, expression of a mutant *Aro A* gene from *Salmonella typhimurium* (which encodes EPSPS) in transgenic tobacco resulted in even higher tolerance to the herbicide than overexpression of the wild-type petunia EPSPS gene (for review see Tsaftaris 1996). This allows farmers to control weeds in transgenic cultivars by spraying with glyphosate alone.

A different approach has been applied for development of resistance to the herbicide phosphinothricin (basta). The *bar* gene from *Streptomyces hygroscopicus* or *S. uiridochromogenes* encodes the enzyme phosphinothricin acetyl transferase (PAT), which converts the herbicide to a nontoxic acetylated form. Expression of the *bar* gene in transgenic tobacco, potato, and tomato plants conferred phosphinothricin resistance at up to ten times the normal application rate of the herbicide in the field (Wohlleben et al. 1988).

Product Quality

Transgenic technologies have been used to modify other important characteristics of plants such as starch composition in potato (Lorberth et al. 1998; Takaha et al. 1998), ripening in tomato (Smith et al. 1990; Klee et al. 1991), lignin content in arabidopsis (Ni et al. 1994), flower vase-life in carnation (Bovy et al. 1995), and to explore many new possibilities for uses in agriculture as well as in industry.

Future Prospects

Major achievements in plant biotechnology are presently limited to traits involving one or a few genes. It will probably require more research before we can manipulate complex traits (such as yield) that are influenced by many genes. However, with newly developed techniques we can now incorporate multiple genes in plant genomes integrating multiple traits.

In addition, advances in structural and functional analysis of higher plant genomes will provide substantial knowledge on important biochemical pathways that are involved in the regulation of more complex characters. This could even enable scientists to identify and transfer entire biochemical pathways from one species to another and incorporate them into new hosts for the benefit of agriculture and/or industry. Eventually it may also be possible to develop crops for nonfood uses by modifying traits to make them more suitable for industrial purposes, or to use plants rather than animals to make antibodies for medical and agricultural diagnostic purposes, and delivering vaccines with food in developing countries.

Current research will see the improvement and development of crops for specific purposes. Plants that require less water could be developed for countries with arid climates. Crop plants engineered to be tolerant to salt could be farmed in salt-damaged farmland or could be irrigated with salty water. Crops with higher yields and higher protein values are also possible. Much current research focuses on understanding and developing useful promoter sequences to control transgenes, and establishing precise methods to insert and place the transgene at specific locations in the recipient chromosomes. Much still needs to be done to improve our knowledge of specific genes and their actions, the potential side effects of adding foreign DNA and of manipulating genes within an organism and the problems associated with transgene silencing.

Phenotypic Stability of Transgenic Varieties

Transgenic crop plants will only be of value if their phenotype is stable in the field and transmitted faithfully in subsequent generations. Although it is possible to study transgenes with a high level of precision, there is often uncertainty related to inactivation and structural instability of the transgenes. This inactivation is well documented and is most frequently correlated with gene silencing, and not loss of the transgene (for review see Tsaftaris and Polidoros 1999). Gene silencing in transgenic plants has been identified as a major obstacle in transgenic technology. Reversal to herbicide sensitivity, for example of transgenic plants bred for tolerance to herbicides, could lead to significant losses. From the applied side, gene silencing has come as an unwelcome surprise and is turning out to be a substantial problem. According to Finnegan and McElroy (1994), of 30 companies polled, nearly all reported some problems with unwanted silencing of transgenes. It has been shown that this unwelcome sensitivity due to inactivation of the transgene mediated by methylation, is triggered by stresses like the common agronomic practice of seedling transplantation in the field (Brandle et al. 1995). Thus, it requires closer attention since many crops need transplantation in the field and this imposes a severe stress on the young plantlet. It is clear that some steps that were taken for granted may need to be further investigated for successful commercialization of transgenic crops.

Risk Assessment and the Regulatory Framework

Risk Assessment

Advances in transgenic technology bring new responsibilities for safe use of transgenic plants for the benefit of humanity and the environment. The objective in risk assessment is to develop safety procedures, proactively rather than reactively. Safety issues are scale-dependent and are probably different in small-scale experimental field trials than in large-scale commercial releases. Long-term effects of transgenic plants and their products may only be detectable after large-scale or even commercial production of transgenic crops. Thus, one major challenge concerning safety of transgenic organisms is to develop procedures to assess long-term effects on human health and the environment.

Other issues have also emerged that could be considered in risk assessment. Competitiveness, jobs, and investment are thought to be at risk if the technology is not adopted, and potential benefits lost. Another dimension is the issue of consumer choice and rights to reject the technology at the point of sale. To a great extent, this has been the focus of the growing consumer movement in all parts of the world, and especially in Europe.

In addition, much of the commercialization of the new technology relies on a few international companies with capital and power to dominate in the forming market of transgenic plants. The future may well find just a few key multinational industries active in producing recombinant plants to manage plant diseases and to produce agricultural and industrial products leaving the developing countries (which could benefit the most from the new technology) well behind. This can be alleviated if laboratories from developing countries along with advanced ag-biotech companies share knowledge and technology in a network, aiming to the development of improved transgenic varieties of crops that may be of minor commercial value for the companies, but critical as source of food or agricultural income for the developing countries.

A summary of possible risks for human health and the environment, associated with transgenic plants is given below.

Risks for Human Health

- Formation of new allergens from the novel proteins expressed in the transgenic organism, which could trigger allergic reactions at some stage.
- Creation of new toxins through unexpected interactions between the product of the genetic modification and other endogenous constituents of the organism.
- Dispersion of antibiotic resistance genes used as markers from the genetically modified organism derived food to gut microorganisms and intensification of problems with antibiotic-resistant pathogens.

Risks for the Environment

- Gene transfer from the transgenic plant to related species as a result of hybridization that could lead to new pests.
- The transgenic plant escapes its intended use and becomes an invader to the natural environment.
- Harmful effects on nontarget species with the expression, for example, of insecticide toxins that can kill beneficial as well as targeted insects.
- Development of resistance from the continuous use of the same agent on the target organism.
- Harmful effects on ecosystems when transgenic plant products interfere with natural biochemical cycles.
- Harmful effects on biodiversity if a transgene offers an adaptive advantage in transgenic plants escaped in the area of cultivation or in wild relatives, where it could be transferred by cross-fertilization. This is of practical importance if occurring at the centers of genetic variation of cultivated plants. In addition, biodiversity concerns have been raised for current cultivation systems including many locally adopted varieties if they are replaced by a few new transgenics.

The process of examining the above risks from the release of a transgenic plant to the environment can provide a framework for risk assessment. Of course, enumeration and listing all the above major questions or possible risks expressed by scientists, consumers, and ecological groups for field testing and commercialization of transgenic plants does not imply that all the above are a concern for all the different transgenics and in different environments. For example, a risk for possible new allergenicity to the consumer could be meaningful, thus requiring testing prior to release, in cases where new genes coding for possibly new allergenic compounds have been cloned into plants. This question should not concern transgenics without such genes being cloned.

The Regulatory Framework

The criteria and factors that determine biosafety assessment of transgenic plants vary in different countries. In the EU all plants produced by genetic modification must be assessed (technology-based assessment), whereas in the USA and Canada only plants modified with particular genes are regulated (product-based assessment). There is considerable debate on the safety guaranteed by the two approaches. The use of *Agrobacterium* as a vector for the transformation process implies that the transgenic plants produced will be regulated under both approaches. However, as other methods avoid *Agrobacterium* and sequences derived from plant pathogens, there is considerable difference in the regulatory requirements between North America and Europe. Time and scale differences in commercialization of transgenic crops will also have an effect, making one system more "experienced" than the other.

In the USA, three agencies share the primary responsibility for regulating the genetically modified organisms, whether they be designed for closed systems or for environmental uses. These are the USEPA, the FDA, and the USDA. In addition to federal regulation, several states and municipalities have enacted biotechnology-related legislation, including provisions related to the environmental release of genetically modified organisms. Each of the federal agencies regulating biotechnology is guided in its analysis and decision-making criteria by its specific legislation, i.e., the laws passed by Congress charging each agency with specific responsibilities. These laws differ in their mandate as to what populations to consider with regard to adverse effects (e.g., humans, crops, the environment), as well as in their mandate as to how to strike a balance between risks and benefits. In addition to its specific legislation, each agency must also adhere to the National Environmental Policy Act (NEPA), which is binding for all federal agencies.

In the EU, the regulatory framework on agricultural biotechnology is made of a few European Directives and Regulations (Vega et al. 1999). Directive 90/219 issues the regulations covering the contained use of genetically modified organisms and Directive 90/220 issues the regulations covering the deliberate release into the environment of genetically modified organisms. Directive 90/219 has recently been totally revised. The European Commission also presented a proposal (COM/98/0085) for amending the Directive 90/220 so as to harmonize European approaches to the issue. Several Member States have refused to approve commercialization of transgenic plants approved in other Member States in their territories, and others have called for a moratorium. The European Parliament's Committee on the Environment, Public Health and Consumers proposed a Europe-wide moratorium on all transgenic crops awaiting authorization to be placed on the market. The Council Regulation (EC) No. 258/97 regulates the compulsory indication of the labeling of certain foodstuffs produced from genetically modified organisms. The Directive on the legal protection of biotechnological inventions 98/44/EEC regulates issues of intellectual property. For more details on risk assessment and the regulatory framework pertaining to transgenic plants, see the chapter by Dr. John Beringer, in this Volume.

Risk assessment needs to have an international dimension and extent beyond the primary country of release or a shared international boundary or a market. International harmonization of regulations and procedures for production, testing, and handling transgenic plants must be a major challenge. Obtaining good scientific data for the long-term effects on the environment will be critical for passing these products from the regulatory framework. This will be even more critical for plants tolerant to different biotic and abiotic stresses. Progress in meeting this challenge will be highly dependent on:

1. how the questions are formulated, and
2. the amount and kind of effort that will be devoted towards this goal.

The role of international organizations in archieving this goal can be critical, and meetings and discussions to facilitate this are of immense importance.

References

Barcelo P, Lazzeri P (1998) Direct gene transfer: chemical, electrical and physical methods. In: Lindsey K (ed) Transgenic plant research. Harwood Academic Publ, Chur, Switzerland

Bechtold N, Ellis J, Pelletier G (1993) In planta transformation mediated gene transfer by infiltration of adult *Arabidopsis* plants. C R Acad Sci Ser III. Sci Vie 316:1194-1199

Bovy AG, van Altvorst AC, Angenent GC, Dons JJM (1995) Genetic modification of the vase-life of carnation. Acta Hortic 405:179-189

Brandle JE, McHugh SG, James L, Labbe H, Miki BL (1995) Instability of transgene expression in field grown tobacco carrying the *csr1-1* gene for sulfonylurea herbicide resistance. Biotechnology 13: 994-998

Caboche M (1990) Liposome-mediated transfer of nucleic acids in plant protoplasts. Physiol Plant 79:173-176

Carozzi NB, Warren GW, Desai N, Jayne SM, Lotstein R, Rice DA, Evola S, Koziel MG (1992) Expression of a chimeric CaMV 35S *Bacillus thuringiensis* insecticidal protein gene in transgenic tobacco. Plant Mol Biol 20:539-548

Chen L, Marmey P, Taylor NJ, Brizard JP, Espinoza CD, Cruz P, Huet H, Zhang S, de Kochko A, Beachy RN, Fauquet CM (1998) Expression and inheritance of multiple transgenes in rice plants. Nat Biotechnol 16:1060-1064

Christou P (1995) Strategies for variety-independent genetic transformation of important cereals, legumes and woody species utilizing particle bombardment. Euphytica 85:13-27

Clark WG, Fitchen JH, Beachy RN (1995) Studies of coat protein-mediated resistance to TMV. I. The PM2 assembly defective mutant conifers resistance to TMV. Virology 208:485-491

Clough SJ, Bent AF (1998) Floral dip: a simplified method for *Agrobacterium*-mediated transformation of *Arabidopsis thaliana.* Plant J 16(6):735-743

Dale PJ, Irwin JA (1998) Environmental impact of transgenic plants. In: Lindsey K (ed) Transgenic plant research. Harwood Academic Publ, Chur, Switzerland

De Cleene M, De Ley J (1976) The host range of crown gall. Bot Rev 42:389-466

D'Halluin K, Bonne E, Bossut M, De Beuckleer M, Leemans J (1992) Transgenic maize plants by tissue electroporation. Plant Cell 4:1495-1505

Feldmann KA, Marks MD (1987) Agrobacterium-mediated transformation of germinating seeds of *Arabidopsis thalliana.* A non-tissue culture approach. Mol Gen Genet 208:1-9

Finnegan J, McElroy D (1994) Transgene inactivation: plants fight back! Biotechnology 12: 883-888

Frame BR, Drayton PR, Bagnall SV, Lewnau CJ, Bullock WP, Wilson HM, et al. 1994. Production of fertile transgenic maize plants by silicon carbide fiber-mediated transformation. Plant J 6:941-948

Gheysen G, Angenon G, Van Montagu M (1998) *Agrobacterium*-mediated plant transformation: a scientifically intriguing story with significant applications.

In: Lindsey K (ed) Transgenic plant research. Harwood Academic Publ, Chur, Switzerland

James C (1998) Global review of commercialized transgenic crops. ISAAA Briefs 8. ISAAA, Ithaca, New York

Joersbo M (1990) A method for introducing molecules, particularly genetic material into plant cell. Danish patent application 3251/89

Joersbo M, Brunstedt J (1990) Direct gene transfer to plant protoplasts by mild sonication. Plant Cell Rep 9:207-210

Klee HJ, Hayford MB, Kretzmer KA, Barry GF, Kishore GM (1991) Control of ethylene synthesis by expression of a bacterial enzyme in transgenic tomato plants. Plant Cell 3:1187-1193

Krens FA, Molendijk L, Wullems GJ, Schilperoort RA (1982) In vitro transformation of plant protoplasts with Ti-plasmid DNA. Nature 296:72-74

Li B, Xu X, Shi H, Ke X (1991) Introduction of foreign genes into the seed embryo cells of rice by electro-injection and the regeneration of transgenic rice plants. Sci China 34 (8):923-931

Liang X, Zhu Y, Mi J, Chen Z (1994) Production of virus-resistant and insect-tolerant transgenic tobacco plants. Plant Cell Rep 14:141-144

Lorberth R, Ritte G, Willmitzer L, Kossmann J (1998) Inhibition of a starch-granule-bound protein leads to modified starch and repression of cold sweetening. Nat Biotechnol 16:473-477

May GD, Afza R, Mason HS, Wiecko A, Novak FJ, Arntzen CJ (1995) Generation of transgenic banana (*Musa acuminata*) plants via *Agrobacterium*-mediated transformation. Bio/Technology 13:486-492

Ni WT, Paiva NL, Dixon RA (1994) Reduced lignin in transgenic plants containing a caffeic acid O-methyltransferase antisense gene. Transgen Res 3:120-126

Park SH, Pinson SRM, Smith RH (1996) T-DNA integration into genomic DNA of rice following *Agrobacterium* inoculation of isolated shoot apices. Plant Mol Biol 32:1135-1148

Schnorf M, Neuhaus-Url G, Galli A, Iida S, Potrykus I, Neuhaus G (1991) An improved approach for transformation of plant cells by microinjection: molecular and genetic analysis. Transgen Res 1:23-30

Smith CJS, Watson CF, Morris PC, Bird CR, Seymour GB, Gray JE, Arnold C, Tucker GA, Schuch W, Harding S (1990) Inheritance and effect on ripening of antisense polygalacturonase genes in transgenic tomatoes. Plant Mol Biol 14:369-379

Takaha T, Critchley J, Okada S, Smith SM (1998) Normal starch content and composition in tubers of antisense potato plants lacking D-enzyme (4-alpha-glucanotransferase). Planta 205:445-451

Tang X, Xie M, Kim YJ, Zhou J, Klessing D, Martin GB (1999) Overexpression of *Pto* activates defense responses and confers broad resistance. Plant Cell 11:15-29

Tsaftaris A (1996) The development of herbicide-tolerant transgenic crops. Field Crops Res 45:115-123

Tsaftaris AS, Polidoros AN (1999) DNA methylation and plant breeding. Plant Breed Rev (in press)

Vega M, Bontoux L, Llobell A (1999) Biotechnology for environmentally safe agriculture. IPTS Rep 31:9-13

Walden R, Schell J (1990) Techniques in plant molecular biology – progress and problems. Eur J Biochem 192:563-576

Weber G, Monajembashi S, Greulish KO, Wolfrum J (1989) Uptake of DNA in chloroplasts of *Brassica napus* (L.) by means of a UV-laser microbeam. Eur J Cell Biol 49:73-79

Wohlleben W, Arnold W, Broer I, Hillemann D, Strauch E, Puhler A (1988) Nucleotide sequence of the phosphinothricin N-acetyltransferase gene from *Streptomyces uiridochromogenes* Tu 494 and its expression in *Nicotiana tabacum.* Gene 70:25-37

Zhang L, Cheng L, Xu N, Zhao N, Li C, Yuan J, Jia S (1991) Efficient transformation of tobacco by ultrasonication. Bio/Technology 9:996-997

Recombinant Proteins from Domestic Animals

LM. HOUDEBINE[1]

Proteins are the direct product of genes. They play a major role in living organisms. Some of them are enzymes, others are hormones, blood factors etc. For many years the only source of proteins was living organisms themselves. Proteins have a too complex structure to be synthesized chemically at a reasonable cost. They must therefore be biosynthesized. The techniques of genetic engineering have offered the possibility to reprogram living organisms by transferring isolated genes coding for proteins. The proteins synthesized in this manner are named recombinant proteins. The preparation of recombinant proteins is currently one of the major successes of biotechnology. Human insulin, human growth hormone etc. synthesized by bacteria are now used as pharmaceuticals. An additional advantage of recombinant proteins is that various forms of the molecules can be prepared by mutating the corresponding genes.

Why Use Animals to Prepare Recombinant Proteins?

The first recombinant proteins were prepared from bacteria. It is relatively easy to modify a human gene to be expressed in bacteria. The cDNA of the gene (devoid of introns) must be added between a promoter and a transcription terminator from bacteria. In some cases, the sequence of a certain number of codons must be modified to optimize the expression of human genes in bacteria. Although quite efficient and commonly used, this technique has intrisic limitations. Some of the recombinant proteins are toxic for the bacteria. Others may aggregate in inclusion bodies and become difficult to extract in a biologically active form. A large proportion of proteins are to be used as pharmaceuticals and biochemically modified after their synthesis. These modifications are relatively complex and some of them such as specific cleavage, glycosylation, γ-carboxylation etc., cannot be performed by bacteria. In these cases, the proteins can be synthesized only by eukaryotic cells.

Yeast shares many advantages with bacteria. It can be easily transformed by foreign genes and be cultured in large amount in fermentors. The proteins synthesized by yeast may be secreted in the culture medium. However, yeasts are too different

1. Biologie Cellulaire et Moléculaire, Institut National de la Recherche Agronomique, 78352 Jouy-en-Josas Cedex France

from animal cells and they are not in all cases able to post-translationally process recombinant proteins in an appropriate manner.

Various animal cells are used to prepare recombinant proteins. Insect cells infected by recombinant baculovirus-harbouring the genes of interest can synthesize quite significant amounts of the corrresponding proteins at a reasonable cost. However, insect cells are not always able to modify the recombinant proteins, after their biosynthesis, in a correct fashion.

Animal cell lines are used on an industrial scale to prepare recombinant proteins. The CHO cells (Chinese hamster ovary) are one of these lines. The proteins secreted from animals cells are usually quite functional. Human erythropoïetin, which is too rare to be extracted from blood and which is glycosylated, is prepared from cultured animal cells and currently used in hospitals. The techniques of animal cell culture in fermentors are not new and they can be used successfully on an industrial scale. However, this process is rather complex and costly; it also has a limited flexibility. On the other hand, purification of the recombinant proteins from proteins from the culture medium is not simple in all cases.

For all these reasons, the preparation of recombinant proteins from transgenic animals appeared an attractive alternative. This approach has several theorical advantages. The cells are as numerous as needed in the animals and they are in an ideal metabolic situation. Several quite different cell types can be used in the animals and the recombinant proteins can be recovered from various biological fluids.

Which Animal Organs to Use

Blood

Blood may be an attractive source of recombinant proteins. Indeed, blood is a low-cost by-product of slaughterhouses, and it is the part where many secreted proteins are found. Several years ago, it was shown that transgenic rabbits harbouring the human α_1-antitrypsin gene synthesized the protein in their liver and the protein was found at a concentration as high as 1 mg ml^{-1} in blood. The recombinant α_1-antitrypsin was correctly glycosylated and perfectly active (Massoud et al. 1991). Although attractive, this process has intrisic limitations. It may not be easy to separate the human recombinant proteins from the homologous animal molecules. On the other hand, many proteins have only a short half-life in blood and cannot accumulate at a high concentration. Morever, the recombinant proteins in blood are in an ideal situation to interact with the cells of the animals and alter their health. For all these reasons, blood has not been retained as a universal source of recombinant proteins. This process may still prove useful in some particular cases.

Reticulocytes are cells synthesizing large amounts of hemoglobin. Transgenic pigs are experimentally being used to synthesize human hemoglobin (Swanson et al. 1992). As long as the endogenous pig hemoglobin is synthesized, this approach, leading usually to the generation of pig and human hybrid hemoglobin, cannot be used on an industrial scale. The specific knockout of pig globin genes might become feasible in the coming years and solve this problem.

Reticulocytes might be the source of recombinant peptides stored rather than secreted. The transcription regulatory mechanism of α- and β-globin genes is being extensively studied and it is possible to express specifically foreign genes in red blood cells. The isolated reticulocytes collected in slaughterhouses might thus be used to isolate recombinant non-secreted peptides or proteins.

Milk

As early as 1982, when the first functional transgenic mice were born, milk appeared as one of the best potential sources of recombinant proteins. Pioneer experiments published in 1987 (Gordon et al. 1987; Simons et al. 1987) demonstrated that foreign proteins can be secreted in the milk of transgenic mice in a biologically active form.

Milk has several theorical advantages and drawbacks. Milk is a very abundant biological fluid, although available only from lactating mammalian females. It is secreted out of the body and its proteins are not expected to interact with the physiology of the animals. Milk has a relatively simple biochemical composition, and it does not contain proteases potentially capable of degrading the recombinant proteins. Morever, milk from domestic ruminants is very abundantly used for human consumption.

Milk protein genes are extensively studied and their promoters can be fused to foreign genes to direct the synthesis of corresponding proteins (Mercier and Vilotte 1997; Houdebine 1991). Several promoters have been selected for their high potency to drive the expression of associated foreign genes. Those which seem the most appropriate are from the ruminant αs_1- and -β casein genes, from the ruminant β-lactoglobulin gene, from mouse and rabbit whey acidic protein genes and from ruminant and human α-lactalbumin genes (Mercier and Vilotte 1997).

The association of a milk gene promoter with a foreign gene does not always lead to efficient expression of the transgenes (Houdebine 1991). This is obviously due to our lack of knowledge of the mechanism of gene expression. A few rules have emerged during the past decade to generate gene constructs capable of expressing transgenes in a reliable manner. Genomic fragments containing native genes are generally well expressed. The same is not true for cDNA, which are relatively long DNA fragments not interrupted by introns. Addition of foreign introns to cDNAs improves transgene expression quite significantly. Yet the expression of transgenes with the currently used constructs is often not fully satisfactory. Many transgenes remain silent or are expressed non-specifically at a low level in different tissues of the animals. It is admitted that this phenomenon is due to a position effect. The transgenes are generally less readily expressed when they are integrated closer to the centromeres. It seems more and more likely that cells of the transgenic animals recognize the foreign DNA. This is particularly true when multiple copies of the transgenes are integrated. These transgenes provoke the formation of heterochromatin, which extinguishes their expression. Long genome DNA fragments are generally well-expressed transgenes. Experiments carried out recently with different genes led to the conclusion that they contain DNA sequences capable of preventing the formation of heterochromatin. Some of these sequences, named insulators or chromatin openers, have been identified. Most of these sequences are specific to a cell type and to a gene. One of them, at least, seems capable of strongly favouring the expression of quite different transgenes in most cell types (Taboit-Dameron et al. 1999). The addition of such sequences to a gene construct allows the transgenes to be expressed at a high level in most animals. This may contribute quite significantly to reduce the number of transgenic animals to be generated.

Gene transfer into mammals relies essentially on the direct micro-injection of DNA into the pronuclei of embryos at the one-cell stage. The efficiency of this technique is low and its cost is very high in large domestic animals. To reduce the number of donors and recipients, embryos at the one-cell stage can be prepared in vitro by the maturation of oocytes collected in the slaughterhouse, followed by fertilization. The microinjected embryos can be cultured until the blastocyst stage. In the best cases, only the embryos which survived to microinjection and in which the transgene was found are transferred into recipients (Houdebine 1991).

Another approach has been recently proposed. Foetal cells can be cultured and receive a foreign gene by classical transfection. The cells harbouring the foreign gene can be used to regenerate living organisms by the cloning technique based on the transfer of nuclei into enucleated oocytes. This protocol proved to reduce quite significantly the number of animals needed to generate transgenic sheep (Schnieke et al. 1997). The same technique is being extended to cow and goat.

The amounts of recombinant proteins which are needed for pharmaceutical use are from a few grams to several tons per year. The species which are currently used to produce recombinant proteins in milk are mouse, rabbit, sheep, pig and cow. Mouse is used essentially to test the gene constructs. Rabbit can produce up to 1 kg of a recombinant protein per year (Stinnakre et al. 1997). Transgenic rabbits can be obtained and bred at a relatively low cost (Strömqvist et al. 1997; Viglietta et al. 1997). Milking in this species is, however, laborious and relatively costly. Sheep (Wright and Coleman 1997), goat (Di Tullio et al. 1997) and pig (Drews et al. 1995) are considered as suitable to produce no more than a few tons of recombinant proteins per year. Cow is being used to produce the largest amounts of proteins (Krimpfort et al. 1991; Eyestone 1999).

The Alternative Systems

Other biological fluids are candidates for use instead of milk. A recent study demonstrated that cells of the urinary bladder epithelium can synthesize human growth hormone which is secreted into the urine. The hormone was found at a concentration of a few $\mu g\ ml^{-1}$ of urine and it was biologically active. Urine, which is abundant and already the source of pharmaceutical proteins, might become a valuable alternative to milk in some particular cases (Kerr et al. 1998).

Other studies carried out recently demonstrated that seminal plasma can contain significant amounts of biologically active recombinant proteins when appropriate promoters are fused to the genes of interest (Pothier et al. 1999).

A quite attractive system to produce recombinant proteins is chicken egg white. Eggs contain huge amounts of ovalbumin, which can be easily purified. This system has not been used for technical reasons. Indeed, it remains a difficult task to generate transgenic birds. Totipotent cell lines have been recently isolated. Chimeric transgenic chickens have been obtained with these cells (Pain et al. 1996). Homologous recombination leading to the replacement of the ovalbumin gene by a gene of interest has been performed successfully (Kunita et al. 1998). Other methods to generate transgenic chicken, namely microinjection or infection by retroviral vectors, may also contribute to using egg white as a source of recombinant proteins.

The sericigen gland of the silk worm might be used to synthesize some particular recombinant proteins. Transgenesis remains a difficult task in this species.

In some cases, the recombinant proteins must be very specifically modified to be active. Their modifications may then occur in only a few cell types. In these cases, it is conceivable to extract the recombinant proteins from the organs of transgenic animals and not from a biological fluid.

Conclusion

Up to 100 proteins have been produced in milk for experimental purposes. There is little doubt that the reproduction of recombinant proteins from domestic animals is becoming a new branch of the pharmaceutical industry. The first protein, human antithrombin III, obtained from goat milk, should be on the market in 2001. It should be followed rapidly by many others. The proteins which will be produced in this way belong to quite different categories. Some will be enzymes or structural proteins. Others will be hormones, growth factors, blood factors, antigens or antibodies. A few hundred proteins could be prepared in milk. The number of antibodies which might be prepared from milk appears particularly high. Indeed, the diversity of antibodies is very broad. They may be used for diagnosis but also for immunotherapy. They may be specific carriers of toxins to destroy tumor cells. The production of monoclonal antibodies from hybridomas is relatively costly and the use of ascites is less and less accepted for biosafety and ethical reasons. Milk is therefore considered as a valuable alternative to prepare recombinant antibodies.

In a certain number of cases, the mammary cells showed astonishing capacity to synthesize foreign protein of complex structure. The simultaneous microinjection into cow embryos of the gene constructs coding for the three fibrinogen subunits led to the abundant secretion of fibronogen containing the three proteins associated in a quite satisfactory fashion (Prunkard et al. 1996). Transgenic rabbits were able to secrete human EC-superoxide dismutase at a concentration as high as 3 mg ml^{-1} of milk. This protein was active, glycosylated and correctly assembled as dimer containing copper ion. The mammary cell was thus able to capture large amounts of copper in the blood circulation although milk from non-transgenic rabbits contains only a low amount of this metal (Strömqvist et al. 1997).

All the production systems have advantages and drawbacks. Milk appears for the next decade the best in vivo production system. A certain number of proteins cannot be prepared easily in this way. This was the case for human erythropoïetin, which seriously altered the health of the transgenic animals (Massoud et al. 1996). For unknown reasons, some recombinant proteins are secreted only at a very low concentration in milk or they severely impair the growth or the activity of mammary cells.

The biochemical purification of recombinant proteins from milk is generally not a difficult task (Wright and Coleman 1997). The contamination of the purified proteins by pathogens remains a concern. It is generally admitted that rabbit, sheep, goat, pig and cow can be bred under conditions eliminating the classical pathogens. Contamination by some viruses or by prions may be more difficult to detect. Specific protocols must be defined and followed to eliminate the biorisks. The Food and Drug Administration in the USA has published guidelines for the preparation of proteins from animals for human use (Miller and Matheson 1997). The biorisks do not appear currently too high to dissuade companies from using this method of pharmaceutical production.

It seems at present likely that milk will be an essential method to produce recombinant proteins. The alternative methods, including the use of transgenic plants, will be most likely used simultaneously. Various criteria will be considered to choose the most appropriate producing system, one of which is the cost. The toxicity of a protein for animals may justify the use of yeasts or plants or perhaps chicken.

All the available data indicate that the use of recombinant proteins produced by animals will be a common practice in the 21st century and possibly later. The techniques defined to prepare pharmaceuticals in milk can be used as well to transform milk into a neutriceutical. Indeed, various proteins or oligosaccharides (Prieto et al. 1995) having antibacterial activity may be added in milk by this process. Recombinant antibodies can thus protect against infection of the digestive tract (Castilla et al. 1998). Antigens might also be secreted to be used as very cheap and efficient vaccines. The same techniques are also being used to optimize milk composition for human consumption (Mercier and Vilotte 1997). Several projects are aimed at reducing the concentration of lactose (Stinnakre et al. 1994; Stacey et al. 1995; L'huillier et al. 1998; Jost et al. 1999) and β-lactoglobulin in cow milk and at replacing some of the casein genes. Thanks to these techniques, milk is expected to be better tolerated by consumers and more easily utilized by the dairy industry.

References

Castilla J, Pintado B, Sola I, Sanchez-Morgado J, Enjuanes L (1998) Engineering passive immunity in transgenic mice secreting virus-neutralizing antibodies in milk. Nat Biotechnol 16:349-354

Di Tullio P, Ebert KM, Pollock J, Edmunds T, Meade HM (1997) Production of complex human pharmaceuticals in the milk of transgenic goat using the goat beta casein promoter. In: Houdebine LM (ed) Transgenic animals. Generation and use. Harwood Academic Publ, Amsterdam, pp 465-468

Drews R, Pal Eyanda RK, Lee TK, Chang RR, Rehemtulla A, Kaufman RJ, Drohan WN, Lubon H (1995) Proteolytic maturation of protein C upon engineering the mouse mammary gland to express furin. Proc Natl Acad Sci USA 92:10462-10466

Eyestone WH (1999) Production and breeding of transgenic cattle using in vitro embryo production technology. Theriogenology 51:509-517

Gordon K, Lee E, Vitale JA, Smith AE, Westphal H, Hennighausen L (1987) Production of human tissue plasminogen activator in transgenic mouse milk. Bio/Technology 5:1183-1187

Houdebine LM (1991) Production of recombinant protein in transgenic animals. J Biotechnol, 34:269-287

Jost B, Vilotte JL, Duluc I, Rodeau JL, Freund JN (1999) Production of low-lactose milk by ectopic expression of intestinal lactase in the mouse mammary gland. Nat Biotechnol 17:160-164

Kerr DE, Liang F, Bondioli KR, Zhao H, Kreibich G, Wall RJ, Sun TT (1998) The bladder as bioreactor: urothelium production and secretion of growth hormone into urine. Nat Biotechnol 16:75-79

Krimperfort P, Rademakers A, Eyestone W, vander Schans A, van der Broek S, Kooiman P, Kootwijk E, Platenburg G, Pieper F, Strijker R (1991) Generation of transgenic dairy cattle using 'in vitro' embryo production. Bio/Technology 9:844-847

Kunita R, Samarut J, Pain B (1998) Establishment of the chicken gene-targeting disruption system. French-Japanese Workshop. Genes and early development, June 4-5, 1998

L'huillier P, Soulier S, Stinnakre MG, Lepoury L, Davis SR, Mercier JC, Vilotte JL (1998) Efficient and specific ribozyme-mediated reduction of bovine α-lactalbumin expression in double transgenic mice. Proc Natl Acad Sci USA 93:6698-6703

Masoud M, Bischoff R, Dalemans W, Pointu H, Attal J, Schultz H, Clesse D, Stinnakre MG, Pavirani A, Houdebine LM (1991) Expression of active recombinant human α1-antitrypsin in transgenic rabbits. J Biotechnol, 18:193-204

Massoud M, Attal J, Thépot D, Pointu H, Stinnakre MG, Théron MC, Lopez C, Houdebine LM (1996) The deleterious effects of human erythropoietin gene driven by the rabbit whey acidic protein gene promoter in transgenic rabbits. Reprod Nutr Dev 36:555-563

Mercier JC, Vilotte JL (1997) The modification of milk protein composition through transgenesis: progress and problems. In: Houdebine LM (ed) Transgenic animals generation and use. Harwood Academic Publ, Amsterdam pp 473-482

Miller MA, Matheson III JC (1997) Food safety evaluation of transgenic animals. In: Houdebine LM (ed) Transgenic animals. Generation and use. Harwood Academic Publ, Amsterdam, pp 563-568

Pain B, Clark ME, Shen M, Nakazawa H, Sakurai M, Samarut J, Etches RJ (1996). Development 122:2339-2348

Long-term in vitro culture and characterization of avian embryonic stem cells with multiple morphogenetic potentialities

Pothier F, Dyck MK, Ouellet M, Gagné D, Sénéchal JF, Bélanger E (1999) The production and secretion of human growth hormone into seminal fluid of transgenic mice. Theriogenology 51, 424

Prieto PA, Mukerji P, Kelder B, Erney R, Gonzal Ez D, Yun JS, Smith DF, Mormen KW, Nardelli C, Pierce M, Li Y, Chen X, Wagner TE, Cumings RD, Kopchick JJ (1995) Remodeling of mouse milk glucoconjugates by transgenic expression of a human glycosyltransferse. J Biol Chem 270:29515-29519

Prunkard D, Cotinngham I, Garner I, Bruce S, Dalrymple M, Lasser G, Bishop P, Foster D (1996) High-level expression of recombinant human fibrogen in the milk of transgenic mice. Nat Biotechnol 14:867-871

Schnieke A, Kind AJ, Ritchie WA, Mycok K, Scott AR, Ritchie M, Wilmut I, Colman A, Campbell KHS (1997) Human factor IX transgenic sheep produced by transfer of nuclei from transfected fetal fibroblasts. Science 278:2130-2133

Simons JP, McClenaghan M, Clark AJ (1987) Alteration of the quality of milk by expression of sheep beta-lactoglobulin in transgenic mouse. Nature 328:530-532

Stacey A, Schnieke A, Kerr M et al. (1995) Lactation is disrupted by α-lactalbumin deficiency and can be restored by human α-lactalbumin gene replacement in mice. Proc Natl Acad Sci USA 92:2835-2839

Stinnakre MG, Vilotte JL, Soulier S, Mercier JC (1994) Creation and phenotype analysis of α-lactalbumin deficient mice. Proc Natl Acad Sci USA 91:6544-6548

Stinnakre MG, Massoud M, Viglietta C, Houdebine LM (1997) The preparation of recombinant proteins from mouse and rabbit milk for biomedical and pharmaceutical studies. In: Houdebine LM (ed) Transgenic animals. Generation and use. Harwood Academic Publ, Amsterdam, pp 461-464

Strömqvist M, Houdebine LM, Andersson JO, Edlund A, Johansson T, Viglietta C, Puissant C, Hansson L (1997) Recombinant human extracellular superoxide dismutase produced in milk of transgenic rabbits. Transgen Res 6:271-278

Swanson ME, Martin MJ, O'Donnell JK, Hoover K, Lago W, Huntress V, Parsons CT, Pinkert CA, Pilder S, Logan JS (1992) Production of functional human hemoglobin in transgenic swine. Bio/Technology 10:557-559

Taboit-Dameron F, Malassagne B, Viglietta C, Puissant C, Leroux-Coyau M, Chéreau C, Attal J, Weill B, Houdebine LM (1999) Association of the 5'HS4 sequence of the chicken β-globin locus control region with human EF1α gene promoter induces ubiquitous and high expression of human DAF and CD59 cDNAs in transgenic rabbits. Transgen Res 8:223-235

Viglietta C, Massoud M, Houdebine LM (1997) The generation of transgenic rabbits. In: Houdebine LM (ed) Transgenic animals. Generation and use. Harwood Academic Publ, Amsterdam, pp 11-14

Wright G, Coleman A (1997) Purification of recombinant proteins from sheep's milk. In: Houdebine LM (ed) Transgenic animals. Generation and use. Harwood Academic Publ, Amsterdam, pp 469-472

Cryopreservation of Mammalian Embryos

H. NAGASHIMA[1]

Embryo cryopreservation is now routine for major experimental and domestic animals. The absence of effective embryo cryopreservation technologies for other animals may result in major loss of genetic diversity. "Delipation" enables cryopreservation of the cryosensitive embryos. Cryobanking of transgenic and cloned embryos is important fort the future security of animal production.

Introduction

The technology for cryopreserving mammalian embryos has been developed in the past three decades, and it has important implications in the long-term storage, propagation and transport of valuable genotypes of agricultural or zoological interest. The birth of live offspring from embryos frozen in liquid nitrogen, however, is still limited to the major experimental and domestic animals and a few other species including primates and antelopes. The absence of effective embryo cryopreservation technologies for other animals, particularly endangered species, may result in a major loss of genetic diversity. The lack of cryopreservation procedures for porcine embryos has been a special challenge in this regard, as pig meat represents 40% of total animal meat consumption by man, and, due to the extreme sensitivity of porcine embryos to low temperatures, no procedure has been available to maintain the rapidly narrowing pool of genetic diversity in this species. We have recently demonstrated that the sensitivity of porcine embryos to low temperature can be decreased by reducing their lipid content, and developing a practical cryopreservation method for pig embryos is well on its way to solution. This breakthrough would also have an impact on conservation biology. Embryos from different species vary considerably in the conditions required to achieve survival after cryopreservation. The development of protocols for cryopreserving mammalian embryos has concentrated on optimizing conditions such as cryoprotectants, freezing media and cooling/warming rates according to the species. The inability to collect sufficient embryos to carry out quantitative experiments is the major factor responsible for the absence of procedures for cryopreserving embryos of

1. Division of Organ Transplantation, Biomedical Research Center, Osaka University Medical School
Present address: Laboratory of Reproduction Engineerimg, Meiji University 1-1-1 Higashimita, Tama, Kawasaki 214-8571, Japan

scarce and endangered species. The method we have developed for early cleavage stage porcine embryos represents a new approach to cryopreservation. In this regard, our experiments may herald new opportunities for the cryopreservation of oocytes and early-stage embryos from species, including several ungulates and carnivores which contain large amounts of cytoplasmic lipid. Furthermore, embryo cryopreservation would facilitate practical application of other reproductive biotechnologies, such as cloning, in vitro fertilization and gene transfer.

The production of live animals from frozen embryos was first successful in mice in 1972 (Table 1). Then live offspring were born from cryopreserved embryos in most species of major experimental and domestic animals and man, and a few species of non-domestic animals. Exceptionally, in the major domestic animals, cryopreservation of pig embryos remained unsuccessful for many years, due to the extremely high sensitivity of pig embryos to low temperatures.

Table 1. History of embryo cryopreservation in mammals. (Mahmoundzadeh 1994)

Production of the first offspring from frozen embryos			
Mouse	1972	Baboon	1984
Cow	1973	Bongo	1985
Rabbit	1974	Hamster	1985
Sheep	1974	Cynomolgus	1986
Rat	1975	Marmoset	1987
Goat	1976	Cat	1988
Horse	1982	Pig	1989
Eland	1983	Rhesus	1989
Man	1983		

There are two major embryo cryopreservation technologies; the slow freezing method and the vitrification method (Fig. 1). In the freezing method, embryos are exposed to the cryoprotectants such as glycerol, DMSO, propanediol or ethylene glycol with relatively low concentrations. Then the embryos are frozen slowly to dehydrated the free water in the cytoplasm. In this method, ice crystals are formed in the freezing solution which holds the embryos. In the other method, vtrification, embryos are cooled very rapidly with cryoprotectants of high concentration, so that embryos will vitrify without icecrystal formation.

Embryo cryopreservation is now routinely applied for reproduction of experimental and domestic animals (Fig. 2). By establishing an embryo cryobank, the cost of maintaining live animals is dramatically reduced, and the genetics of valuable experimental animals can be distributed worldwide for use in scientific researches. Embryo cryopreservation technology has also been changing the form of international trades of domestic animals, particularly in the cattle industry. It provides a hygienic and economical way of transporting animals.

Application of embryo cryopreservation for conserving scarce breeds of domestic animals is also very important (Fig. 3). Embryos collected from scarce breeds and conserved in a cryobank can be transferred in future to recipient animals of a common breed for proliferation.

Cryopreservation procedures for porcine embryos have been undeveloped for many years, and this has emerged as a special challenge in the pig industry (Fig. 4).

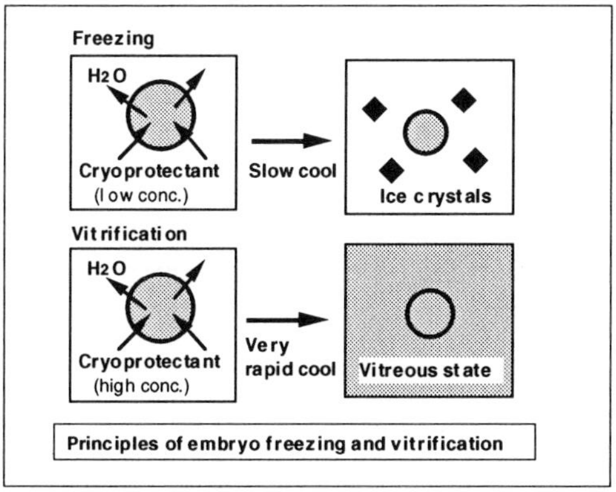

Fig. 1: Principles of embryo freezing and vitrification

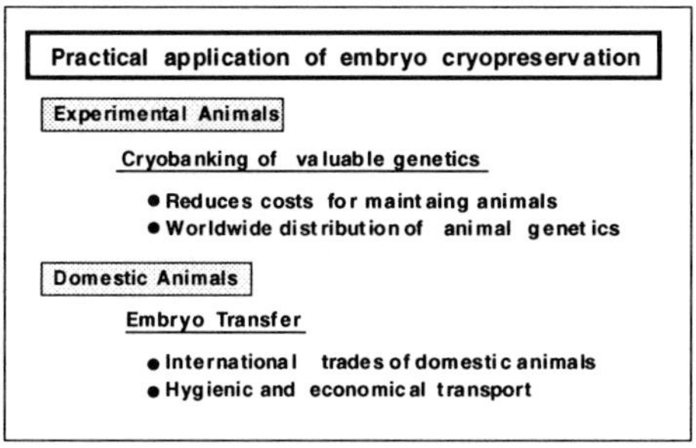

Fig. 2: Practical application of embryo cryopreservation

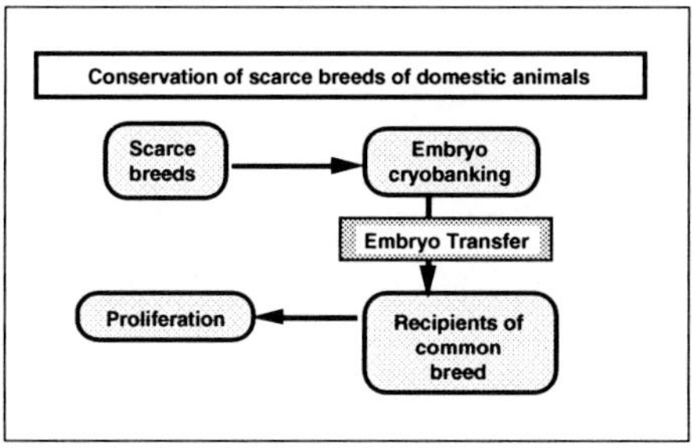

Fig. 3: Conservation of scarce breeds of domestic animals

Despite the fact that pig meat represents nearly 40% of total animal meat consumption by man, the pool of genetic diversity in this species has been rapidly narrowing, due to the productivity-oriented breeding program in most of the industrialized countries and lack of effective cryopreservation procedures for porcine embryos.

However, recent development in technology may provide a solution to this problem. This slide shows a new technology which was developed recently. We demonstrated that the torelance of porcine embryos to cryopreservation can be increased by removing their cytoplasmic lipid droplets. Pig embryos contain very high amount of cytoplasmic lipid droplets. These lipid droplets can be polarised by centrifugation, and removed by microsuction. The lipid-removed embryos or "delipated" embryos are very tolerant to cryopreservation, and very efficient production of live piglets has now become possible.

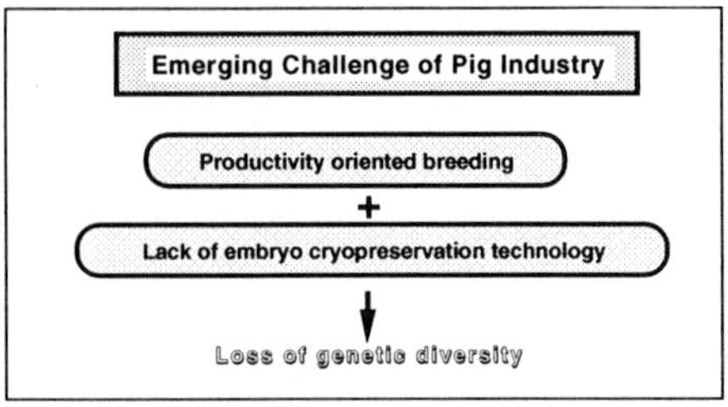

Fig. 4: Emerging challenge of the pig industry

The method developed for pig embryos would represent a new approach to embryo cryopreservation in scarce and endangered species. The development of protocols for cryopreserving mammalian embryos has concentrated on optimizing conditions such as cryoprotectant, freezing media and cooling/warming rates according to the species (Fig. 5). The inability to collect sufficient embryos for carrying out quantitative experiments is a major factor responsible for the absence of procedures for cryopreserving embryos of scarce and endangered species. In this regard, the method we have developed may herald new opportunities for the cryopreservation of early-stage embryos from species including several ungulates and carnivores which contain large amounts of cytoplasmic lipid.

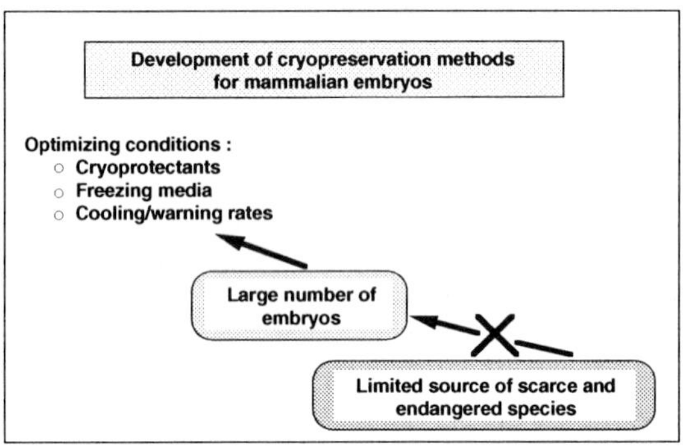

Fig. 5: Development of cryopreservation methods for mammalian embryos

Procedures for producing embryos by in vitro maturation and fertilization of the oocytes collected from abattoir ovaries have been developed in cattle, sheep and pigs. The in vitro production system for cattle was used as a model to demonstrate that those in vitro-matured fertilized embryos can also be cryopreserved after removal of their cytoplasmic lipid. Development of assisted reproductive technology (Fig. 6) such as in vitro embryo production combined with embryo cryopreservation, would provide a novel means for the conservation of wildlife animals.

One of the most important applications of embryo cryopreservation in future will be the cryobanking of the embryos of transgenic animals (Fig. 7). Variety of transgenic animals have been considered so far and some of the genetically modified animals in domestic species have already been produced; for example animals with enhanced performance in growth or meat/milk production, animals which produce valuable pharmaceutical proteins in the milk, and animals which produce organs or tissues for medical treatment of humans. In future, we may need to rely on the genetically engineered animals, due to the increase in population, or violent changes of climate or environment. To provide for such possible demands in future, cryobanking embryos of various transgenic animals will be important.

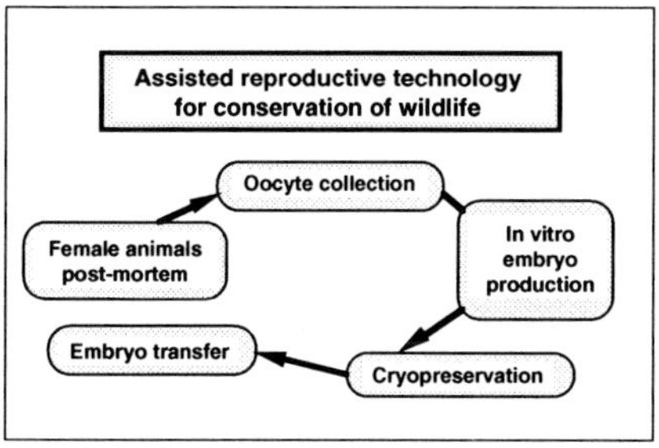

Fig. 6: Assisted reproductive technology for conservation of wild life

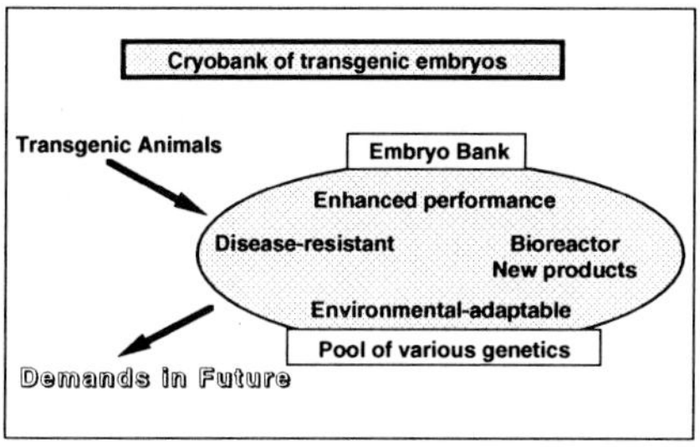

Fig. 7: Cryobank of transgenic embryos

Cryobanking of cloned embryos (Fig. 8) would also have great impact on animal production. One of the aims of cloning is to replicate currently exsisting superior animals. The worldwide distribution of superior animals through embryo cloning and cryopreservation would enable us to improve the efficiency of animal production systems, and hence reduce the negative impact of agriculture on the environment of the planet.

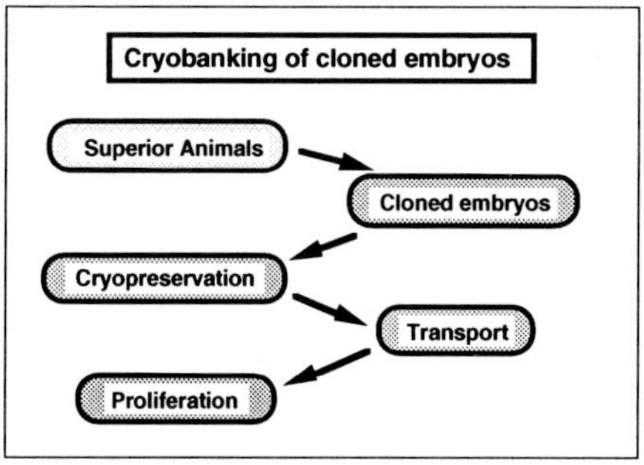

Fig. 8: Cryobanking of Cloned Embryos

Fungal Molecular Biology

CP. KUBICEK[1]

Filamentous fungi play a considerable role in modern biotechnology due to their metabolic versatility, ability to excrete high concentrations of extracellular protein but also because of their damage done to important crop plants. This article highlights the current stage of recombinant technology of filamentous fungi, thereby particularly emphasizing (a) progress in gene cloning and transformation, (b) gene regulation, (c) molecular biotechnology, and (d) the status of genome sequencing projects.

Introduction

It has been estimated that there are well over one million species of filamentous fungi (compared with around 250 000 bacterial species). Such organisms range from the simple water moulds through the Zygomycetes and Ascomycetes to the complex Basidiomycetes. They exhibit tremendous diversity in habitat and general physiology, and frequently possess exceptional metabolic capabilities which, e.g., allow the detoxification of potentially hazardous components by biotransformation, excretion of large amounts of enzymes capable of degrading plant polymers or production of substances which enable successful competition with other organisms. Their metabolic versatility is illustrated by the copious amounts and extensive range of extracellular products produced, such as metabolic intermediates (e.g. citric, malic, *cis*-itaconic acid) or molecules which bring about biological control of other organisms such as bacteria (e.g. ß-lactam antibiotic, which has had an enormous beneficial effect on the health and well-being of humans and domestic animals) and plants (e.g. gibberellins or alkaloids) (Table 1). Other fungi, such as certain Deuteromycetes, are useful in the development of mycoinsecticides and mycofungicides against several insect and fungal pests, respectively, of agricultural crops.

The biochemistry and molecular genetics of fungi has been and still is strongly dominated by research on *Saccharomyces cerevisiae*, which culminated in its complete genomic sequencing and current elucidation of the function of the genes found (Oliver *et al.* 1998). However, there are a number of serious arguments why

1. Institut für Biochemische Technologie und Mikrobiologie, TU Wien, Getreidemarkt 9/172-5, A-1060 Wien (Vienna), Austria

scientific findings on *S. cerevisiae* should not be uncritically transferred to other fungi: first, *S. cerevisiae* has been domesticated for about 4000 years towards an exceptionally efficient fermentative metabolism, which is illustrated by the proliferation of genes associated with glucose uptake and glycolysis; second, there is evidence of an ancient duplication of the entire yeast genome, which probably took place 108 years ago (Wolfe and Shields 1997); third, there are an accumulating number of fungal genes, encoding gene regulators and proteins involved in signal transduction, for which *S. cerevisiae* counterparts are missing. These points have provided a late justification that the filamentous fungi have to be studied separately, and findings obtained with *S. cerevisiae* must thus be applied to filamentous fungi with utmost caution.

Here I will focus on recent achievements and developments in recombinant DNA technology and research of filamentous fungi.

Table 1. "Modern" examples of mycotechnology

	Primary Metabolites
Citric acid	*Aspergillus niger*
Itaconic acid	*Aspergillus terreus*
Gluconic acid	*Aspergillus niger*
	Secondary metabolites
Penicillin	*Penicillium chrysogenum*
Cephalosporin	*Acremonium chrysogenum*
Cyclosporin	*Tolypocladium inflatum*
Ergot alkaloids	*Claviceps purpurea*
Griseofulvin	*Penicillium griseofulvum*
Mevalonin	*Aspergillus terreus*
Enzymesα-Amylases	*A. niger, A. awamori, A. oryzae*
Glucoamylase	*Aspergillus spp., Rhizopus spp.*
Glucose Oxidase	*A. niger*
Invertase	*A. niger*
Laccase	*Coriolus versicolor*
Proteinases	*Aspergillus spp.; Rhizopus spp.*
Cellulases	*Trichoderma reesei*
Hemicellulases	*Trichoderma reesei, Aspergillus pp.*

Gene Cloning and Transformation

The availability of the desired genes is paramount to any further recombinant approaches. The development of polymerase chain reaction (PCR) technology has revolutionized gene cloning, and the use of PCR-generated probes (derived from protein or foreign gene sequences) is now standard. However, not all genes are accessible from information of the encoded proteins or their similarity to genes already known in other organisms. Many genes, notably those encoding regulatory proteins (transcription factors), or proteins involved in signal transduction became accessible only because of the availability of the respective mutants.

Classic genetic analysis using mutagens such as chemicals or ultraviolet light has yielded a wealth of knowledge on fungal physiology and development. These mutagens mainly generate base pair deletions or substitutions, which can result in the loss or alteration of gene function, and their relative lack of specificity allows saturation of a genome with mutations. Subsequent genetic analysis of mutant strains can, however, be time-consuming because it usually involves isolation of the mutated gene by complementation with genomic DNA library from the wild-type strain. The most obvious problem is the relative size of fungal genomes. Assuming that mutagenesis is completely random, the theoretical number of mutants necessary for saturation mutagenesis can be estimated from the size of the genome and the mean gene density. The genome size of *A. nidulans* is approximately 31 megabases with a mean gene density of 1 per 3.2-3.8 kb. The mean length of 50 randomly chosen genes whose complete sequences are available in gene databases is about 2 kb. A mutagenesis of *A. nidulans* will therefore require about 15 000 mutants to have a mean mutation density of one per gene, and will require several times that number to be fully comprehensive.

For this reason, techniques for the introduction of DNA (=transformation) into fungi (to complement mutants with gene libraries and so clone the gene) became an important topic right from the start. Already during the 1980s, the development of genetic engineering procedures for filamentous fungi, in particular for *Aspergillus* and *Neurospora* spp., took place. Although today *de facto* all scientifically and industrially relevant filamentous fungi have been successfully transformed, fungal frequencies are normally relatively low compared with yeast (10 to 50 vs. 20 000 transformants per microgram of plasmid DNA), and moreover integrate the transformed DNA into genomic DNA. Such low frequencies make it difficult to clone genes by transformation and functional complementation of mutant strains, as well as (in the positive case) to recover the desired DNA fragment.

However, in recent years, significant progress has been made in this direction: plasmid replicators were isolated from an *A. nidulans* genomic library by screening transformants for irregular morphology of colonies, which was expected to result from sectorial plasmid loss. This led to the isolation and characterization of a single sequence, which proved to be an effective plasmid replicator (AMA1; Gems et al. 1991). AMAl-bearing plasmids transformed the fungus at frequencies 2000-fold higher than integrative vectors and were able to persist in mycelium for an indefinite number of asexual generations without rearrangement or integration into chromosomal DNA (Aleksenko and Clutterbuck, 1995). Analysis of the genome of laboratory strains of *A. nidulans*, which are all derived from a single wild isolate, has shown that AMA1 is an inverted duplication of a dispersed repeat (mobile *Aspergillus* transformation enhancer, or MATE), several single copies of which occur elsewhere in the genome. The MATE element comprises two long perfect inverted repeats flanking a short, unique, central spacer. Subclone analysis indicates that the complete inverted duplication, but not the unique central spacer, is necessary for efficient plasmid replication. Two or more copies of the repeat in any relative orientation are able to perform the replicator function, whereas a single copy of a MATE element increases transformation frequency only to a modest extent and leads to multiple rearrangement, unstable integration or concatenation of vector molecules. Plasmid vectors which contain the MATE element have consequently been constructed and shown to be effective with a number of biotechnologally important fungi.

Table 2. Techniques for DNA-transformation in filamentous fungi

Protoplasts
Particle bombardment
Electroporation
Agrobacterium tumefaciens-mediated

Methods to transform filamentous fungi are listed in Table 2 and usually necessitate the removal of the cell wall and the generation of protoplasts. However, de Groot et al. (1998) recently reported a promising alternative, using *Agrobacterium tumefaciens* –T-DNA-mediated transformation. This technique is routinely used for the genetic modification of a wide range of plant species. The authors showed that *A. tumefaciens* can also transfer its T-DNA efficiently to the filamentous fungus *Aspergillus awamori*, thereby demonstrating also for the first time DNA transfer between a prokaryote and a filamentous fungus. The transformation frequency varied from approximately 300 to 7200 transformants per 10^7 protoplasts, which is up to 600-times higher than the values obtained for transformation of protoplasts. When conidia were used, the transformation frequency varied from approximately 1000 to 9000 transformants per 10^7 conidia, which is in the same range as the frequency with protoplasts. The majority of the *A. awamori* transformants contained a single T-DNA copy randomly integrated at a chromosomal locus, in a manner similar to that described for plants. The authors were also able to transform a variety of other filamentous fungi, including *Aspergillus niger*, *Fusarium venenatum*, *Trichoderma reesei*, *Colletotrichum gloeospodoides*, *Neurospora crassa*, and the mushroom *Agaricus bisporus* (Table 3).

Table 3. Fungi transformed using *Agrobacterium tumefaciens*

No. of transformants/10^7 recipients	
Fungus	9000 (conidia)
Aspergillus awamori	7000 (protoplasts)
Aspergillus niger	5
Fusarium venenatum	25
Trichoderma reesei	1200
Neurospora crassa	5000
Agaricus bisporus	5

For some fungi, transformation with heterologous plasmid DNA carrying a selectable marker results in its single copy insertion. Although the transformation frequency is generally low, this method's simplicity is attractive, and has proved useful for the identification of virulence genes from the plant pathogens *Colletotrichum lindemuthianum* and *Ustilago maydis* (Table 4). For most fungi, however, multiple copy integration, either at different genomic sites or in tandem at one genomic site, is the rule, and alternative methods of insertional mutagenesis are necessary. This problem can now be circumvented by two promising strategies: restriction enzyme-mediated integration (REMI) and transposon tagging (Brown and Holden 1998). REMI, which facilitates single copy integration of heterologous transforming DNA, requires the addition of a restriction enzyme with a single site in the transforming plasmid to the transforming reaction. The restriction enzyme presumably cleaves genomie DNA at some of its recognition sites, a proportion of which then provide targets for integration of the linear transforming plasmid. This results in genomic DNA containing a copy of the transforming plasmid flanked by restriction sites for the REMI restriction enzyme. Comprehensive mutagenesis of fungi using REMI requires the use of at least two 6-bp recognition sequence restriction enzymes. This technique has meanwhile been used successfully to isolate virulence genes for several fungal phytopathogens, including *U. maydis*, *Magnaportha grisea* and *Cochliobolus heterostrophus* (Table 4).

Table 4. Gene cloning in pathogenic fungi by insertional mutagenesis

Fungus	Mutagenesis	Screening	Gene
Colletotrichum lindemutanium	PT	Excised leaves	clk1
Magnaportha grisea	REMI	*In planta acropetal*	con4, con7,
Ustilago mayidis	REMI	*In planta*	Genes for reduced virulence
Cochliobolus heterostrophus	REMI	T-toxin assay	toxin-deficient

Bacterial transposons derived from Tn5, Tn10 and Tn917 have been used extensively for insertional mutagenesis of bacteria. Numerous fungal transposons have been identified, including some which show evidence of active transposase function (Kempken and Kück 1998). In *Fusarium oxysporum*, the naturally occurring fungal transposon *impala* has been used to generate insertional mutants. This type of transposon would appear to have considerable promise as a tool for insertional mutagenesis of fungi, although the incidence of homologous recombination for many pathogenic fungi appears to be too low for this method to be generally feasible.

Gene Regulation

The ease by which genes and their nt-sequences nowadays become available has opened the way to concentrating more strongly on their function and regulation. The respective techniques used (Northern analysis, footprinting assays, EMSA) are standard procedures. However, there is a growing interest to study the regulation of genes *in vivo* (i.e. without having to extract components from the fungus). Towards this end, Wolschek *et al.* (1998) have recently described an efficient method to carry out *in vivo* footprinting analysis of fungal promoters, using ligation-mediated PCR. This method has since then been used by our lab and others to identify protein-DNA interactions in *Trichoderma* in vivo (Zeilinger et al. 1998), and may, upon optimization, be transferable to any other fungal promoter.

An already long-practiced compromise to study gene expression in vivo was the development of reporter systems, in which the 5'-regulatory regions of the gene of interest were fused to a foreign gene, which was not present in the fungus itself, and whose mRNA was reasonably stable to not interfere with the regulation to be investigated (Timberlake 1991). While this still required the assay of the gene product, it enabled the analysis of regulation of genes whose product was difficult to quantify. To this end, the β-galactosidase (*lacZ*) and β-glucuronidase genes of the bacterium *Escherichia coli* have been traditional favorites owing to the ease and sensitivity with which they can be assayed either in situ or in vitro (Soll and Srikantha 1998). However, certain filamentous fungi, notably plant pathogens, may contain endogenous activies of these enzymes which thus interfere with the assay; and more importantly, the demonstration of their activity still requires enzymatic assays to be carried out.

A breakthrough in this field was the demonstration that a mutant version of the green fluorescent protein (GFP) from the jellyfish *Aeqorea victoria*, which has been used with success in green plants, can be functionally expressed in filamentous fungi (Spellig et al. 1996). This system has the advantage that no activity measurements are necessary since the expressed GFP can be visualized by its fluoresence. Also, the construction of gene fusions does not affect the emission of fluorescence, and this system can therefore also be used to monitor the localization of the gene product *in situ*. The GFP system has therefore been used numerous times in the past 3 years, including investigations on the expression of fungal genes during plant infection, during mycoparasitism and on the localization of the final gene product in organelles such as vacuoles or mitochondria (Table 5). A still prevailing drawback, however, is the low intensity of the fluorescent signal, which has so far limited its use to the investigation of rather strongly expressed genes.

Table 5. Applications of GFP in filamentous fungi

Fungus	Topic
Aspergillus nidulans	Protein localization Nuclear migration
Aspergillus flavus *Cochliobolus heterostrophus* *Trichoderma harzianum* *Ustilago mayidis*	Plant-fungus interaction
Trichoderma harzianum	Fungus-fungus interaction

Molecular Biotechnology

Recombinant DNA and its cognate technologies have, of course, revolutionized industrial mycology. Some examples of the developments coming out of this hybrid mix are listed in Table 6. With the development of fungal genetic transformation systems and gene isolation technology, there has been widespread interest in the application of the technology for commercial purposes. The primary goal was to improve existing fungal industrial processes. This has led to the isolation of genes of commercial importance such as genes encoding industrial enzymes, e.g. amylases, cellulases, glucoamylases, ligninases, pectinases, proteases. The goals, of course, were to increase production levels or to change their physical properties conferring upon them more advantagous properties such as more convenient temperature, pH optima or to extend the substrate range. In addition, the yield of metabolites such as antibiotics has been successfully improved by gene cloning and amplification of the biosynthetic genes involved.

A second, long-term goal was to generate new fungal industries. It was hoped that fungi may produce potentially useful enzymes or metabolites, the concentration of which to date was too low for commercial exploitation, such as mammalian proteins of pharmaceutical importance, whose availability in addition, may be erratic, so that expression in a microbial host can fulfill the demand. Some of these, such as human interferon-γ or the human tissue plasminogen activator, are otherwise only obtained from biological samples such as serum, and their production can thus be profitable even at very low levels. *Aspergillus* spp. and *Trichoderma reesei* have been particularly studied as host for the production of mammalian proteins, since they have the capacity to produce and secrete impressive amounts of their own proteins (e.g. as much as < 40 g l^{-1}). It further helps that certain mammalian proteins have been found to be difficult to express in bacteria or yeast. Before successful expression of a foreign gene by a fungal cell can take place, however, the foreign gene requires to be inserted between an efficient fungal transcriptional promoter and terminator system. The choice of promoter may be important as, e.g. the production of certain heterologous proteins may be detrimental to the host and thus an inducible promoter being preferable. Case situations of foreign genes expressed thus far in fungi are given in Table 7.

Table 6. Current trends in modern fungal biotechnology

Development	Examples
Expression and secretion of heterologous proteins	mammalian
	plant
	fungal
Tayloring of secondary metabolic pathways	hybrid antibiotics
Large scale genomics	*Aspergillus nidulans*
	Aspergillus fumigatus
	Neurospora crassa
	Magnaporthe grisea
	Phytophtora infestans
DNA chips	High-density DNA arrays
	for screening gene expression
Mining fungal biodiversity	DNA-sequence based environmental sampling

Table 7. Examples of proteins which have successfully been expressed in filamentous fungi

Chymosin
Preprocathepsin D
Thauamatin
Avian pathogenic surface antigen
Corticosteroid binding globulin
E. coli enterotoxin subunit B
Epidermal growth factor
Human growth hormone
Interleiukin 6
Interferon α_2
Human parathyroid hormone
Tissue plasminogen activator
Hen egg white lysozyme

Despite their promising secretory capacity, filamentous fungi have so far not proven to be capable of producing mammalian proteins in significantly higher amounts as, e.g. some yeasts. A major breakthrough was the finding that a translational fusion between a fungal protein and the mammalian protein of interest considerably improves secretion (Ward et al. 1990), which suggests that correct protein folding may be a limiting factor to heterologous protein secretion. However, expression of heterologous proteins by fungi apparently also suffers from a number of other factors such as incomplete glycosylation, which can be of importance in immunogenicity, function and stability of proteins. Fortunately, extensive cellular hyperglycosylation, which is often observed in yeast, is not a problem in filamentous fungi. Other factors potentially limiting secretion are reviewed by Gouka *et al.* (1997). The molecular biology of the fungal secretory pathway has recently been a major focus of interest in the European fungal research community.

Secondary metabolic pathways are also amenable to molecular analyses (Radzio and Kück 1997). Predictably, the penicillin family of antibiotics was the first group of compounds to benefit from the new methodologies. During cloning of the genes functionally involved in the penicillin biosynthetic pathway, it was discovered that certain high-yielding strains contained multiple copies of these genes. Researchers at Panlabs Inc. (Crawford et al. 1995) were the first to demonstrate the use of selected genes to modify the existing β-lactam-producing pathways: the *Cephalosporium acremonium* expandase-hydroxylase-encoding gene was expressed in a penicillin-producing strain of *Penicillium chrysogenum*. The recombinant strains, when cultivated in the presence of adipinic acid, produced adipyl-cephalosporins, thus demonstrating the use of genetic engineering to produce novel antibiotics not occurring in nature.

Polyketides are another group of secondary metabolites that have benefited from molecular analyses. To date, most of the research on mixing-and-matching polyketide synthases has been conducted in actinomycetes only. The permutations generated by the use of different starter units, adjustment of oxidation and stereochemistry during elongation and the introduction of various post-polyketide modifications led to an almost astronomical number of theoretically possible different molecules. It is to be expected that these strategies will soon be applied in fungi as well to create, "unnatural natural products".

Fungal Genomics

The acceleration and automation of DNA sequencing has made large-scale genomics possible. The first complete DNA sequence of an organism was for the bacterium *Haemophilus influenzae* Rd (Fleischmann et al. 1995); since then several additional genomes have been completed and it is estimated that about 150 species are currently under investigation. Microbial genomics promises to revolutionize not only basic biology but also to accelerate the drug discovery processes by guiding clinical trials. Correlations between sets of genes and drug activities will open new avenues for screening. The first fungal genome which was fully sequenced is, of course that of *S. cerevisiae* (Dujon 1996), which consists of 16 chromosomes encompassing 12 067 266 base pairs (excluding repeats). The results showed that many yeast genes have mammalian homologues, whose functional compatibility has been exploited in studying cancers and other human diseases (Botstein et al. 1997). On the other hand, most of the open reading frames identified by the sequencing project had gone unnoticed, i.e. they had no known phenotypes. The challenge is now to discover what each of the gene products does and how they interact in a living yeast cell. The systematic and comprehensive approaches currently ongoing are reviewed by Oliver *et al.* (1998).

In the meantime, genomics projects have also been initiated for the yeasts *Schizosaccharomyces pombe* and *Candida albicans.* With respect to the filamentous fungi, genomic sequencing has been completed for the genetic model fungus *Aspergillus nidulans*, and is underway for the medically important species *Aspergillus fumigatus* and the closest competitor for the model fungus status, *Neurospora crassa* (Bennett 1997, 1998; Hamer 1997). The large amount of classical genetic information available for these species will make it easier to correlate sequence data with phenotypes.

The human genome project has led to the development of high-throughput technologies that provide powerful approaches to analyze complex systems. Among these, a central question was the comprehensive analysis of the multitude of genes discovered. Using photolithography and technology derived from the silica chip industry, immobilized oligonucleotides encompassing the entire yeast genome can be arrayed on a small surface creating a DNA microchip. Oligonucleotide arrays representing all 6000 genes can be used to interrogate how gene expression patterns change when specific biological activities are triggered (e.g. DNA replication, DNA repair, protein secretion, and change in yeast mating type) or particular genes are destroyed. Indeed, Shoemaker et al. (1996) demonstrated the elegant use of oligonucleotide markers and oligonucleotide arrays to monitor large numbers of yeast deletion mutants under differing selective conditions. The DNA array technology is particularly exciting because it offers the potential of using hundreds of thousands of differing oligonudeotides to analyze changing mRNA populations. Using probes to measure hybridization, the chip can be used as a reagent for measuring the expression of any number of genes. DNA chips can also be used to identify important molecular targets for drug discovery, and to search for bioactive products using environmental DNA samples from organisms that are difficult to culture such as marine fungi or basidiomycetes (Blanchard and Hood 1996).

Conclusions

Despite their industrial, agricultural, and ecological importance, filamentous fungi have for most of the past been scientific victims of the methodically advantageous yeast *Saccharomyces cerevisiae.* However, most of the biological and genetic bottlenecks seem now to have been overcome. The complete sequencing of the *A. nidulans* genome has been the logical last step in this development and therefore opens new avenues towards an understanding of fungal biology. One can anticipate that this development will ultimately lead to a picture of fungal genetics which is considerably different from that of *S. cerevisiae.*

References

Aleksenko A, Clutterbuck AJ (1996) *Mol Microbiol* 19:565-574

Bennett JW (1997) *Fungal Genet Biol* 21:3-7

Bennett JW (1998) *J Biotechnol* 66:101-107

Blanchard AP, Hood L (1996) *Nature Biotechnol* 14:1649

Botstein D, Cherfitz SA, Cherry JM (1997) *Science* 277:1259-1260

Brown S, Holden DW (1998) *Curr Opin Microbiol* 1:390-394

Crawford L, Stepan AM, McAda PC, Rambosek JA, Conder MJ, Vinci VA, Reeves CD (1995) *Bio/Technol* 13:58-62

De Groot MJA, Bundock P, Hooylkaas PJJ, Beijerbergen AGM (1998) *Nature Biotechnol* 16:839-842

Dujon B (1996) *Trends Genet* 12:263-270

Fleischmann RD, Adams MD, White O, Clayton RA, Kirkness EF et al. (1995) *Science* 269:496-512

Gems D, Johnstone IL, Clutterbuck AJ (1991) *Gene* 98:61-67

Gouka RJ, Punt PJ, Van den Hondel CAMJJ (1997) *Appl Microbiol Biotechnol* 47:1-11

Hamer L (1997) *Fungal Genet Biol* 21:8-10

Kempken F, Kück U (1998) *BioEssays* 20:652-659

Oliver SG, Winson MK, Kell DB, Baganz F (1998) *Trends in Biochem Sci* 16:373-378

Radzio R, Kück U (1997) *Proc Biochem* 32:529-539

Shoemaker DD, Lashkari DA, Morris D, Mittmannn M, Davies RW (1996) *Nature Genet* 14:450-456

Soll DR, Srikantha T (1998) *Curr Opin Microbiol* 1:400-405

Spellig T, Bottin A, Kahman R (1996) *Mol Gen Genet* 252:503-509

Timberlake, W.E. (1991) In: Bennett JW, Lasure LL (eds) *More gene manipulations in fungi.* pp. 51-85. Academic Press Inc., San Diego, CA

Ward M, Wilson LJ, Kodama KH, Rey MW, Berka RM (1990) *Bio/Technol* 8:435-440

Wolfe KH, Shields DC (1997) *Nature* 387:708-713

Wolschek M, Narendja F, Karlseder J, Kubicek CP, Scazzocchio C, Strauss J (1998) *Nucl Acids Res* 26:3862-3864

Zeilinger S, Mach RL, Kubicek CP (1998) *J Biol Chem* 273:34464-34471

Panel on Production Systems

Chair: Senator Kerrey (US Senator, Agriculture Appropriation Committee)
Experts: Josef Fulka, Allan King, Lazló Solti, Mark Tepfer, José-Antonio Vazquez-Boland

Introduction by Senator Kerrey

Public trust and confidence are keys to acceptance in any arena. In general, the public does not trust scientists but they trust politicians even less. In terms of food and agricultural issues, public expectations about food safety, health, and the environment are seeped in both emotional and rational, scientific issues. However, public acceptance in the United States of genetically modified organimsms, or GMOs, has been nearly universal, with little or no public outcry. This is compared to the outcry that is being experienced in Europe, where recent experiences have perhaps led to a breach in public faith. In Europe, ongoing mistrust may be tied to, or be an extension of, the mad-cow disease issue. The public received assurances from scientists that the disease was not a human threat, but when it proved in fact to be a threat, public confidence was jolted. Now the credibility of scientists and policy-makers is being questioned. Similar problems with the acceptance of hormones in beef are being experienced. But is this really just an issue of education and information, or the lack thereof? Is it a matter of explaining the facts and the science behind these new technologies, either through the media or some other method? New technologies can benefit us all, now and in the future, but any new technologies must be accepted by the public, or there will not be a market. Without consumer acceptance and confidence, any technologies, regardless of their potential benefit, will fail.

Vazquez-Boland José-Antonio (Complutense University of Madrid, Spain)

I will contribute to the discussions by making five general comments.

1. My first point deals with the subject of the conference itself, i.e., connecting science and policy. A sine qua non requisite for such a connection is that scientists can interact face to face, in an effective way, with the political sector. The OECD provides an ideal forum for such an interaction, and I encourage the organization to proceed further with the Cooperative Research Programme in the future, organizing new conferences with the increasing involvement of policy-makers to improve interactions with scientists.

2. Clearly, scientific and technological research is tightly coupled to economic development. Countries must therefore protect and promote basic research in the academic sector as a source of ideas for new biotechnological developments. In Europe, it is particularly important for member states to promote the creation of small and medium-sized enterprises to carry out applied research and to develop new technologies.
3. In the context of rapidly growing economic interdependence, it is of fundamental importance to promote research cooperation at the international level. In the European Union, we have transnational cooperative research programs aimed at increasing our R+D capabilities so that we can compete successfully with the USA and Japan in the technological field. I believe that, for sustainable global economic development, we must promote a third level of international scientific cooperation. The OECD could play a key role in such an initiative by implementing a program for funding cooperative research projects involving several countries from different geographical areas.
4. The release onto the market of modern biotechnological developments may contribute to sustainable economic development. The exploitation of such developments depends on political decisions, which, in a democracy, are not strictly based on advice from scientists, and depend heavily on public opinion. Therefore, scientists and policy-makers must make every effort to provide society with adequate, objective, comprehensible information about the applications of biotechnology. This is the only way to allay many of the fears associated with the use of genetically manipulated organisms (GMOs).
5. For the OECD's new Cooperative Research Programme, which will hopefully be launched at the beginning of the new century, I suggest genomic and postgenomic research into organisms relevant to agriculture as a major new theme to be addressed. Indeed, knowledge obtained through basic research is the only means of overcoming the fear of the unknown in the general population. Only the comprehensive understanding of the biology of an organism, which requires a profound knowledge about its genome (and how it functions), may enable us to predict its behavior. This is of fundamental importance in the case of GMOs. Genomic research also has enormous potential value in agriculture for the quantitative and qualitative improvement of crop production, via the identification of useful plant traits that can be incorporated into new plant varieties by transgenic technology or traditional selection methods. Genomic research may also be instrumental in the development of new vaccines and live antigen delivery vectors and in the identification of targets for new antibiotic substances or antimicrobial technologies aimed at preventing the proliferation of pathogenic microorganisms in food. Finally, I would also suggest that emphasis be placed on another area of research: the development of genetic tools for the biosafety and containment of GMOs, confining them to a particular niche and preventing their uncontrolled spread.

Allan King (University of Guelph, Canada)

The full potential and impact of biotechnology has yet to be realized in livestock production. The current animal biotechnology has its roots in introduction of artificial insemination using frozen semen in the 1950s and the development of superovulation and embryo-transfer procedures in cattle in the 1970s. While these techniques are widely used in beef and milk production, their use in other sectors of the livestock industry has been limited. In the 1980s techniques such as in vitro fertilization, embryo sexing, and genetic-marker-assisted selection were at the forefront of animal biotechnology research. The incorporation of these practices has been on an even more limited scale due to technical inefficiencies, added production cost with limited financial or production gain, and lack of national or regional strategies for their incorporation into farming practices. These techniques are, however, the basis for stem cell culture, transgenesis, and nuclear transfer (cloning) currently under intensive development in most OECD member countries. The emerging animal biotechnologies are faced with many societal and technical challenges. The technical challenges are due to less than complete knowledge of the fundamental processes involved, cumbersome techniques, and a low level of efficiency. Cloning and transgenesis have not reached a level of development that allows practical agricultural application. Nevertheless, developments in the areas of cloning and transgenesis are of utmost importance, not only because of their promise for more efficient and targeted production of animal proteins and new and safe food products, but for production of pharmaceutically relevant macromolecules, reduction of the environmental impact of animal waste, and retaining rare breeds and endangered species.

With the rush to move from the laboratory to commercial practice, the transfer of technology into the public domain and the dissemination of information to producers, user groups, and public stake holders is often overlooked. Hence, strategies for the implementation of production systems and the testing and marketing of altered products are either poorly developed or non-existent. Coupled with the increasing sophistication of emerging technologies and the potential to permanently alter the basic genetic makeup of production animals as well as changes in the way that animal husbandry is practiced, public concern over the direction of this fundamental research has increased. Considering the inefficiencies of these techniques, the biological differences between species, and the relatively long generation interval in farm animals, it is not surprising that the impact of biotechnology on production is slow in comparison to other agrifood sectors. The slow progress does, however, provide an excellent opportunity to rectify some of the deficiencies in the area of public and consumer product safety, maintenance of genetic integrity of existing species, and development of international standards.

The OECD fellowship program has contributed to the acquisition of knowledge in the areas of animal biotechnology and international transfer of technology. In addition, it has fostered an appreciation of socially and culturally different issues surrounding research and development in these areas. The OECD as an organization is well suited to bridge research and societal challenges facing biotechnology.

Josef Fulka (Institute of Animal Production, Prague, Czech Republic)

There can be no doubt that all those techniques and approaches we have heard about will play an important role in the future. They are potentially very useful but, on the other hand, they may be very dangerous. This was evident from the lecture given by Prof. Aguzzi. Somebody said that you can do basicallly everything in the former Eastern countries. That is true. Why? We are not yet fully incorporated into EC structures; the flow of information from West to East is still limited; the information is not always correctly translated and some information sometimes disappears somewhere. As we are not yet fully incorporated into EC structures, and as this will perhaps be the case only after another 5 years, the OECD must substitute for EC structures in Eastern countries at least during this transitory period. In this field I can see the further role of this organisation.

Laszló Solti (Veterinary University, Budapest, Hungary)

Regarding the use of biotechnology in agricultural (or medical) production systems, two main aspects should be differentiated and evaluated before deciding whether or not it can be used.

Technical problems are evoked by the research that has been performed before. Generally speaking they must be solved by professionals working in that field. The solution to these problems these problems is a matter of time only. An incomplete list of these tasks is as follows:

- Cryopreservation of gametes (semen, oocyte, embryo) by simple, inexpensive, and user-friendly freezing methods resulting in high survival rate. While acceptable protocols for in vivo bovine embryos and semen of several domestic species are available, freezing methods for oocytes and embryos of wild and some domestic animals are still missing. Without effective cryopreservation, gene banking is not possible.
- The efficiency of in vitro embryo production, transgenesis and cloning, along with long-term storage of these embryos must increase significantly... The success rate of current cloning and gene transfer methods in farm animals is far below 1%, which is inacceptable for their routine use.
- Clearing up and solving the problem of large-offspring syndrome after cultivation of embryos in vitro, by optimizing the culture conditions
- Finding a method for targeted gene integration. Current methods used for gene transfer (microinjection, cloning after transfection of nucleus-donor cells, etc.) result in a random integration of genes.

Avoiding the risk of disease transmission by transfer of in vivo or in vitro embryos.

Society issues mean that to gain public perception and acceptance of biotechnology is more difficult than solving the technical problems. Society needs time to accept ideas that are entirely new and unconventional.

- Continuous dialogue between scientists and the public is required in order to educate ordinary people as to what biotechnology is all about, what the methods are good for, what are the risks and benefits.
- Analyzing potential risks of the new methods and remaining on the safe side (e.g., in release of GMOs or in risking an outbreak of disease caused by viruses from other species, etc.).
- Evaluating how far we may interfere with nature: avoiding "playing God" by tailoring or creating new organisms.
- Long-term consequences of consuming genetically manipulated food: food safety.

Can we afford to stop biotechnology experiments and ban research? My personal view is **no**. Looking back in history, revolutionary new methods always provoked rejection by the church, society or politics. In the last century the use of artificial insemination was threatened with excommunication. About 30 years ago the first heart transplantation of Barnard was also heavily criticized, the test-tube child of Edwards and Steptoe was discussed 20 years ago, the use of GMOs was stopped in Asilomar, and the first lamb cloned by Wilmut and colleagues two years ago from somatic cells generated a very big debate. If banned, research would go on behind the scenes in an uncontrolled manner that could give rise to misuse of the new technologies. Science cannot be stopped by banning it. Mankind cannot say no to research aiming at raising pharmacologically important proteins produced by GMOs, or at producing transgenic animals that may donate organs for xenotransplantation, or increasing the yield of agriculture with less detrimental impact on the environment.

Mark Tepfer (INRA Versailles, France)

One of the areas where plant biotechnology can be a source of important progress in agriculture is that of the control of plant pests and pathogens. It is clear that if the world population continues to increase as expected, then there is a serious possibility of food shortages within a few decades. Preventing losses due to pests and pathogens through careful use of biotechology could contribute in a significant fashion to increasing global productivity in an ecologically sound fashion. Even in regions such as Europe, where it is generally perceived that agricultural overproduction is a problem, plant pathogens such as viruses can have an extremely negative impact on the quality of plant products, and thus here, too, plant biotechnology can be expected to bring significant positive benefits.

Since the very early stages of plant biotechnology, the OECD has played a particularly significant role as a forum for creating international consensus on the scientific basis for regulation in this new field. This is evident from the continuing influence of the OECD "Blue Book" of 1986 (Recombinant DNA Safety Considerations). Nonetheless, over the past few years, there has been a clear degradation of international consensus, with a strikingly different situation in North America and in Europe. The differences concerning virus-resistant transgenic (VRT) plants can be cited as an example. Five VRT crop varieties have been authorized for unrestricted release in the US, with many others in the process of authorization. In Europe, at least one VRT crop was withdrawn from the authorization process because of ethical concerns. In fact, the new difficulties in Europe concerning GMOs in general can, to a large extent, be attributed to a breakdown in public trust in the ability of national and EU agencies to guarantee food safety. The handling of the outbreak of bovine spongiform encephalopathy has had particularly unfortunate effects in this regard. Although the issues of ethics and public acceptability have not been addressed extensively in the OECD context, concerns in this area can only be resolved if the scientific and technical issues are resolved as completely as possible, and here the OECD has an essential role to play in preparing the future worldwide acceptance that is necessary to allow reaping the enormous potential benefits of plant biotechnology for humanity.

General Discussion

In the general discussion the question was raised whether the lack of trust that surrounds biotechnology and GMOs in Europe, especially as compared to the rather silent, blanket acceptance in the United States, is due solely to cultural differences? Or is there something else at work, something that is not obvious to consumers in the US. Why, in Europe, are emotions so high and tension so great about GMO crops that the public would want to ban biotechnology research, or destroy experimental crop plots? This was compared to the relatively low level of outcry and opposition, even in Europe, that accompanied the pronouncement by Dr. Wilmut and others that they had cloned Dolly, the sheep. Dr. Wilmut was asked whether or what was different about their work compared to that of GMO crops, for instance? Dr. Wilmut explained that, as their work neared completion, and especially when it became apparent that word of their work was becoming known, he and others worked very closely with the media to lay the groundwork for their story to be told. This generated discussion about the role of the media, and of media methods, in communicating new technologies as well as new potential dangers, and whether GMOs could have been handled differently in the media.

Other topics for discussion included the fact that science and research will continue with or without public support, and that a lack of confidence or trust is not a valid reason to suggest halting research and progress. In the past, agricultural advances in the US in the form of traditional breeding practices led to the development of corn hybrids with yields 600% higher than their parent crops. Rapid change nearly always elicits negative and sometimes dramatic responses, but is not to be abandoned. However, public acceptance, as has been noted, is necessary to make new technologies and new products viable in the market place. A lack of consumer acceptance will not stop research, but can block sales, use, and potential benefits.

Cautionary measures on behalf of European regulators were also discussed, and the issue of whether or not this was more prudent than what may be seen as less cautionary regulatory procedures in the United States. Long-term environmental risks, as well as potential human health safety issues, are perhaps not adequately addressed without longer-term, more deliberative regulatory scrutiny. The issue of how far, scientifically and ethically, we could or should go, was also raised, without apparent resolution.

A general agreement was found to explain research to the public even if it is often difficult: 90% of the population even ignore what a gene is and therefore cannot understand what biotechnologies are. One proposed solution is to open scientific meetings to the media and to convince researchers to communicate with the public. The conclusion by **Ian Wilmut** was that there is no other option than to wait for (and to help promote) increasing acceptability.

Part V
Beneficial Organisms

James LYNCH

P. NANNIPIERI et al.
C. ALABOUVETTE
D. ANDOW
J. BERINGER

Management of Soil Microbiota

P. Nannipieri, L. Falchini, L. Landi, G. Pietramellara[1]

The manipulation of soil microflora can involve changes in microbial diversity and microbial activities and it is carried out for practical purposes. The microbial diversity can be modified by the inoculation of beneficial microorganisms into soil or by agricultural management practices. Indigenous soil microorganisms show a remarkable range of catabolic and anabolic activities. In addition, they are able to degrade xenobiotics whose molecular structure does not resemble that of naturally occurring compounds. It is unknown how the development of cluster of genes encoding the enzymes involved in the degradation of these compounds occured. Another way of manipulating soil microflora is to stimulate or inhibit its activities and these approaches may involve changes in the composition of soil microflora due to both target and non-target effects. Enzymes immobilized on supports resembling soil colloids can be added to soil so as to increase the activity of microbial reactions.

Introduction

Microorganisms, mainly bacteria and fungi, are considered to be largely responsible for the transformations occurring in soil which are essential for the completion of the global nutrient cycles and for the degradation or inactivation of pollutants (Alexander 1977). Despite the plethora of books, reviews and papers dealing with soil microbiology, understanding of the activity, ecology and population dynamics of microorganisms is still problematic, because soil is the most complex of all microbial habitats (Alexander 1977; Stotzky 1986). The predominance of a solid phase, composed of inorganic and organic particulates, distinguishes soil from most other microbial habitats. Of the soil particulates, clays, iron and aluminium oxides and hydroxides, and organic particles, for their high surface area and exchange capacity, retain water films and nutrients and are essential for microbial life. Microorganisms are aquatic organisms and in soil living microbial cells are also physiologically restricted to the liquid phase (Stotzky 1986).

1. Dipartimento della Scienza del Suolo e Nutrizione della Pianta, P.le delle Cascine 28, 50144 Florence, Italy

The soil microflora is characterized by a large variety of species. The determination of the microbial diversity in soil has been problematic due to nonculturability of the majority of microbial species inhabiting soil. DNA reassociation kinetics have given information about the absolute level of diversity of bacterial population in soil. By hypothesizing that the total number of bacteria (1.5×10^{10}) per gram of soil was distributed among 4 000 clones, Torsvik et al. (1990) calculated a mean number of about 3.8×10^6 bacteria clone^{-1} g of dry soil^{-1}. It was also hypothesized that most diversity was associated to the unculturable microorganisms. Microorganisms are not evenly distributed in soil. Most of them are heterotrophic and thus their number and activity are higher near organic debris or in the rhizosphere, where organic C concentration is higher than in the bulk soil due to rhizodepositions (Lynch 1990). However, other constraining factors of the microbiological activity in soil include N, P, and S, water potential, soil aeration and pH, substrate quality, clay type and osmotic pressure (Alexander 1977; Stotzky 1986).

The management of soil microflora involves an efficient use of the activities of soil microorganisms for practical purposes such as: (1) the improvement of the nutrients supply to crops; (2) the stimulation of plant growth, e.g. through the production of plant hormones; (3) the control or inhibition of the activity of plant pathogens; (4) the improvement of soil structure; (5) the bioremediation of polluted soils (Wilson and Lindow 1993; Burns 1995; van Veen et al. 1997). This has been a primary scope of research since the birth of soil microbiology. Of the seven questions posed by Waksman (1927) in his classic monograph on soil microbiology, one was related to the manipulation of soil microflora: how can one modify soil population and to what ends? (McLaren 1977).

The manipulation of soil microflora can involve change in microbial diversity and microbial activities. The microbial diversity can be modified by the inoculation of beneficial microorganisms into soil or by agricultural management practices. Today, there is a great interest in manipulating soil microflora through the introduction of beneficial microorganisms, mainly bacteria, in soil favoured by the vast possibilities for strain improvement offered by modern biotechnology. Less attention has been paid to the potential applications of changes in the activity, size and composition of the indigenous soil microflora by agricultural management practices. Another way of manipulating soil microflora is to stimulate or inhibit its activities and these approaches may involve changes in the composition of soil microflora due to both target and non-target effects. Indeed, the activity of metabolic processes or enzyme reactions can be decreased by using specific inhibitors or enhanced by inducing or repressing the synthesis of microbial enzymes in soil (Nannipieri 1994). Eventually, specific enzymes can be added to soil so as to increase the activity of microbial reactions (Burns 1987; Nannipieri and Bollag 1991).

The aim of this chapter is to review the main ways employed to manipulate soil microflora, underlining the present limits and discussing, when possible, future realistic and productive research. Constraints in space and time do not allow an exhaustive review of the various subject areas that impact this topic, a task that would exceed the limit of this chapter. Consequently, relevant reviews are cited more than the original literature that is the foundation for these and the present chapter.

Addition of Beneficial Microorganisms and Specific Enzymes to Soil

Soil Inoculation with Specific Microorganisms

Several bacteria have been used as soil inoculants to improve the supply of nutrients to crop plants. An effective establishment of the dinitrogen-fixing symbiosis with leguminous plants has been successfully obtained worldwide by inoculating *Rhizobium* species (Eaglesham 1989). Non symbiotic nitrogen-fixing bacteria, such as *Azotobacter, Azospirillum, Bacillus* and *Klebsiella* spp., and phosphate-solubilizing bacteria, such as *Bacillus* and *Pseudobacillus,* have been applied to soils to improve N and P availability to plant, respectively (Lynch 1983). The increase in plant growth following inoculation is not always the result of a direct response to the introduced microorganisms. For example, the increase in germination and crop yield, following inoculation of soil or seeds with *Azotobacter chroococcum,* can be attributed not only to the increase in N_2-fixation but also to other factors such as the physical protection from soil pathogens, the production of plant regulators, or, in pot experiments, the possible fertilizer effect due to the addition of large numbers of microbial inoculants with the subsequent release of plant nutrients as the cells are lysed and degraded (Lambert and Joos 1989).

The term plant growth-promoting rhizobacteria (PGPR) was coined to encompass bacteria with plant growth-stimulating activity resulting from several different mechanisms (Kloepper et al. 1980a). The best-known microorganisms with PGPR activity are bacteria belonging to the group of fluorescent *Pseudomonas* species (Kloepper et al. 1980b) and a wealth of literature has been accumulated on the mechanisms underlying their plant growth-promoting activity as well as their ecological performance (van Veen et al. 1997). This term was a misnomer when first coined as it was used to indicate *Pseudomonas fluorescens* and *P. putida,* which indirectly increased the yield when inoculated onto seed pieces of potato through their antagonistic action on "deleterious rhizobacteria" (Burr et al. 1978). The direct stimulation of the plant by PGPR can occur either by increasing the supply of mineral nutrients, such as P and N, or by the production of phytohormones, such as indole acetic acid (IAA), gibberellin and cytokinin-like substances and vitamins (Brown et al. 1968; Weller and Thomashow 1994).

The interest in controlling soil-borne plant diseases by biological means has increased in recent years with the growing concern for the impact of chemical pesticides on the environment (Burns 1995). Successful applications have included both bacteria and fungi. These antagonistic organisms generally exert their action directly on the pathogen through the production of antibiotics or by decreasing micronutrient (iron or manganese) concentration in the environment. A control strategy can also involve the induction by the antagonistic microorganism of systemic resistance in the host plant against the soil-borne pathogen (Maurhofer et al. 1994). Interactions between the induction of systemic acquired resistance in plants and the involvement of siderophores in direct antagonism of root pathogens have proved to be subtle and difficult to distinguish. Leeman et al. (1996) proved that the induction of systemic resistance in radish (*Raphanus sativa* L.) against *Fusarium* wilt occurred when iron availability in the nutrient solution was low. The synthesis of both pseudobactin and salicylic acid was involved in the suppressive action of pseudomonads towards *Fusarium* wilt. Experimental evidence seemed to prove that both iron chelators directly caused an induced systemic resistance in the host plant (Leeman et al. 1996).

Microorganisms contribute to the formation of soil structure by enmeshing inorganic particles within filamentous structures and their adhesion by extracellular polysaccharides (Lynch 1983; Oades 1984; Ladd et al. 1996). Thus, the inoculation of polysaccharide-producing microorganisms, if these become established and function, may improve soil structure due to the aggregation action of polysaccharides. Generally, phototrophic microorganisms are preferred to heterotrophs due to nutrient deficiencies of many soils (Metting 1991; Burns 1995). Under laboratory conditions, *Nostoc* spp. inoculation of two clay soils, representative of two physically degraded areas of central Italy, increased the percentage of transmission pores and promoted aggregation due to their secretion of polysaccharides but they did not increase soil stability (Falchini et al. 1996).

In recent years, there has been a development of studies in soil bioremediation due to the desire to reclaim polluted soil for agricultural and recreational purposes. The directed mutation or genetic manipulation of bacteria has provided a broad number of strains capable of degrading recalcitrant organic pollutants as well as causing bioaccumulation or mobilization of inorganic compounds. However, the success of these strains has been generally assessed in vitro, whereas their use is problematic in the field (Drahos et al. 1986; Rojo et al. 1987). As discussed later, several biotic and abiotic factors (Fig. 1) may limit the efficacy of any inoculant in vivo. In the case of soil bioremedation, inoculants have to face other factors. The limited mobility of the cells within the soil could restrict the use of microorganisms to detoxify organic pollutants. Inoculated cells might use other and more desirable C sources, especially when the concentration of the organic pollutant is low (Nannipieri and Bollag 1991). The sorption of the pollutant and the uneven distribution of the contaminant may also reduce the bioavailability. Another important aspect to be considered in soil bioremedation is that contaminated soils usually contain a mixture of organic pollutants, and the biodegradation of these compounds is likely to require an interactive community of microbial species. This is made more complicated by the inoculation approach. This problem can be overcome by using genetically engineered bacteria expressing all the relevant biochemical properties needed to degrade a mixture of contaminants (Timmis et al. 1985).

Many failures or inconsistencies in achieving the objective have been reported, and this has raised concerns about the perspective of the great practical potential offered by microbial releases into soils (van Elsas and Heijnen 1990; Akkermans 1994). Not all inoculated species survived in soil, either because they are not indigeneous (like *Escherichia coli*) or because they respond differently to diverse soil types. The inoculant is effective in soil when the number of active and inoculated microorganisms reaches a certain size (van Elsas and Heijnen 1990; Akkermans 1994). The problem of inoculum size and activity can be less important when the introduced microorganisms have a selective advantage in the soil system from the process in which they are involved (van Elsas and Heijnen 1990; Akkermans 1994). This is the case in the symbiosis of *Rhizobium* species with leguminous plants, and generally a small number of active cells is initially necessary for the application to be effective. Some applications can be successful if spatial and temporal aspects are carefully considered for the release. Thus, the time of inoculation is important in the control of soil-borne plant pathogens because they are active at certain stages of plant development or under certain climatic conditions (van Elsas and Heijnen 1990; van Veen et al. 1997). In addition, an effective plant pathogen control strategy is dependent on inoculum site; for example, the active antagonistic microorganisms can be effective if inoculated in the rhizosphere rather than in the bulk soil (van Overbeek and van Elsas 1997). On the other hand, other applications, such as the use of microorganisms in the bioremedation of soil, are relatively independent of time and space.

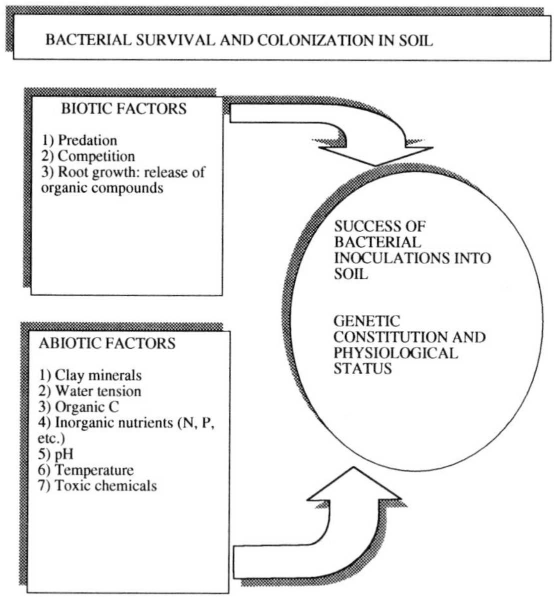

Fig. 1: Abiotic and biotic factors affecting the bacterial survival and colonization in soil

Several abiotic and biotic factors influence the survival and colonization of inoculated bacteria (Fig. 1). Biotic factors, such as competition with the indigenous soil microflora, presence of toxins, predation and presence of roots, are particularly important as testified by the fact that increases in inoculant populations are observed at low inoculum levels in sterile soils whereas a decline is observed in the respective non sterilized soil (van Veen et al. 1997). Abiotic factors such as texture, pH, temperature, moisture content and substrate availability (Fig. 1) determine the survival and activity of the inoculated microorganisms by imposing stresses of various nature in the inoculants, and indirectly, by affecting the activity of the indigenous microflora (Berry and Hagedorn 1991; van Elsas et al. 1991). These abiotic factors need to be critically assessed for each type of inoculum (van Elsas and Heijnen 1990; Akkermans 1994). The response of the inoculant to the prevailing soil conditions depends on its genetic and physiological constitution. Thus, the effects of soil factors on the inoculated microorganism will differ with the ability of the inoculant to cope with adverse and fluctuating conditions to survive and to remain active. Different species will show different responses in terms of survival and activity and these responses will depend on the intrinsic physiological characteristics of the inoculant strains used.

Colonization of soil particles is essential for the survival of the inoculum in soil (Hattori and Hattori 1976), and the effect of clay surfaces in conferring protection to the microflora has been proved (Marshall 1975; Stotzky 1986; Ladd et al. 1996). Higher survival occurred for inoculated bacteria localized in the interior parts of 1- to 2-mm soil aggregates than those present on the external surfaces which were easily grazed by protozoa (Vargas and Hattori 1986). The access of inoculated bacteria to the protective sites, such as pores with small necks (diameter < 6 μm) can be facilitated by adding water solution containing the inoculant to dry soil (Heijnen and van Veen 1991; van Veen et al. 1997). In this case, the bacteria will reach the protective pores with the water of the inoculum. However, the number of bacteria per unit of soil is much lower than that of cells which theoretically occupy the volume of the space potentially accessible to the inoculated cells (van Veen et al. 1997). As mentioned above, the availability of organic carbon is restricted to sites containing organic debris, rhizodepositions or to pores where water soluble compounds have been transported by water. Some of these sites are not easily reached by microorganisms (van Veen et al. 1997). In addition, some of the organic matter is relatively recalcitrant and it is metabolized slowly. Each soil type is characterized by a determined ecological space available for maintenance and persistence of the inoculated microorganism. Nannipieri et al. (1983) hypothesized that each soil ecosystem has its own distinctive biological space with respect to the level of microbial biomass and enzyme activity. This hypothesis can be challenged, since it did not consider changes in microbial diversity in response to changes in environmental conditions including differences in substrate preference between different groups.

Addition of Specific Enzymes to Soil

Many enzymes that are usually active within the cell can also function extracellularly if they do not require cofactors for their activity. For this reason, cell-free enzymes may be used in soil for increasing the supply of nutrients such as phosphorus to plants, reclaiming pesticide-polluted soils, controlling soil-borne pathogens and for rapidly degrading waste residues (Burns 1987; Nannipieri and Bollag 1991).

Soluble phosphate is essential for plant growth. Synthetic polyphenol-phosphatase complexes added as seed coatings increased the alkaline phosphatase activity of soil and this effect lasted for 51 days (Burns 1987).

The use of polyphenolic-β-glucosidase complexes was hypothesized to increase the degradation rate of the lignocellulosic material in soil due to higher hydrolysis rates of cellobiose whose accumulation repressed the cellulase enzymes (Sarkar and Burns 1984).

Both *Trichoderma* and *Penicillium* isolates synthetize antibiotics and enzymes, such as proteases, cellulases and β-gluconases, which are active in controlling damping-off diseases promoted by the fungus *Pythium ultimum* (Thompson and Burns 1989). Proteases hydrolyze pathogen-secreted enzymes involved in the degradation of the plant cell wall whereas cellulases and β-gluconases degrade the pathogen *Pythium* cell wall. The treatment of seedlings with antagonistic enzymes from *Penicillium claviforme*, entrapped in alginate or coated by clays, decreased the percentage of seedlings infected by the pathogen fungus from 84 (the control) to 46%.

Enzymes can be used to detoxify organic polluted soils if the product of the reaction is less toxic than the pollutant molecule (Nannipieri and Bollag 1991). Since the product of pesticide hydrolysis is frequently less toxic than the parent compound, and hydrolases do not require cofactors or coenzymes, these enzymes can have promising applications in the detoxification of pesticides in the environment. Extracellular and intracellular hydrolases, such as esterases, acylamidases, phosphatases, lyases etc., are common in the microbial metabolism of pesticides. Both technical and formulated diazinon added to an autoclaved soil to reach a concentration of 1% (w/w) were almost completely hydrolyzed if a parathion hydrolase from *Pseudomonas* was added to the sterile soil (Barik and Munnecke, 1982). Complete hydrolysis by the enzyme occurred at all various concentrations (from 0.05 to 0.5%) but the highest (0.5%) concentration of diazinon was in a non sterile soil (Honeycutt et al. 1984). The proteolytic degradation of the parathion hydrolase by soil microflora may have shortened the persistence of the enzyme in soil, thus preventing the hydrolysis of diazinon at the highest concentration (Nannipieri and Bollag 1991). It is well established that free enzymes in the aqueous soil solution are short-lived due to the combination of adsorption, denaturation and degradation (Nannipieri 1994). Some extracellular enzymes, however, do persist because they are inglobated and protected by soil colloids (Burns 1982; Nannipieri 1994). Thus, a prerequisite for using enzymes in soil is their immobilization on supports protecting the enzyme molecules against the proteolytic degradation by soil microflora (Burns 1987; Nannipieri and Bollag 1991). For this reason, enzymes, such as hydrolases, have been inglobated into synthetic polyphenolic polymers prepared from aromatic compounds typically found in humates (Burns 1987). These phenolic compounds respond to enzymes, such as peroxidases and laccases, to form radicals and quinones which subsequently polymerize through covalent bonds formation or nucleophilic addition. The enzyme to be immobilized is added to the phenolic mixture after the formation of radicals and quinones (Sarkar and Burns 1984). Chemical and structural properties of synthetic polyphenol-enzyme complexes resemble those of naturally occurring humic-enzyme complexes extracted from soil (Sarkar and Burns, 1984). The type of support should be carefully considered, not only for its property of preserving the activity of the enzyme to be immobilized but also for its possible effects on no-target organisms of soil. The matrix on which the enzyme is immobilized might have either stimulatory or inhibitory effects on non-target soil microorganisms.

In addition to the immobilization of the enzyme on a support, other factors need to be considered for a successful use of enzymes in soil. The immobilized enzyme must be capable of expressing its activity at a high level and for an extended period of time. The site of application has to be carefully considered because the efficacy of the inoculated enzyme can depend on its location on the soil matrix. For example, when the activity of the enzyme is required in the rhizosphere, probably the best way of application is as a seed coating. In this case there are good chances that the enzyme will cover the rhizoplane when roots will develop from the seedling. A comparison with microbial inoculants selected for the same purpose of enzyme preparation should be carried out under laboratory conditions so as to determine what is the most effective. The efficacy of the enzyme preparation should be investigated not only in the laboratory but also under field conditions. Finally, an economic analysis of the enzyme application should be made and compared with alternative methods available for the purpose for which the enzyme preparation is used.

Enzymes present several advantages on microorganisms (Nannipieri and Bollag 1991). They can often withstand environmental extremes that would be lethal to microbial cells. For example, whereas parathion hydrolase is still active at 50 °C, at a salt concentration of 10% and at solvent concentrations of 10 g l^{-1}, *Pseudomonas* cells, from which the enzyme is extracted and then purified, are negatively affected by these conditions (Bollag and Liu 1990). In addition, enzymes are not affected by predators and toxins, and may be effective even at low substrate concentration. Under these conditions, as reported above, microorganisms generally prefer more easily degradable compounds. The enzyme technology is now so sophisticated that it is possible to immobilize two enzymes acting sequentially and one cofactor can be also included if required for the activity of one of the two enzymes (Gu and Chu 1988). This achievement enlarges the possibility of using enzyme preparations when the production of the less toxic by-product than the parent pollutant requires two instead of one single reaction. On the other hand, there are several factors that may limit the use of the enzyme. It may be problematic to extract and purify the enzyme. Even if the enzyme is easily and cheaply extracted and purified, it may be unstable and this may restrict its use (Nannipieri and Bollag 1991).

Induced Changes in the Activities of the Indigenous Soil Microflora

Stimulation of Soil Microbial Activities

It is well established that different practices, such as tillage methods, cropping systems or manure applications can affect the size, composition and activities of soil microorganisms. By comparing the two extremes of tillage (conventional tillage versus no-tillage), it was observed the greatest microbial biomass content, dehydrogenase and potentially mineralizable nitrogen activities occurred in the surface (0 to 7.5 cm) soil layer of no-tillage; in the case of the respective ploughed soil the greatest values were observed at the 7.5-15 cm depth (Doran 1980, 1987). This is not surprising, because the no-till practice results in crop residue concentrations at the soil surface that stimulate a greater microbial activity. Reduced tillage has a positive effect on the activity of hydrolases (Angers et al. 1993; Nannipieri 1994). This may depend on the higher level of intracellular hydrolase activities due to the higher microbial biomass and/or the higher immobilization of extracellular hydrolases by recalcitrant humic moietes. Reduced tillage increased microbial diversity, as determined by differences between soil isolates in the decomposition of carbon-rich compounds, metabolism of nitrogen compounds and antagonistic activity against selected microorganisms (Hassink et al. 1991). This effect was confirmed by specific patterns of bacterial substrate utilization; the effect of tillage on microbial diversity was more pronounced when soil was sampled under wheat at flag-leaf stage that at planting time (Lupwayl et al. 1998).

Cropping regimes influence both microbial activity and microbial biomass (Beuchamp and Hume, 1997). A higher microbial biomass C content (256 and 192 mg kg^{-1} soil, respectively) was observed in a pearl millet-wheat-green manure rotation than in a pearl millet-wheat-fallow rotation (Chander et al. 1997). The content of microbial biomass C decreased when an oilseed crop, such as sunflower or mustard, was included in the rotation. The fallow phase of a crop rotation reduces soil microbial diversity as evaluated by fatty acid analysis of soil extracts (Zelles et al. 1992). The substrate utilization patterns of the bacterial communities isolated from soil revealed that biodiversity was higher under wheat preceded by red clover green manure or field peas than under continuous wheat or wheat-fallow rotation

(Lupwayl et al. 1998). Constitutive rather than repressible or inducible enzyme activities of soil can discriminate between soil management treatments probably because they are related to microbial biomass content, which depends on such treatments (Nannipieri 1994).

Soil has an inherent capacity to degrade synthetic pollutants, even those with molecular structures not resembling those of natural compounds. However, synthetic pollutants that differ considerably from those found in nature persist for longer periods of time than those with structures resembling natural compounds (Bollag and Liu 1990). It has been speculated that genetic manipulation based on random mutations and selection may explain the appearance of microbial strains capable of degrading organic pollutants with structures resembling natural compounds. On the other hand, acquisition of novel pathways is needed to degrade more dissimilar pollutants. This acquisition requires a cluster of "degradative" genes on a single genetic unit, such as a plasmid or bacteriophage. Numerous genes involved in the degradation of organic compounds are plasmid-encoded and this may help to explain the remarkable capacity for the adaptation of microbial enzymes to xenobiotic substances. Indeed, plasmids show unique properties: (1) they can replicate independently of their host chromosomes; (2) they possess transposable elements with terminal repeated DNA sequences which can be removed as whole blocks; (3) they can accept insertion sequences, including transposons. For these reasons, plasmids possess a degree of natural exchanges of genetic information between different strains, species, or genera of microorganisms, and this exchange may promote the evolution of new degradative capacities (Williams 1981). It is necessary to monitor these processes in soil and determine the effects of environmental conditions on the rate of evolution of clusters of degradative genes.

The broad metabolic activity of soil can also depend on the fact that some enzymatic systems are non-specific (so-called enzymatic combustion). The ligninolytic system of white rot fungi also attacks a large variety of aromatic pollutants, such as polycyclic aromatic hydrocarbons and their derivatives (polychlorinated biphenyl and even dioxin), in high efficiency and through mineralization (Jian and Tso 1996).

The degradation of hydrocarbons occurs through oxidation pathways which feed into central metabolism (Williams 1981). Generally, these oxidative pathways are inducible and include (1) an upper pathway, where enzymes with specialized activities convert the initial hydrocarbon to a substrate for (2) a lower pathway characterized by more generalized enzyme activities for the conversion of this substrate to intermediates of central metabolism. The location of genes encoding the relative enzymes is plasmids (Timmis et al. 1985). By genetic manipulation it is possible to extend the substrate range of a bacterial strain. This can be obtained by constructing strains that carry a number of degradative plasmids. Thus, a constructed strain of *Pseudomonas*, called superbug, harboured a hybrid CAM-OCT, an NAH and a TOL plasmid encoding the oxidation of camphor-octane (also hexane and decane), naphtalene and toluene (also meta- and para-xylene), respectively (Friello et al. 1976). This strain was more efficient in degrading the crude oil that was a mixture of individual strains each carrying a single plasmid. However, it is questionable if this multiplasmid strain, which is capable of degrading various different hydrocarbons, is more efficient in degrading oil in situ than a mixed microbial population. Indeed, the indigenous soil microflora is capable of decomposing petroleum, which is a complex mixture of non-aromatic and aromatic hydrocarbons, including heterocyclic compounds (McGill et al. 1981). Petroleum is degraded by first-order reaction rates, if the supply of oxygen and nutrients are not present in limiting concentrations. During petroleum degradation, N is immobilized at higher rates. An empirical relationship was developed, where the amount of N to be added to soil for reducing the oil contamination at the half value (half-life) in 1 year can be calculated as follows:

kg N ha^{-1} = (% oil in the 0-15 cm soil layer) (190) - 140.

For a contamination of 7% oil by weight in soil, 1170 kg N ha^{-1} are needed to reduce the contamination to 3.5% after 1 year (McGill et al. 1981). Agricultural practices involving the adjustment of the C: N ratio to 50: 1 by applying NPK fertilizers with rototilling and irrigation of soil every week and the presence of the plant cover (alfalfa inoculated with rhizobium coltures), after partial remediation of the soil, were effective in the decontamination of petroleum-polluted soils (alkaline and calcareous) of Kuwait (El Nawawy et al. 1996). The inoculation of the polluted soils with indigenous hydrocarbons utilizing bacteria lead to the degradation of 90 and 75% of the total petroleum hydrocarbon content from light (3% oil in soil) and heavy (8% oil in soil) polluted soils, respectively. The content of polycyclic aromatic hydrocarbons with three to five rings was also significantly reduced in reclaimed soils. Addition of fertilizers, composts or degrading bacteria have been reported to stimulate the microbial degradation of polycyclic aromatic hydrocarbons in soil (Wang et al. 1990). The degradation of petroleum in Canadian and Arabic soils proves that the indigenous microflora is capable of degrading a complex mixture of recalcitrant compounds under different environmental conditions. The potential activity of the indigenous soil microflora versus any recalcitrant compound should be determined under laboratory conditions also in relation to changes in the amount and type of nutrients, O_2 supply, temperature and pH values. This investigation should permit the efficient use of the degradative activity of indigenous soil microflora under field conditions.

Inhibition of Soil Microbial Activities

Inhibitors of specific enzyme reactions, such as those of urease, or inhibitors of metabolic processes, such as those of nitrification, have been commercialized after testing their effects under laboratory and field conditions. Urea is the most used N fertilizer and it is hydrolyzed to NH_4^+ and CO_2 in soil. If the rate of hydrolysis is too high, the N-NH_4^+ concentration can reach high values, and these can promote high N-NH_3 volatilization losses in alkaline soils; in addition toxic effects can occur especially for germinating seedlings and young plants (Fig. 2). For this reason, there is the need to control the urea hydrolysis in soil under certain conditions. Several inhibitors have been tested and some of them are used in agricultural practice (Mulvaney and Bremner 1981). Phosphoroamides seem to be the most effective for inhibition of urease activity in soil (Held et al. 1976). N-(n-butyl) thiophosphoric triamide (NBPT), N-(diaminophosphinyl)-cyclohexylamine (DPCA) and phenylphosphorodiamidate (PPD) are among the most-used phosphoroamides (Bremner and Chai 1989). Usually, the most efficient inhibitor in retarding urea hydrolysis in soil is NBPT because it is rapidly decomposed in N-(n-butyl) phosphoric triamide (BNPO), which shows a greater inhibition than PPD, DPCA and the parent compound (NBPT) (McCarty and Bremner 1989).

Nitrate is the most mobile N compound in the soil-plant system since it does not interact with the negatively charged soil colloids and, although easily taken up by plants, can be leached by water with risks of groundwater contamination. In addition, nitrate, under anaerobic conditions, can be denitrified to N_2O and N_2. Therefore nitrification needs to be controlled if nitrification rates are too high. The first reaction, involving the production of hydroxylamine, is catalyzed by an ammonia mono-oxygenase. The oxidation of hydroxylamine to nitrite, through the formation of an unknown intermediate, is catalyzed by the hydroxylamine oxidoreductase and the final oxidation to nitrate is catalyzed by a nitrite oxidoreductase. Generally nitrification is carried out in soil by chemoautotrophic bacteria even if heterotrophic nitrification has been reported in acid soils (Adams 1986).

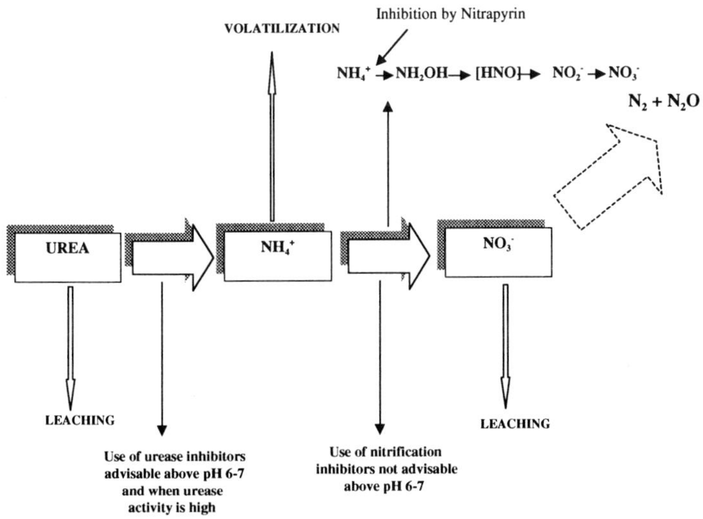

Fig. 2: Ureolysis, nitrification, N losses and inhibition of urease and nitrification

There are many compounds that have been shown to be inhibitory to nitrification (Bundy and Bremner 1973). Dicyan [the commercial name of dicyandiamide] and Nitrapyrin [the commercial name of 2-Chloro-6(trichloromethyl) pyridine] are among the most used in the agricultural practice and both inhibit the oxidation of ammonia to nitrite. Despite their widespread use and commercial interest, very little is known of their mechanism of action or of their effects on the physiology of ammonia oxiders. Probably nitrapyrin acts as a copper chelating agent which is specific for ammonia mono-oxygenase (Campbell and Aleem 1965; Prosser 1990).

The N mineralization and immobilization turnover affects the amount of available N in soil. Both processes occur simultaneously and thus it is difficult to measure the gross rates of the two processes. N immobilization is mediated by soil microflora and can occur through a reaction catalyzed by glutamine synthetase (GS) and involving the reaction of ammonium ions with glutamate in the presence of ATP and formation of glutamine which reacts with α-ketoglutarate producing glutamate through a reaction catalyzed by glutamate synthase, known as glutamine oxoglutarate amino transferase (GOGAT) (Reitzer and Magasanik, 1987). The NH_4^+ is also incorporated as glutamate via glutamate dehydrogenase (GDH) and, in some microbial strains, as asparagine from aspartate through a reaction catalyzed by asparagine synthetase (Reitzer and Magasanik 1987). The K_m for GS is lower than those of GDH and asparagine synthetase and when the NH_4^+ concentration is lower than 0.1 mM, NH_4^+ is incorporated through the reaction catalyzed by GS (Reitzer and Magasanik 1987). Ammonium concentrations are very low in the soil solution and, therefore, it is reasonably to hypothesize that microbial N immobilization in soil occurs through the GS/GOGAT pathway (Nannipieri et al. 1994).

The ^{15}N dilution techniques can be used to determine the gross rates of N mineralization and immobilization (Powlson and Barraclough 1993). Because the methodology for isotope dilution is complicated and requires expensive instrumentation, there is a clear need for a simpler way of determining gross rates of mineralization in soil. A method based on the use of L-methionine-DL-sulphoximine (MSX), an inhibitor of GS activity, for easily and accurately determining gross N mineralization rate was proposed by Nannipieri et al. (1994). The premise for this approach was that N mineralization in MSX-treated soil would equal gross mineralization in the untreated soil. The approach has also biotechnological implications owing to the possible use of the inhibitor under field conditions to block or reduce the immobilization of fertilizer N so as to increase the amount of available N to the plant. Indeed, a competition for available N exists between plant and soil microorganisms. To optimize the uptake of available N by crops, it is essential to inhibit microbial N immobilization when the plant needs N. Each crop is characterized by a particular period of the growing season where N uptake is essential for an optimal yield.

In order to test the efficacy of the inhibitor, the isotopic dilution technique was used to determine gross N mineralization and immobilization rates of two soils (Pistoia, sand-clay-loam, and Romola, sandy) in the presence of MSX (Landi et al. 1999). The inhibitor was only partially effective at blocking immobilization because MSX was itself mineralized releasing N, as evidenced by the increase in gross N mineralization rates in both soils in the presence of the inhibitor and by the decreases in the MSX recoveries. In addition, soil microorganisms appeared to switch to NO_3-N as an N source.

The effectiveness of any inhibitor in soil depends on many factors. To be effective, an inhibitor must be able to maintain its activity for the required period of time. Biotic and abiotic degradation may trasform the inhibitor to inactive by-products. For example, dicyandiamide (DCD) can be used as an N source by soil bacteria (Hauser and Haselwandter 1990). Other environmental factors such as adsorption and volatilization can decrease the bioactivity of the inhibitors in soil. In addition, the inhibitor does not always show only target effects. Nitrapyrin was very efficient in inhibiting the nitrification process (up to 90% inhibition for 12 days) but also slightly inhibited gross rates of N immobilization and mineralization, as determined by the isotope dilution technique (Chalk et al. 1990).

Conclusions

Soil microorganisms can be manipulated so as to increase specific microbial activities which can promote plant growth (by increasing the supply of nutrients, controlling soil-borne pathogens or, directly, by synthetizing phytohormones), improve soil structure or reclaim polluted soils (Lynch 1983; Burns 1995; van Veen et al. 1997). These tasks can be obtained by inoculating beneficial microbial strains, adding specific enzymes or, directly, by stimulating or inhibiting activities of the indigenous soil microflora.

Modern genetic manipulation and enzyme technologies are so sophisticated that a broad number of strains and enzyme preparations are available for manipulating activities and composition of soil microflora. However, many failures or inconsistencies in achieving the objectives by inoculating microbial strains have been reported, and this has raised concern about the perspective of the great practical potential offered by microbial releases into soils (van Elsas and Heijnen 1990; Akkermans 1994). To overcome the impact of harsh soil conditions on the inoculated cells, the delivery of microbial cells to soil or rhizosphere can be carried out as inoculant formulations with carrier materials. These carriers may provide a physical and nutritional protective niche to inoculated cells in soil by physically preventing the contact between inoculants and predators and providing nutrients to the inoculated cells that may facilitate their competition against the indigenous soil microflora. Van Veen et al. (1997) have discussed the characteristics required of the carrier materials to be used to formulate inoculant bacteria for delivery in soil. It can be concluded that a successful application of beneficial microorganisms to soil depends on the proper characterization of target soil and rhizosphere as habitats for introducing microorganisms, as well as the choice of the proper carrier material, of time and site of application.

The use of enzyme preparations has been discussed, but the relative research under laboratory conditions is still rare. Only a few enzyme preparations have been experimentally tested in soil for degrading pesticides, improving the amount of plant available P and controlling soil-borne pathogens. Further investigations are needed for assessing the efficacy of enzyme preparations in different types of soil and different environmental conditions. A wider range of enzymes than that used so far needs to be investigated. Eventually, the use of each enzyme preparation should be compared with the microbial inoculants available for the same purpose for which the enzyme is used.

A better knowledge of the activity and composition of soil microflora is required for a successful exploitation of metabolic activities of the indigenous soil microflora. Tillage, type of crop, amount and type of agrochemical can affect size, composition and activity of soil microflora. Current methods to determine microbial biomass and activity are quite accurate, whereas plate counts only give information on culturable microorganisms (Nannipieri et al. 1990). The extraction and characterization of soil DNA might permit monitoring the effects of agricultural practices on the composition of soil microflora more accurately than plate counts.

Soil microorganisms show a remarkable range of catabolic and anabolic activities. Some of them are able to degrade xenobiotics whose molecular structure does not resemble that of naturally occurring compounds. The development of a cluster of genes encoding the enzymes involved in the degradation of these compounds may require time and the presence of suitable environmental conditions. It is unknown how this process occurs in situ. Future research is needed to know which microbial strains are involved in the degradation, to determine the activity of these strains in soil and to understand the type of gene exchanges and mutations at the basis of the development of new clusters of gene encoding enzymes degrading the xenobiotics in soil. Further insights are also needed on the role of non-specific enzymatic systems, such as the ligninolytic one, on the degradation of recalcitrant pollutants in soil and how the activity of these systems is affected by environmental conditions. As reported above in the case of petroleum degradation (McGill et al. 1981), today it is possible to determine the optimal environmental conditions favouring the degrading activity of the indigenous soil microflora against a mixture of recalcitrant compounds and efficiently use these conditions for reclaiming polluted soils.

References

Adams JA (1986) Nitrification and ammonification in acid forest litter and humus as affected by peptone and ammonium-N amendment. Soil Biol Biochem 18:45-51

Akkermans ADL (1994) Application of bacteria in soils, problems and pitfalls. FEMS Microbiol Rev 15:185-194

Alexander M (1977) Soil microbiology John Wiley, New-York

Angers DA, Bissonnette N, Lègère A, Sanson N (1993) Microbial and biochemical changes induced by rotation and tillage in a soil under barley production. Can J Soil Sci 73:39-50

Barik S, Munnecke DM (1982) Enzymatic hydrolysis of concentrated diazinon in soil. Bull Environ Contam Toxicol 29:235-239

Berry DF, Hagedorn C (1991) Soil and ground water transport of microorganisms. In: Ginsburg LR (ed) Assessing ecological risks of biotechnology. Butterworth-Heinemann, Stoneham, Massachusetts, pp 57-73

Beuchamp EG, Hume DJ (1997) Agricultural soil manipulation: the use of bacteria, manuring and plowing. In: van Elsas JD, Trevors JT, Wellington EMH (eds) Modern soil microbiology. Marcel Dekker, New York, pp 643-664

Bollag JM, Liu SY (1990) Biological transformation processes of pesticides. In: Cheng HH (ed) Pesticides in the soil environment: processes, impacts, and modeling. Soil Science Society of America Book Ser 2. Soil Science Society of America, Madison, Wisconsin, pp 169-209

Bremner JM, Chai HS (1989) Effects of phosphoroamides on ammonia volatilization and nitrite accumulation in soils treated with urea. Biol Fertil Soils 8:227-230

Brown ME, Jackson RM, Burlingham JK (1968) Effects of produced on tomato plants, *Lycopersicum esculentum* by seed or root treatment with gibberellic acid and indol-3yl-acetic acid. J Exp Bot 19:544-552

Bundy LG, Bremner JM (1973) Inhibition of nitrification in soils. Soil Sci Soc Am Proc 37:396-398

Burns RG (1982) Enzyme activity in soil: location and a possible role in microbial ecology. Soil Biol Biochem 14:423-427

Burns RG (1987) Interactions of humic substances with microbes and enzymes in soil and possible implications for soil fertility. An Edafol Agrobiol 46:1247-1259

Burns RG (1995) Enumeration, survival, and beneficial activities of microorganisms introduced into soil. In: Huang PM, Berthelin J, Bollag JM, McGill WB, Page AL (eds) Environmental impact of soil component interactions. Lewis, Boca Raton, Florida, pp 145-164

Burr TJ, Schroth MN, Suslaow T (1978) Increased potato yields by treatment of seed pieces with specific strains of *Pseudomonas fluorescens* and *P. Putida*. Phytopathology 68:1377-1383

Campbell NER, Aleem MIH (1965) The effect of 2-chloro-6-(trichloro-methyl) pyridine on the chemoautotrophic metabolism of nitrifying bacteria. I. Nitrite oxidation by *N. Antonie*. J Microbiol Serol 31:124-136

Chalk PM, Victoria RL, Muraoka T, Piccolo MC (1990) Effect of a nitrification inhibitor on immobilization and mineralization of soil and fertilizer nitrogen. Soil Biol Biochem 22:533-538

Chander K, Goyal S, Mundra MC, Kapoor KK (1997) Organic matter, microbial biomass and enzyme activity of soils under different crop rotations in the tropics. Biol Fertil Soils 24:306-310

Doran JW (1980) Soil microbial and biochemical changes associated with reduced tillage. Soil Sci Soc Am J 44:765-774

Doran JW (1987) Microbial biomass and mineralizable nitrogen distributions in no-tillage and plowed soils. Biol Fertil Soils 5:68-75

Drahos DJ, Hemming BC, McPherson S (1986) Tracking recombinant organisms in the environment: β-galactosidase as a selectable non-antibiotic marker for fluorescent pseudomonads. Biotechnology 4:439-444

Eaglesham ARJ (1989) Global importance of *Rhizobium* as an inoculant. In: Campbell R, Macdonald R (eds) Microbial inoculation of crop plants. Oxford University Press, Oxford, pp 29-48

El Nawawy AS, Al-Daher R, Yateem A, Al-Awadhi N (1996) Bioremediation of oil contaminated soil in Kuwait II. Enhanced landfarming for bioremediation of oil-contaminated soil. In: Moo-Young M, Anderson WA, Chakrabarky AM (eds) Environmental biotechnology: principles and applications. Kluwer, Dordrecht, pp 249-258

Falchini L, Sparvoli E, Tomaselli L (1996) Effect of *Nostoc* (Cyanobacteria) inoculation on the structure and stability of clay soils. Biol Fertil Soils 23:346-352

Friello dA, Mylroie JR, Chakrabarty AM (1976) Use of genetically engineered multi-plasmid microorganisms for rapid degradation of fuel hydrocarbons. In: Sharpley JM, Kaplan A (eds) Proc 3rd Int Biodegradation Symp. Applied Science Publ, London, pp 205-214

Grego S, D'Annibale A, Luna M, Badalucco L, Nannipieri P (1990) Multiple forms of synthetic pronase-phenolic copolymers. Soil Biol Biochem 22:721-724

Gu KF, Chu TMS (1988) Immobilization of a multienzymic system and dextran-NAD^+ in semipermeable microcapsules for use in a bioreactor to convert urea into L-glutamic acid. In: Moo-Young M (ed) Bioreactor immobilized enzymes and cells: fundamentals and applications. Elsevier, London, pp 59-62

Hassink J, Oude Voshaar JH, Nijhuis EH, van Veen JA (1991) Dynamics of the microbial population of a reclaimed-polder soil under a conventional and reduced-input farming system. Soil Biol Biochem 23:515-524

Hattori T, Hattori R (1976) The physical environment in soil microbiology: an attempt to extend principles of microbiology to soil microorganisms. CRC Crit Rev Microbiol 4:423-461

Hauser M, Haselwandter K (1990) Degradation of dicyandiamide by soil bacteria. Soil Biol Biochem 22:113-114

Heijnen CE, van Veen JA (1991) A determination of protective microhabitats for bacteria introduced into soil. FEMS Microbiol Ecol 85:73-80

Held P, Lang S, Tradler E, Klepel M, Drohne D, Hartbrich HJ, Rothe G, Scheler H, Grundmeier S, Trautman A (1976) Agent for reducing the loss of plant-available nitrogen in cultivated soil. East German Patent 122 177 (Chem Abstr 87:67315w).

Honeycutt R, Ballantine L, LeBaron H, Paulson D, Seim V, Ganz C, Milad G (1984) Degradation of higher concentrations of a phosphorothoic ester by hydrolase. In: Krueger RF, Seiber N (eds) Treatment and disposal of pesticide wastes. ACS Symp Ser 259:343-352

Hussain A, Vancura V (1970) Formation of biologically active substances by rhizosphere bacteria and their effect on plant growth. Folia Microbiol 11:468

Jian H, Tso WW (1996) DNA recombination, plasmid dissemination, enzymatic "combustion" and cell immobilization: their potential effect on environmental biotechnology. In: Moo-Young M, Anderson WA, Chakrabarty AM (eds) Environmental biotechnology: principles and applications, Klumer, Dordrecht, pp 28-37

Kloepper JW, Leong J, Teintze M, Schroth MN (1980a) Enhanced plant growth by siderophores produced by plant growth-promoting rhizobacteria. Nature 286:885-886

Kloepper JW, Lifshitz R, Schroth MN (1980b) *Pseudomonas* inoculants to benefit plant production. In: ISI atlas of science. Institute for Scientific Information, Philadelphia 1988, pp 60-64

Ladd JN, Forster RC, Nannipieri P, Oades JM (1996) Soil structure and biological activity. In: Stotzky G, Bollag JM (eds) Soil biochemistry vol 9. Marcel Dekker, New York, pp 23-78

Lambert B, Joos H (1989) Fundamental aspects of rhizobacterial plant growth promotion research. Tibetechnology 7:215

Landi L, Barraclough D, Badalucco L, Gelsomino A, Nannipieri P (1999) L-methionine-sulphoximine affects N mineralization-immobilizatiom in soil. Soil Biol Biochem 31:253-259

Leeman M, Den Ouden FM, Van Pelt JA, Dirkx FPM, Steijl H, Bakker PAHM, Schippers B (1996) Iron availability affects induction of systemic resistance to fusarium wilt of radish by *Pseudomonas fluorescens*. Phytopathology 86:149

Lynch JM (1983) Soil biotechnology. Blackwell, Oxford

Lynch JM (1990) The rhizosphere. John Wiley, Chichester

Lupwayl NZ, Rice WA, Clayton GW (1998) Soil microbial diversity and community structure under wheat as influenced by tillage and crop rotation. Soil Biol Biochem 13:1733-1741

Marshall K (1975) Clay mineralogy in relation to survival of soil bacteria. Annu Rev Phytophatol 13:157-373

Maurhofer M, Hase C, Meuwly P, Metraus JP, Defago G (1994) Induction of systemic resistance of tobacco necrosis virus by the root colonizing *Pseudomonas fluorescens* strains CHAO: influence of the gacA gene and of pyoverdine production. Phytophatology 84:139-146

Metting B (1991) Biological surface features of semiarid lands and deserts. In: Skujins J (ed) Semiarid lands and deserts: Soil resources and reclamation. Marcel Dekker, New York, pp 257-293

McCarty GW, Bremner JM (1989) Formation of phosphoryl triamide by decomposition of thiophosphoryl triamide in soil. Biol Fertil Soils 8:290-292

McGill WB, Rowell MJ, Westlake DWS (1981) Biochemistry, ecology, and microbiology of petroleum components in soil. In: Paul EA, Ladd JN (eds) Soil biochemistry, vol 5. Marcel Dekker, New York, pp 229-296

McLaren AD (1977) The seven questions of Selman A. Waksman. Soil Biol Biochem 9:375-376

McLaren AD, Peterson GH (1967) Introduction to the biochemistry of terrestrial soils. In: McLaren AD, Peterson GH (eds) Soil biochemistry vol 1. Marcel Dekker, New York, pp 1-13

Mulvaney RI, Bremner JM (1981) Control of urea transformations in soil. In: Paul EA, Ladd JN (eds) Soil biochemistry, vol 5. Marcel Dekker, New York, pp 153-196

Nannipieri P (1994) The potential use of soil enzymes as indicators of productivity, sustainability and pollution. In: Pankhurst CE, Doube BM, Gupta VVSR, Grace PR (eds) Soil biota. Management in sustainable farming systems. CSIRO, East Melbourne, Victoria, Australia, pp 144-155

Nannipieri P, Bollag JM (1991) Use of enzymes to detoxify pesticide-polluted soils and waters. J Environ Qual 20:510-517

Nannipieri P, Muccini L, Ciardi C (1983) Microbial biomass and enzyme activity: production and persistence. Soil Biol Biochem 15:679-685

Nannipieri P, Grego S, Ceccanti B (1990) Ecological significance of the biological activity in soil. In: Bollag JM, Stotzky G (eds) Soil biochemistry vol 6. Marcel Dekker, New York, pp 293-355

Nannipieri P, Badalucco L, Landi L (1994) Holistic approach to the study of populations, nutrient pools and fluxes: limits and future research needs. In: Ritz K, Dighton J, Giller KE (eds) Beyond the biomass. John Wiley and Sons, New York, pp 231-238

Oades JM (1984) Soil organic matter and structural stability: mechanisms and implications for management. Plant Soil 76:319-337

Powlson DS, Barraclough D (1993) Mineralization and assimilation in soil-plant systems. In: Knowles R (ed) Nitrogen isotope techniques. Academic Press, San Diego, pp 209-242

Prosser JI (1990) Autotrophic nitrification in bacteria. In: Rose AH, Tempest DW (eds) Advances in microbial physiology. Academic Press, London, pp 125-181

Reitzer LJ, Magasanik B (1987) Ammonia assimilation and the biosynthesis of glutamine, glutamate, aspartate, asparagine, L-alanine and D-alanine. In: Neidhardt FC, Ingraham JL, Low KB, Magasanik B, Schaechter M, Umbarger HE (eds) *Escherichia coli* and *Salmonella typhimurium*, cellular and molecular biology. American Society of Microbiology, Washington, DC, pp 302-320

Rojo F, Pieper D, Engesser KH, Knackmuss HJ, Timmis KN (1987) Assemblage of ortho cleavage routes for simultaneous degradation of chloro- and methylaromatics. Science 238:1395-1398

Sarkar JM, Burns RG (1984) Synthesis and properties of β-D-glucosidase-phenolic copolymers as analogues of soil humic-enzyme complexes. Soil Biol Biochem 16:619-625

Stotzky G (1986) Influence of soil mineral colloids on metabolic processes, growth, adhesion and ecology of microbes and viruses. In: Huang PM, Schnitzer M (eds) Interactions of soil minerals with natural organics and microbes. Special Pub 17. Soil Science Society of America, Madison, Wisconsin, pp 305-428

Thompson RJ, Burns RG (1989) Control for *Phytium ultimum* with antagonistic fungal metabolites incorporated into sugar beet seed pellets. Soil Biol Biochem 21:745-748

Timmis KN, Lehrbach PR, Harayama S, Don RH, Mermod N, Bas S, Leppik R, Weightman AJ, Reinecke W, Knackmuss HJ (1985) Analysis and manipulation of plasmid-encoded pathways for the catabolism of aromatic compounds by soil bacteria. In: Helinski DR, Cohen SN, Clewell DB, Jackson DA, Hollaender A (eds) Plasmid in biology. Basic Life Sciences, vol 30. Plenium Press, New York, pp 719-739

Torsvik V, Goksoyr J, Daee FL (1990) High diversity in DNA of soil bacteria. Applied and Environ Microbiol 56:782-787

van Elsas JD, Heijnen CE (1990) Methods for the introduction of bacteria in soil: a review. Biol Fertil Soils 10:127-133

van Elsas JD, Heijnen CE, van Veen JA (1991) The fate of introduced genetically engineered microorganisms in soil, in microcosms and the field impact of soil textural aspects. In: MacKenzie DR, Henry SC (eds) Biological monitoring of genetically engineered plants and microbes. Agricultural Research Institute, Bethesda, Maryland, pp 67-79

van Overbeek LS, van Elsas JD (1997) Adaptation of bacteria to soil conditions: applications of molecular physiology in soil microbiology. In: van Elsas JD, Trevors JT, Wellington EMH (eds) Modern soil microbiology. Marcel Dekker, New York, pp 441-477

van Veen JA, van Overbeek LS, van Elsas JD (1997) Fate and activity of microorganisms into soil. Microbiol Mol Biol Rev 61:121-135

Vargas R, Hattori T (1986) Protozoan predation of bacteria cells in soil aggregates. FEMS Microb Ecol 38:233-242

Waksman SA (1927) Principles of soil microbiology. Williams & Wilkins, Baltimore, USA

Wang X, Yu X, Bartha R (1990) Effect of bioremediation on polycyclic aromatic hydrocarbon residues in soil. Environ Sci Technol 24:1086-1089

Weller D, Thomashow LS (1994) Current challenges in introducing beneficial microorganisms into rhizosphere. In: O'Gara F, Dowling DN, Boesten B (eds) Molecular ecology of rhizosphere microorganisms. VCH Weinheim, New-York, pp 1-18

Williams PA (1981) Catabolic plasmids. Trends Biochem Sci 6:23-60

Wilson M, Lindow SE (1993) Release of recombinant microorganisms. Ann Rev Microbiol 47:913-944

Zelles L, Bai QY, Beck T, Besse F (1992) Signature fatty acids in phospholipis and lipopolysaccharides as indicators of microbial biomass and community structure in agricultural soils. Soil Biol Biochem 24:317-323

Biological Control of Plant Diseases

C. ALABOUVETTE[1]

The present chapter reviews the question of developing a microbial control method of plant diseases. All problems which may occur, from the initial screening of an efficient biocontrol agent to the final step of integration of the biological control method into the crop management system are briefly examined.

Introduction

Biological control of plant pathogens has been studied for many years; Baker and Cook published the first book entirely dedicated to biological control in 1973. Since that time, numerous research papers, several books and review articles have been produced, but practical application of biological control agents remains limited.

Despite great progress made in the understanding of the modes of action of biological control agents, biological control products have difficulty in finding a market. Many reasons can be evoked to explain this apparent failure of biological control, and the OECD cooperative research programme Biological Resource Management for Sustainable Agriculture has supported the organisation of several workshops in order to identify scientific and technological bottlenecks and propose new research and development strategies to achieve success in microbial control of plant diseases.

This chapter briefly reviews the questions and problems to be solved throughout the process of developing a biological control method, from the initial screening of an efficient biocontrol agent to the final step of integration of the biological control method into the crop management system.

This chapter is mainly based on the discussions and conclusions drawn during an OECD workshop held in Dijon in August 1998 and entitled Strategies of Microbial Inoculation for Sustainable Agriculture and Forestry. Indeed, there is a large similarity in approach in developing microbial products whatever their intended use: growth promotion, nitrogen fixation, and resistance to stress or disease control. Therefore the microbial inoculation of legumes with nitrogen-fixing bacteria can serve as an example of successful application of beneficial microorganisms in agriculture.

1. I.N.R.A.-C.M.S.E., Flore pathogène, BV 1540, F 21034 Dijon, France

Screening of Effective Strains of Biological Control Agents

The first step in developing a biocontrol method is the screening of an effective strain of biological control agent. Two different approaches can be followed.

The first, traditional approach, is based on a random screening among many strains owing to a standardised method where the antagonist is confronted with the pathogen, in the soil environment and in the presence of the host plant. Several levels of bioassays are conducted, enabling the number of strains tested to be progressively decreased. At the beginning, with the largest number of strains, the bioassay is conducted under artificial conditions, sometimes in vitro, most often in a sterile substratum such as sand or peat to grow the plant. At the end of the process, a very limited number of strains are evaluated for their biocontrol capacity under conditions similar to that of their application in the targeted crop (Hökeberg et al. 1997). This approach needs no preexisting knowledge of the modes of action of the antagonists, that are most often chosen at random. However, in order to improve the probability of finding effective strains, it is possible to isolate the potential antagonists from specific places, such as suppressive soils, where one can assume that a natural level of biocontrol does exist. This approach is space – and time-consuming, but enables biocontrol agents to be detected that fit with the environment where they will be applied.

In contrast, the second approach is based on the preexisting knowledge that a given function, for example antibiotic production, plays a major role in the antagonism expressed by a microbial species against the pathogen. Then, the strategy consists in screening for this function by in vitro assays. In fact, when the genes coding for this function are known, it is possible to base the screening procedure on the tagging of these genes among a large population of microorganisms. For example, in the case of the fluorescent *Pseudomonas* spp., most of the genes coding for antibiotic production, such as phenazine or 2-4 diacetylphloroglucinol, are characterised. Therefore it is possible to screen among a large collection of bacteria for the presence of these genes. It is, however, then necessary to study the expression of these genes, since the presence of the genomic sequence does not necessarily imply the production of the given metabolite in the environment where the biocontrol agent will be used.

Scientists in favour of the first approach argue that, to be effective, a biocontrol agent must not only possess the required modes of action but also be well adapted to the environment where it must express these functions (rhizosphere, phyllosphere, spermosphere). Until now, only a few teams have been involved in the study of the genes coding for the "fitness" of the biocontrol agents. Therefore there is a risk of selecting potentially very active antagonists that will be not able to survive or to express their beneficial properties in the environment. Scientists in favour of the second approach argue that knowing the most important functions will enable the manipulation of the biocontrol agents in order to add several modes of action in a single strain or to deregulate the production of an important metabolite in order to produce it in greater quantity or at the right time. The last difference between the two approaches is that the second one, allowing progress on the understanding of the modes of action of the biocontrol agents and in the regulation of the microbial functions leads to publication of good scientific papers while the first approach cannot be evaluated by scientific papers, but only by the commercial success of the biocontrol agent. This last difference might explain why so many academic teams in the world are now involved in the second approach, attracted more by knowledge than by practical application.

Modes of Action of Biological Control Agents

Antagonistic effects responsible for disease suppression result either from microbial interactions directed against the pathogen, mainly during its saprophytic phase, or from an indirect action through induced resistance of the host plant.

Microbial antagonism implies direct interactions between two microorganisms sharing the same ecological niche. Three main types of direct interactions may be characterised: parasitism, competition for nutrients and antibiosis.

Parasitism of plant pathogens by other microorganisms including viruses is a well-distributed phenomenon. The parasitic activity of strains of *Trichoderma* spp. towards pathogens such as *Rhizoctonia solani* has been extensively studied (Chet and Baker 1981) and other mycoparasites such as *Coniothyrium minitans* and *Sporidesmium sclerotivorum* are efficient in controlling diseases caused by *Sclerotinia* spp. and other sclerotia-forming fungi (Whipps and Lewis 1980; Adams and Fravel 1993). Parasitism is the only documented mode of action of *Ampelomyces quisqualis* a biological control agent already used in several countries to control powdery mildew in grapes and vegetables (Sundheim and Krekling 1982).

Competition for nutrients is a general phenomenon regulating population dynamics of microorganisms sharing the same ecological niche and having the same physiological requirements. Competition for carbon in soil is considered responsible for the well-know phenomenon of fungistasis (Lockwood 1977), describing the inhibition of fungal spore germination in soil. Energy deprivation in soil is also partly responsible for "general suppression of a pathogen which is directly related to the total amount of microbial activity at a time critical to the pathogen" (Cook and Baker 1983). This general suppression has been demonstrated to play a role in the determinism of the suppresiveness of soils to *Fusarium* wilts, where it controls competition for carbon between pathogenic and non-pathogenic *Fusarium oxysporum* (Alabouvette et al. 1985). Some strains of non-pathogenic *F. oxyxporum* are more competitive than others and should be selected for biological control (Couteaudier and Alabouvette 1990). Competition for carbon has also been involved in determining the antagonism expressed by yeasts against several plant pathogens at the leaf surface (Blakeman and Fokkema 1982). Competition for minor elements also frequently occurs in soil; for example, competition for iron is one of the modes of action by which fluorescent pseudomonads limit the growth of pathogenic fungi and reduce disease incidence or severity (Schippers et al. 1987; Bakker et al. 1991; Lemanceau and Alabouvette 1993).

Antibiosis is the antagonism resulting from the production by one microorganism of secondary metabolites toxic for other microorganisms. Antibiosis is a very common phenomenon responsible for the biocontrol activity of many biological control agents such as fluorescent *Pseudomonas* spp., *Bacillus* spp., *Streptomyces* spp. or *Trichoderma* spp. A large diversity of antibiotics, bacteriocines, enzymes and volatile compounds have been described and their role in suppression of several plant pathogens has been documented (Fravel 1988; Loper and Lindow 1993; Weller and Thomashow 1993, Alabouvette et al. 1996). A given strain of biocontrol agent may produce several types of antifungal compounds effective against certain species of fungal pathogens. For example, the strain CHAO of *Pseudomonas fluorescens* produces siderophores, phenazines, 2.4-diacetylphloroglucinol and cyanide, a different combination of these metabolites being responsible of the antagonism expressed against *Gaeumannomyces graminis* var. tritici and *Chalara elegans* (Defago and Haas 1990). It is important to emphasise that one single antifungal metabolite generally does not account for all the antagonistic activity of a biocontrol agent.

Induced systemic resistance classically occurs when an inducing agent is applied prior to inoculation with a pathogen, resulting in reduced disease in comparison to the non-induced control. More and more studies are devoted to induced resistance of the host plant after application of biocontrol agents. Kuc (1987) reported the first evidence of systemic protection of cucumber against *Colletotricum orbiculare* after preinoculation of the cotyledons of the plant with the same pathogen. It is also well established that the preinoculation of a host plant with an incompatible *forma specialis* of *F. oxysporum* results in reduced disease severity when the plant is inoculated with the compatible pathogen (Biles and Martyn 1989). The fluorescent pseudomonads selected for their plant growth-promoting capacity or for their biocontrol activity have been shown to induce systemic resistance in the plant (Kloepper et al. 1996; Van Loon 1996). Since systemic induced resistance is a general phenomenon that can protect the plant against several pathogens and can be induced by many biological control agents, it receives more attention today that any other mode of action of biological control agents. It also enables the study of plant-microorganism interactions at the cellular and molecular levels, leading to publication in highly rated international journals. However, it must be said that induced systemic resistance does not exclude other modes of actions and may often exert only a complementary effect to microbial antagonism.

More generally, consistency of biological control needs the association of several modes of action, acting simultaneously or successively. As stated above, it is proposed by genetic manipulation to associate several modes of action in a single antagonistic strain and the first improved strains of *Pseudomonas fluorescens* producing phenazine and phloroglucinol are being evaluated for their improved biocontrol activity in the fields (Thomashow and Weller 1996). Another approach consists in associating several strains of biocontrol agents in the same product. It has been well established that association of certain strains of *Pseudomonas fluorescens* with non-pathogenic *Fusarium oxysporum* always improves the control of fusarium diseases. Obviously these associations have to be based on knowledge of the mode of action to create a synergetic effect.

Production and Formulation of the Biocontrol Agents

Production and formulation of the biological control agents, the two last steps before application, probably constitute a bottleneck for the development of biological control strategies. Indeed, too often these steps are not carefully considered by the academic research laboratories, which consider that these technological problems have to be solved by industry. However, producing and formulating an efficient biomass at a low cost needs a scientific approach based on the knowledge of the physiology of the microorganisms. The aim of fermentation is to produce and harvest a viable biomass that will have to express its beneficial properties after some time of storage and application to the crop. To achieve this goal, it is absolutely necessary to study the physiology of the microorganism to determine the fermentation parameters that will enable an effective biomass to be obtained at an affordable cost.

Several review papers have addressed these questions of how to produce and formulate an active biomass (Lumsden et al. 1995; Lewis et al. 1991; Whipps 1997). In most examples, the microorganisms are grown in liquid fermentation and the propagules after harvest are either mixed with a solid substratum, such as clay talc or peat, or embedded in alginate pellets (Fravel et al. 1985; Lumsden and Lewis 1989). The final product must be easy to handle, store and then apply.

An alternative to liquid fermentation is solid state fermentation, where the biocontrol agent is directly produced on solid material that provides nutrients and a substratum that can help to solve the formulation problem. The inoculum, being stored in the substratum on which it has grown, usually shows better survival. Moreover, solid state fermentation enables different types of agricultural waste products to be utilised that are cheap and can be found on the local market, especially in developing countries.

Application of Biological Control Agents

The final step in developing a biocontrol method is to choose a method of application that will enable delivery of the biocontrol agent at the right time and right place where it has to be active. Depending on the target pathogen, the antagonist will be delivered with the seed or in the potting mixture to let it colonise the young roots of the plant or will be applied by spraying the aerial part of the plant in order to let it colonise the stems and leaf surface. Obviously, seed coating is the best approach to introduce a biocontrol agent in the rhizosphere of open field crops, and this is the technique used to apply *Pseudomonas chlororaphis* to wheat seeds in Northern Europe (Hökeberg et al.1997).

Spraying Trichodex on the leaves of grape is the technique used to control *Botrytis cinerea* (Elad and Shtienberg 1996). This example is very interesting because it demonstrates the necessity of epidemiological studies to determine application dates. Indeed, to be effective, *Trichoderma* must be applied when the disease pressure is still not too high. Therefore, the company has developed an epidemiological model to advise growers on the most suitable time for Trichodex application. In some cases, it may be necessary to apply a fungicide alternately with the biological application.

In any case, it is always necessary to study the compatibility of the biological product with the chemical pesticides used in the same crop. Indeed, the biocontrol agent most often targets a single type of pathogen, therefore it has to be integrated into the pest management programme. Much more research is needed to determine the exact use of biological strategies in disease and pest management.

Finally, before marketing the product, it is necessary to register the biological control agent; today, this is not an easy task, since microorganisms are submitted to the same types of requirements as chemical molecules regarding toxicity for man and the environment. The bioassays developed to study the toxicity of chemicals to man and the environment are not necessarily relevant to assess the toxicity of microorganisms; therefore the actual guidelines for registration of pesticides are difficult to apply to microbials. Moreover, microorganisms, being living entities, there is a large concern about their capacity to survive and multiply in the environment. To address this important question, it is necessary to study the population dynamics of the biocontrol agent in the environment where it will be applied. However, as the biocontrol agent is most often a natural microorganism isolated from the same environment, it is not possible to easily recognise the introduced strain from preexisting wild strains. This requires utilising marked strains (antibiotic-resistant, GFP marked strains, etc.) which are no longer identical to the wild-type strain used as biological agent. Today, there is much debate among national and international organisations to propose a new legislation better adapted to the registration of microbial products.

Conclusion

As stated by Fravel (1999) "there are still very few people worlwide whose primary job is to discover and develop biocontrol system for plant diseases". Obviously, more funding is needed not only for research but also for implementation of biological control. Too often, potential beneficial microorganisms are studied from an academic point of view, nobody being really interested in the applied research needed to develop a microbial product and determine its conditions of application to attain consistent biocontrol activity.

The present concern of the public for healthy and safe food products and for the preservation of the environment should promote a renewed interest in biological methods of disease control, especially in fruit and vegetable production. At the same time, several pesticides including fungicides will be removed from the market, inviting scientists and growers to look for alternative strategies. For example, methyl bromide application, which has already been reduced, will be totally banned in the near future. This offers a good opportunity to develop biological procedures to control soil-borne plant pathogens and nematodes. Obviously, biological control will not solve all the problems and has to be considered as one method to be used in association with other control methods such as crop rotation, tolerant cultivars and adequate chemical control. Considering the very important effort made worldwide by private companies and public institutions to create new resistant cultivars by introduction of microbial genes into the plant genome, it must be recalled that biological control is an easy way to deliver microbial genes of interest into a crop without transforming the plant. If only a small percentage of the funds dedicated to the study of systemic induced resistant and to the creation of transgenic plants had been dedicated to biological control studies, more products would be on the market today.

References

Adams PB, Fravel DR (1993) Dynamics of *Sporidesmium*, a naturally occuring fungal mycoparasite. In: Lumsden RD, Vaughn JL (eds) Pest management: biologically based technologies. American Chemical Society, Washington, DC, pp 189-195

Alabouvette C, Couteaudier Y, Louvet J (1985) Soils suppressive to *Fusarium* wilt: Mechanisms and management of suppressiveness. In: Parker CA, Rovira AD, Moore KJ, Wong PTW, Kollmorgen JF (eds) Ecology and management of soil-borne plant pathogens. American Phytopathological Society, St Paul, Minnesota, pp 101-106

Alabouvette C, Hoeper H, Lemanceau P, Steinberg C (1996a) Soil suppressiveness to diseases induced by soil-borne plant-pathogens. In: Stotzky G, Bollag JM (eds) Soil biochemistry, Vol 9. Marcel Dekker, New York, pp 371-413

Bakker PAHM, Van Peer R, Schippers B (1991) Suppression of soil-borne plant pathogens by fluorescent pseudomonads: mechanisms and prospects. In: Beemster ABR, Bollen GJ, Gerlach M, Ruissen MA, Schippers B, Tempel A (eds) Development in agriculturally managed-forest ecology. Vol. 23. Elsevier, Amsterdam, pp 217-230

Biles CL, Martyn RD (1989) Local and systemic resistance induced in waterme-lons by formae speciales of *Fusarium oxysporum*. Phytopathology 79:856-860

Blakeman JP, Fokkema NJ (1982) Potential for biological control of plant diseases on the phylloplane. Annu Rev Phytopathol 20:167-192

Chet I, Baker R (1981) Isolation and biocontrol potential of *Trichoderma harmatum* from soil naturally suppressive to *Rhizoctonia solani*. Phytopathology 71:286-290

Cook RJ, Baker KF (1983) The nature and practice of biological control of plant pathogens. American Phytopathological Soc., St Paul, Minnesota

Couteaudier Y, Alabouvette C (1990) Quantitative comparison of *Fusarium oxysporum* competitiveness in relation with carbon utilization. FEMS Microbiol Ecol 74:261-268

Defago G, Haas D, (1990) Pseudomonads as antagonists of soilborne plant pathogens: mode of action and genetic analysis. In: Bollag JM, Stotzky G (eds) Soil biochemistry, vol. 6. Marcel Dekker, New York, pp 249-291

Elad Y, Shtienberg D (1996) Trichoderma harzianum T39 (Trichodex) integrated with fungicides for the control of grey mould of strawberry, vegetable greenhouse crops and grapes. In: Wenhua T, Cook RJ, Rovira A (eds) Advances in biological control of plant diseases. China Agricultural University Press. Beijing, pp 310-319

Fravel DR (1988) Role of antibiosis in the biocontrol of plant diseases. Annu Rev Phytopathol 26:75-91

Fravel DR (1999) Hurdles and bottlenecks on the road to biocontrol of plant pathogens. Australas Plant Pathol 28:53-56

Fravel DR, Marois JJ, Lumsden RD, Connick WJ (1985) Encapsulation of potential biocontrol agents in an alginate-clay matrix. Phytopathology 75:774-777

Hökeberg M, Gerhardson B, Johnsson L (1997) Biological control of cereal seed-borne diseases by seed bacterization with greenhouse-selected bacteria. Eur J Plant Pathol 103:25-33

Kloepper JW, Zehnder GW, Tuzun S, Murphy JF, Wei G, Yao C, Raupach G (1996) Toward agricultural implementation of PGPR-mediated induced syste-mic resistance against crop pests. In: Wenhua T, Cook RJ, Rovira A (eds) Advances in biological control of plant diseases. China Agricultural University Press, Beijing 100094, pp 165-174

Kuc J (1987) Plant immunization and its applicability for disease control. In: Chet I (ed) Innovative approaches to plant disease control. John Wiley, New York, pp 255-274

Lemanceau P, Alabouvette C (1993) Suppression of fusarium wilts by fluorescent pseudomonads: mechanisms and applications (review article). Biocontrol Sci Technol 3:219-234

Lewis JA, Papavizas GC, Lumsden RD (1991) A new formulation system for the application of biocontrol fungi to soil. Biocontrol Sci Technol 1:59-69

Lockwood JL (1977) Fungistasis in soils. Biol Rev 52:1-43

Loper JE, Lindow SE (1993) Roles of competition and antibiosis in suppression of plant diseases by bacterial biological control agents. In: Lumsden RD, Vaughn JL (eds) Pest management: biologically based technologies. American Chemical Society, Washington, DC, pp 144-155

Lumsden RD, Lewis JA (1989) Biological control of soil-borne plant pathogens: problems and progress. In: Whipps JM, Lumsden RD (eds) Biotechnology of fungi for improving plant growth. Cambridge University Press, Cambridge, pp 171-190

Lumsden RD, Lewis JA, Fravel DR (1995) Formulation and delivery of biocontrol agents for use against soilborne plant pathogens. In: Hall FR, Barry JW (eds) Biorational pest control agents-formulation and delivery. American Chemical Society, Washington, DC

Shippers B, Bakker AW, Bakker PAHM (1987) Interactions of deleterious and beneficial rhizosphere microorganisms and the effect of cropping practices. Annu Rev Phytopathol 25:339-358

Sundheim L, Krekling T (1982) Host-parasite relationships of hyperparasite *Ampelomyces quisqualis* and its powdery mildew host *Sphaerotheca fuliginea.* Phytopathol Z 104: 201-210

Thomashow LS, Weller DM (1996) Molecular basis of pathogen suppression by antibiosis in the rhizosphere. In: Hall R (ed) Principles and practice of managing soilborne plant pathogens. American Phytopathological Society, St Paul, Minnesota, pp 80-103

Van Loon LC (1996) Disease-suppressive actions of *Pseudomonas* bacteria: induced resistance. In: Alabouvette C (ed) Biological and Integrated Control of Root Diseases in Soilless Cultures, IOBC/WPRS Bull, Vol 19, pp 53-61

Weller DM, Thomashow LS (1993) Microbial metabolites with biological activity against plant pathogens. In: Lumsden RD, Vaughn JL (eds) Pest management: biologically based technologies. American Chemical Society, Washington, DC, pp 173-180

Whipps JM (1997) Developments in the biological control of soil-borne plant pathogens. Adv Bot Res 26:1-134

Whipps JM, Lewis DH (1980) Methodology of a chitin assay. Trans Br Mycol Soc 74:416-417

Management of Transgenic Pesticidal Crops

DA. ANDOW[1]

Transgenic pesticidal crops have the potential of revolutionizing modern crop production. Commercial transgenic pesticidal varieties of potato, cotton and maize have been used in the United States for several years. Of the three, transgenic *Bt* maize appears to have the greatest potential for transforming crop production in the US. Two scientific concerns, however, need to be addressed by policy makers to ensure that farmers benefit from these technologies over medium to long range time horizons. First, these transgenic plants may harm nontarget species, including butterflies, natural enemies, other crops, and other species. Second, the pests may evolve resistance to the transgenic plants, rendering them useless. Scientific knowledge can contribute substantially to provide sound, science-based strategies for addressing these issues. Small but significant low magnitude effects, however, are very costly to evaluate and have not been tested prior to commercialization. Scientists must be perfectly clear for policy makers that this class of risks cannot readily be evaluated in small-scale experiments, and will become observable only when the plants are grown on a commercial scale.

Use and Potential of Transgenic Insecticidal Crops in the United States

Transgenic pesticidal crops have the potential of revolutionizing modern crop production by providing the opportunity to design plant defenses to kill or repel important insect, disease, and weed pests. Transgenic insecticidal plants have been produced in broccoli (Metz et al. 1995a, b), cabbage (Metz et al. 1995b), canola (Stewart et al. 1996a), cotton (Flint et al. 1995, Jenkins et al. 1997), maize (Koziel et al. 1993, Armstrong et al. 1995), eggplant (Arpaia et al. 1997), potato (Perlak et al. 1993), soybean (Stewart et al. 1996b), tobacco (Carozzi et al. 1992), and tomato (Fischhoff et al. 1987). These insecticidal crops have the potential to reduce the overall use of insecticides, reduce losses to insects, stabilize yield, and increase production of higher-quality foods.

Fulfilling this promise lies still far off in the future. To date, no transgenic disease-resistant crop plant is in widespread use, no weed-resistant crop plant has been

1. Department of Entomology, 219 Hodson Hall, University of Minnesota, St. Paul, MN 55108 USA

commercialized, and the only insect-resistant crop plants in widespread use rely on a narrow range of defenses. All the commercialized transgenic insect resistant crops rely on *Bt* technology. *Bt* technology uses genes from the bacterium *Bacillus thuringiensis* (*Bt*) that code for a crystalline protein toxin called a Cry toxin. Although there are hundreds of Cry toxins that are found in *Bacillus thuringiensis*, only about four different ones are presently used in commercial transgenic crops, creating a potential vulnerability to insects resisting the resistance.

During 1998, 20.1 million acres were planted to transgenic insect resistant crops worldwide (excluding China for which adequate data are not available), nearly doubling from the 10.3 million acres during 1997. Pesticidal crop plants, however, represent a declining proportion of total transgenic crops. During 1997 they were 36% of the worldwide acreage, while in 1998 they were only 28% of the worldwide acreage. Indeed, fully 71% of the area of transgenic crops planted during 1998 worldwide had traits for herbicide tolerance rather than pest resistance. About 74% of the worldwide total acreage was planted in the United States during 1998.

The main transgenic insecticidal crops are *Bt* maize, *Bt* cotton and *Bt* potato. During 1998, approximately 50 000 acres of *Bt* potato were planted, 5.8 million acres of *Bt* cotton, and ~15 million acres of *Bt* maize were planted in the United States. The area of *Bt* potatoes and *Bt* cotton has not been growing very much during the recent past, but the area of *Bt* maize has been limited by the availability of seed, and is expected to increase to more than 20 million acres during 1999.

Bt potato was the first insecticidal plant commercialized, but it has had relatively little impact on potato production in the US. It has had to compete with conventional insecticides, and acceptance by growers has been slow, in part because of its agronomic characteristics. Present acreage is only 50 000 (3.5% of the total potato acreage in the US), and this has not been growing very rapidly.

Bt cotton was the next insecticidal plant commercialized, and it was rapidly adopted in the US to help control *Heliothis virescens* (tobacco or cotton budworm), one of the key cotton pests. All *Bt* cotton varieties use Cry1Ac as the main Cry toxin. Early overpromotion of the efficacy of *Bt* cotton gave way to skepticism, when it became clear that it did not provide very good control of the related pest, *Helicoverpa zea* (cotton bollworm or corn earworm). Indeed, in parts of the world where *Helicoverpa armigera* is the key cotton pest, *Bt* cotton has relatively low efficacy, and has not been highly adopted. This meant that prior scouting expenses could not be eliminated by the *Bt* cotton technology, and cotton farming was not greatly affected operationally.

Bt maize has been rapidly adopted in many areas of the US corn belt. Most varieties use Cry1Ab as the main Cry toxin, but varieties using Cry9 are becoming available. These toxins protect maize primarily from damage by *Ostrinia nubilalis* (European corn borer). In parts of Kansas and Texas where *Diatrea grandiosella* (southwestern corn borer) occurs, *Bt* maize protects against this severe pest. Both insects have been difficult to control using conventional insecticides because the larvae bore into corn stalks where the conventional insecticide cannot penetrate. These transgenic varieties are likely to be combined with transgenic traits that will control other maize pests, such as the corn rootworms. When this occurs, it will create a complex mix of possible varieties that will challenge our ability to develop science-based public policy regarding the risks and benefits of the technology.

This chapter will concentrate on the example of *Bt* maize and explore the scientific evidence for non-target effects and resistance management, and make a case that science-based risk assessment is possible. At the outset, it should be clarified that scientific investigation will not allow policymakers to claim that there is no risk associated with *Bt* maize. Science allows assessment of the type and degree of risk. From this information intelligent policy can be formed.

Nontarget Effects

Previous research on *Bt* Cry toxins had indicated their relative safety to vertebrates, including humans, and their relatively narrow spectrum of action. For example, Cry1A toxins (both Cry1Ab and Cry1Ac) are considered to be active against a subgroup of the Lepidoptera, but not toxic to other insects. The range of species killed by a particular *Bt* toxin is much narrower than the range of species killed by conventional chemical insecticides.

Lepidoptera

To say that the toxicity range was narrow, however, is an overstatement, because the Lepidoptera constitute about 20% of the earth's known biodiversity. Recent work has suggested that the toxic pollen from *Bt* maize might detrimentally affect monarch butterflies (Losey et al. 1999). The magnitude of this potential effect will probably be relatively easy to estimate, but mitigating any detrimental effect may be costly. The use of a scientifically justified worst-case scenario can be used to indicate how science can help estimate the magnitude of the effect. Let W be the proportion of overwintering monarchs that originate from the central US Corn Belt, M be the proportion of milkweed, the most frequent host plant for monarch butterflies, that is close enough to maize so that if it were exposed to *Bt* maize all the monarchs would die, and P be the proportion of maize that is *Bt* maize. Then the projected reduction in overwintering monarch populations would be WMP. Weak evidence suggests that $W \leq 0.5$. If less than between 30-40% of common milkweed were within 20 m of corn in this region, and mortality to monarchs is £100% when within 20 m of *Bt* maize, and the proportion of maize planted to *Bt* maize is 0.2 (presently) or 0.8 (at nearly complete penetration), then the maximum reduction in overwintering monarchs might be between 3 and 4% (presently) to 12 and 16%. This worst case scenario suggests that potential detrimental effect on monarch populations could be considerable.

Several responses may be reasonable. Many individuals have called for additional scientific investigation to better quantify the potential detrimental effect of *Bt* pollen on monarchs. This will be reasonable if there is adequate expectation that additional study will result in an order of magnitude reduction in this worst-case estimate. Clearly, the most uncertain parameter is M, and investigation should focus on its estimation. Mitigation of the potential effect by developing transgenic lines that do not express Cry toxin in pollen would be another reasonable approach.

Bt pollen could also affect endangered or threatened species of Lepidoptera, if they are close enough to agricultural fields. In the United States, this might not be very likely, because most of the endangered and threatened species are associated with native habitats that occur in areas that are not very well suited for arable crops. In many other parts of the world, however, endangered and threatened species are often associated with arable crops, and the impact of *Bt* pollen on these species could be significant.

Non-Lepidoptera

It would be an oversimplification to claim that all of the investigations on the host range and toxicity of *Bt* as an insecticide are directly relevant to host range determination of Cry toxins in *Bt* transgenic crops. In the *Bt* crops, Cry toxin occurs in its activated form while in *Bacillus thuringiensis* the Cry toxins are in inactivated forms. Activation of the toxin in *B. thuringiensis* normally requires a reaction in the insect gut, which is an unnecessary step for the *Bt* crops.

Recent research using sophisticated methodologies has indicated that non-Lepidopteran, nontarget effects of these activated Cry toxins could be more common than we initially believed. Hilbeck et al. (1998a, b) showed that green lacewing larvae, a species very distantly related to Lepidoptera were killed by activated Cry toxin. Previous work had not found significant effects of *B. thuringiensis* on non-Lepidopteran nontargets, although the methodologies in previous studies were not as sophisticated as the Hilbeck et al. studies. Unpublished work by the same group of investigators has shown that the toxins also kill a ladybird beetle, *Coccinella septempunctata*, which is also not closely related to Lepidoptera.

These results raise three related risk issues, one of which will be covered in the next section on resistance evolution. Both of the nontarget species are natural enemies of insect pests of maize, so reductions in their density could have a detrimental effect on natural control, leading to pest resurgence or secondary pest outbreaks in maize or nearby crops. Research on the natural enemies associated with maize in the northern corn belt suggests that lacewings and *C. septempunctata* are not major predators (Andow 1996). Moreover, while it has not been demonstrated conclusively, circumstantial evidence indicates that maize may normally be population sink for lacewings and an incidental habitat for *C. septempunctata*. Consequently, altered predation on nearby crops is not likely. Another ladybird beetle species *Coleomegilla maculata* is a common predator in maize, and if Cry toxin in *Bt* maize were to kill this species, then the assessment of risk would be considerably more difficult.

A more challenging risk to evaluate stems from the broader implications of Hilbeck et al.'s results. A broader range of non-Lepidoteran, non-target effects than had been initially anticipated could occur. A close look at their studies reveals that the activated Cry toxins probably affected the lacewings and beetles by a physiological mode of action that is different from how they affect the target Lepidoptera. Because of this result, it is also possible that other species can be affected. Moreover, research by Tapp and Stotzky (1998) in small-scale soil microcosms suggests that activated toxin will bind to clay and humic substances in soil and remain toxic for 6 months, implying that nontarget exposure could be greater than initially considered. It is not yet possible to predict which species will be affected, so these nontarget effects are not, in principle, scientifically knowable from small-scale experimental tests. These risks fall into the category of low-magnitude effects, and will be discussed more in the section on science-based risk assessment.

Nontarget Crop Loss

Bt maize can cause a novel form of nontarget crop loss. With conventional pesticides, herbicide drift was the primary cause of nontarget crops losses (Pimentel et al. 1980). In a similar way, *Bt* maize pollen can drift off the *Bt* maize field and "contaminate" nearby non-*Bt* maize. Most of a kernel of maize is composed of endosperm tissue, which is triploid, expressing two copies of the maternal genome (the plant) and one copy of the paternal genome (the pollen). If the pollen is *Bt* pollen, then the endosperm will express the *Bt* gene and produce Cry toxin. If the grower of the non-*Bt* maize does not want any Cry toxin in the grain, *Bt* pollen drift will cause contamination of the crop. A small US company producing organic corn chips for export to Europe had $65 000 of its product destroyed during 1998 because the chips were found to contain Cry toxin when they were tested at the port of entry. Although it is not certain, it is likely that the organically produced maize was contaminated by pollen drift from nearby *Bt* maize fields. Management and mitigation of this risk may be costly.

Resistance Evolution

Large-scale use of *Bt* maize will result in resistance to Cry toxin in European corn borer and southwestern corn borer and control failures. Because the ability to control pests is a public common good (Regev et al. 1976), it has become necessary to develop methods to delay the evolution of resistance using resistance management. The primary strategy for resistance management of *Bt* maize is the high-dose plus refuge strategy, which is supported by the USEPA (EPA 1997, 1998), industry (Fishhoff 1996) and public sector scientists (NC205 1998). It involves exposing a portion of the pest population to *Bt* plants with an extremely high concentration of toxin, while maintaining another part of the population in a refuge where the pests do not encounter any *Bt* toxin. By maintaining the refuges near the *Bt* maize, susceptible pests from the refuge are expected to mate with any resistant pests that survive the *Bt* maize. The offspring from these matings are assumed to be unable to survive on *Bt* maize.

Assumptions for Resistance Management

Population genetic theory predicts that this strategy will succeed at delaying resistance evolution if three essential assumptions and two additional assumptions are met (summarized in NC205 1998). (1). Major resistance alleles must be sufficiently rare so that nearly all such alleles will be found in heterozygous individuals. A heterozygous individual has only one copy of the resistance allele and is referred to as an *RS* heterozygote. An allele frequency of ≤ 0.001 is necessary for the strategy to have some success. (2). Major resistance alleles must be nearly recessive. This means that the *RS* heterozygote will have very low survival on *Bt* maize. *RS* survival rates that are less than 5% of the *RR* survival rate on *Bt* maize are needed for the strategy to have some success. (3). Non-*Bt* maize refuges are needed to provide susceptible pests to mate with the *RR*-resistant ones so their offspring will be *RS* heterozygotes. Specifically, there must be random mating between *RR* individuals from *Bt* maize and the susceptible ones from the refuge. While these are the critical assumptions, most of the theoretical models also assume that the pest population exhibits local random mating and no regional genetic isolation. These two additional assumptions are also necessary for the strategy to be successful.

Present scientific information on the frequency of resistance alleles in European corn borer suggests that resistance management may be possible. Although rigorous estimates of resistance frequency have not yet been conducted, present scientific information suggests the frequency of resistance alleles may be $\leq 10^{-3}$ in parts of Minnesota and Iowa (Andow et al. 1998, 1999, Andow and Alstad 1999). Under a worst-case scenario, these allele frequency estimates are low enough to delay resistance for a few generations, but not for very long. Better estimates are needed to more rigorously verify this assumption of the high-dose plus refuge resistance management strategy.

The survival rate of *RS* heterozygotes is unknown because major resistance alleles have not yet been found and characterized. Until resistance alleles are found and characterized, it can only be assumed that *RS* heterozygote survival is low enough (i.e., recessive enough) to enable effective resistance management. Two lines of indirect evidence indicate that resistance may be recessive, or if dominant, extremely rare. Work on other insects has shown that resistance to *B. thuringiensis* is recessive (McGaughey and Beeman 1988; Tabashnik et al. 1992; Gould et al. 1997), and despite searches for resistance in field populations of European corn borer, no resistance has been found. The US-EPA (1998) and Gould and Tabashnik (1998) have proposed an operational criterion for inferring that resistance is recessive, but this is misleading, because it implies that resistance is recessive when there is no direct evidence to substantiate the conclusion. Recent evidence that dominant partial resistance is common in European corn borer (Huang et al. 1999) suggests that the assumption of recessive resistance may not be true. It will be essential to investigate this assumption more thoroughly and to implement a resistance management strategy that can be modified should a dominant resistance be discovered (Andow and Hutchison 1998).

Most European corn borers move short distances from adult emergence to egg laying. This is especially true of the second generation in a year (Andow DA and Alstad DN, unpubl data, Hunt et al. unpubl data). These results and other anecdotal evidence suggest that all *Bt* maize plants should be within 800 m of a non-*Bt* maize refuge so that random mating is likely. Additional research on pre- and postmating movement distances is important to directly confirm this assumption.

European corn borer populations exhibit local nonrandom mating. Nonrandom mating in local populations can lead to faster evolution of resistance because *RR* homozygotes are more likely to mate with each other and fewer *RS* heterozygotes will be produced. Local nonrandom mating can be measured by the F_{is} statistic. When $F_{is} = 0$, local mating is random and $F_{is} > 0$ implies that fewer *RS* heterozygotes are produced than under random mating. Electrophoretic analysis of three variable genetic isozyme markers revealed that 15 of 45 European corn borer samples collected from 40 North American localities have $F_{is} > 0.27$, a significant amount of nonrandom mating (Alstad et al. unpubl data). In addition, this electrophoretic analysis revealed that European corn borer populations exhibit significant regional genetic isolation (Alstad DN et al. unpubl data). These results demonstrate high levels of genetic isolation (less than one migrant exchange per generation) between locations separated on average by 300 km. Both findings suggest that resistance will evolve faster than predicted by most models.

Consensus and Unresolved Scientific Issues

Research by public sector scientists (NC205 1998) has resulted in a policy consensus around using the high-dose plus refuge strategy to manage resistance in *Bt* maize and that the refuge must be at least 20% non-*Bt* maize. Several controversial scientific issues remain to be resolved.

First, it is clear that some of the *Bt* maize varieties do not produce a high dose of Cry toxin throughout the life cycle of European corn borer. Specifically, varieties based on Event 176 and DBT-418 do not express the toxin at high doses during the second corn borer generation (after pollen shedding). For both transformation events, toxin concentration declines sufficiently that *SS*-susceptible insects have a moderately high survival rate a few weeks after pollen shedding. This implies that some time between pollen shedding and a few weeks later, *RS* heterozygotes are likely to have a high survival rate relative to the *SS* types, violating the high dose assumption. Scientifically based strategies for managing resistance for these nonhigh dose events have not been developed.

Another problem is how to manage the non-*Bt* maize refuges. US growers believe that they will suffer yield losses in the refuge, and will usually try to minimize the area of refuge to maximize returns on their investment. Their view, however, is overly simplistic. Mortality to corn borers in the *Bt* maize will create local reductions in pest density so that refuges planted near the *Bt* maize will suffer less damage (Alstad and Andow 1996). This localized effect is called a halo effect, and when added together across many fields in a landscape, should result in an area-wide suppression of pest density. When these effects occur, the cost of the refuge could be reduced substantially, enabling growers to plant larger refuges economically. US growers may also want to spray the refuge with insecticides to reduce pest yield losses. The effect of these sprays on resistance evolution is not well understood. This problem will need additional investigation until a sound science-based policy can be developed.

The scientific basis for monitoring is being developed, but considerable disagreement remains about how to monitor, when to monitor, where to monitor, and how to respond to the monitoring information. The aim of monitoring is to provide timely information to trigger alternative management actions (Andow and Hutchison 1998). Larval-based sampling methods and grower-survey methods are too crude to establish anything but that resistance management has failed and resistance is very prevalent. These methods will be useful for documenting the need to join the pesticide treadmill and switch to a different means of pest control (Andow and Alstad 1998, Andow and Hutchison 1998). Two significantly more sensitive monitoring methods have been developed, the F_2 screen (Andow and Alstad 1998) and in-field sweet corn sentinel plots (Andow and Hutchison 1998, Venette RC, Hutchison WD, Andow DA, submitted). The cost of these methods could be less than about $250 000 per year each in the US (compared to over $4 billion per year maize seed sales in the US). The F_2 screen can estimate recessive resistant allele frequency to about ±0.001 precision, and the sweet corn sentinel plots have a precision of about ±0.005 to 0.01. In contrast, larval sampling methods have a precision of ±0.1-0.15 for recessive resistant alleles. The possible management responses to these monitoring methods are still being developed.

Southwestern corn borer is a severe pest of maize in parts of Kansas and Texas. In these areas, over 80% of the maize area is sprayed annually with insecticides to control this pest. Under these conditions, nearly all growers will want to spray the refuge every year, and resistance is expected to evolve rapidly in both European and southwestern corn borer. The resistance management strategy will need to be modified for this region, but the scientific evidence has not yet become available to guide these modifications.

Natural enemies may accelerate or delay resistance evolution (Gould et al. 1991), depending on whether they concentrate their attack on resistant insects or susceptible ones. Empirical studies in tobacco, potato and cotton suggest that some natural enemies will probably delay and other will accelerate resistance. These effects are presently being studied in *Bt* maize. If natural enemies that help delay resistance are killed directly by Cry toxins (Hilbeck et al. 1998a, b), this would increase the risk of resistance.

In conclusion, resistance management has been strongly influenced by public sector scientific research, and a science-based, adaptive resistance management policy is being developed in the US. Unresolved scientific issues abound, but most of these are under active investigation, and it is likely that new scientific findings will be incorporated in this developing policy during the next few years.

Science-Based Risk Assessment

Scientists do not make policy, but scientific evidence can weigh significantly in the formulation of policy, especially related to risk assessment. The weighing of risk is not done on an objective scale, but is contextualized in real social environments, which determine the scale used to measure risk. Several factors greatly affect the role of science in the measuring of these risks.

Precautionary Principle

The precautionary principle has been stated in many ways, but it fundamentally captures the folk wisdom in the aphorism "look before you leap." In its present form, it says that when an activity raises threats of harm to human health or the environment, precautionary measures should be taken even if some cause and effect relationships are not fully established scientifically. The principle suggests that society should avoid risks that are both irreversible and unmitigable, i.e., risks that cannot be eliminated or mitigated once they first occur. Under the precautionary principle, the burden of proof shifts slightly and it is necessary to show that irreversible, unmitigable risks are acceptably small, rather than having to show that a risk is likely. Scientific evidence is essential for documenting that the risks are not irreversible and unmitigable. Present scientific evidence suggests that resistance evolution is irreversible and unmitigable. Investigations on monitoring are critical to develop the flexibility to manage resistance as it occurs. Nontarget effects on monarchs or rare or endangered species of Lepidoptera may be irreversible and unmitigable, depending on the magnitude of the effect. These potential risks may require additional investigation into methods for mitigating the risk. Nontarget crop risks may require additional regulatory oversight. Nontarget, non-Lepidoptera risks are the most difficult to address because they are likely to be a low magnitude risk.

Low Magnitude Risk

Two types of low-magnitude risk can be distinguished: those that have a small probability of occurrence but potentially large impacts (some nuclear power risks fall into this category), and those that may occur frequently but have a small effect.

Both present similar scientific difficulties because they have a component that is small and difficult to measure. For example, suppose a transgenic crop reduces yield by 1%. For maize, this could cost \$3.00 per acre or \$1500 for a Midwest US farm, which is a significant loss. Because yields frequently vary by as much as 10% due to other environmental factors, statistically detecting a 1% yield drag would require several hundred carefully controlled replicate experiments. For economically important characters such as yield such massive replication is considered economically feasible, indeed necessary; but for environmental effects, such replication is not routinely done. Consequently, scientific evidence usually cannot indicate that small, but significant environmental effects are unlikely.

This issue is particularly acute for transgenic crops such as *Bt* maize. In general, transgenic crops can cause environmental effects related to the direct and indirect effects of the transgenic traits, including the selectable marker alleles. All of the topics discussed in this chapter are direct or indirect environmental effects of the Cry toxin trait. Human health effects have not been discussed. Effects of the selectable marker allele, either herbicide tolerance or antibiotic resistance, have not been discussed. For *Bt* maize in the US, escape of the transgene via hybridization with native plants has also not been considered, because no other plant species occur in the US that can hybridize with maize. Three other classes of effects, however, are theoretically possible, all of which are likely to be small effects when they occur. These are pleiotrophic effects (i.e., effects of the transgene on other phenotypic traits in the plant), epistatic effects (i.e., interactions of the transgene with particular alleles at other loci in genome to alter some phenotypic traits), and genotype by environment interactions (i.e., effects of the transgene that depend on the environment the plant is grown in). Scientists must be perfectly clear for policy makers that these entire classes of risks cannot readily be evaluated in small-scale experiments, and will become observable only when the plants are grown on a commercial scale.

This creates a gap in the scientific evaluation of the potential risks of transgenic crops and presents a paradox. Not all risks can be scientifically evaluated prior to commercialization. Thus, a regulatory decision to allow commercialization of a transgenic crop cannot be construed to be a determination of safety. Given this, how can the safety of these crops be ensured? Clearly, regulatory oversight cannot end with a commercialization decision. Additional oversight will be needed to ensure safety. It remains a significant challenge to develop these systems of postcommercial oversight that balance the costs of such oversight with the need to ensure the safety of the transgenic crops.

Acknowledgements. This work was funded in part by USDA Regional Research Committee NC205 and is publication 99117-0017 of the Minnesota Agricultural Experiment Station.

References

Alstad DN, Andow DA (1996) Implementing management of insect resistance to transgenic crops. AgBiotech News Inf 8:177-181

Andow DA (1996) Augmenting natural enemies in maize using vegetational diversity. In: Kiritani K (ed) Biological pest control in systems of integrated pest management. Food and Fertilizer Technology Center, Taipei, Taiwan, pp 137-153

Andow DA, Alstad DN (1998) The F_2 screen for rare resistance alleles. J Econ Entomol 91:572-578

Andow DA, Alstad DN (1999) Credibility interval for rare resistance allele frequencies. J Econ Entomol, 92:755-758

Andow DA, Hutchison WD (1998) Bt-Corn resistance management. In: Mellon M, Rissler J (Eds) Now or never: serious new plants to save a natural pest control. Union of Concerned Scientists, Washington, DC, pp 19-66

Andow DA, Alstad DN, Pang YH, Bolin PC, Hutchison WD (1998) Using an F_2 screen to search for resistance alleles to *Bacillus thuringiensis* toxin in European corn borer (Lepidoptera: Crambidae). J Econ Entomol 91:579-584

Andow DA, Olson DM, Hellmich RL, Alstad DN, Hutchison WD (1999) Frequency of resistance alleles to *Bacillus thuringiensis* toxin in an Iowa population of European corn borer. J Econ Entomol, in press

Armstrong CL, Parker GB, Pershing JC, Brown SM, Sanders PR, Duncan DR, Stone T, Dean DA, DeBoer DL, Hart J, Howe AR, Morrish FM, Pajeau ME, Petersen WL, Reich BJ, Rodriquez R, Santino CG, Sato SJ, Schuler W, Sims SR, Stehling S, Tarochione LJ, Fromm ME (1995) Field evaluation of European corn borer control in progeny of 173 transgenic corn events expressing an insecticidal protein from *Bacillus thuringiensis*. Crop Sc. 35:550-557

Arpaia S, Mennella G, Onofaro V, Perri E, Sunseri F, Rotino GL (1997) Production of transgenic eggplant (*Solanum melongena* L.) resistant to Colorado potato beetle (*Leptinotarsa decemlineata* Say). Theor Appl Genet 95:329-334

Carozzi NB, Warren GW, Desai N, Jayne SM, Lotstein R, Rice DA, Evola S, Koziel MG (1992) Expression of a chimeric CaMV 35S *Bacillus thuringiensis* insecticidal protein gene in transgenic tobacco. Plant Mol Biol Int J Mol Biol Biochem Genet Eng 20:539-548

Environmental Protection Agency US (1997) Plant pesticides resistance management. Fed Regist 62(36):8242-8244

Environmental Protection Agency US (1998) Final report of the FIFRA Scientific Advisory Subpanel on *Bacillus thuringiensis* (Bt) Plant Pesticides and Resistance Management, Feb. 9-10, 1998. Docket Number: OPP 00231. 60 pp. http://www.epa.gov/pesticides/SAP/finalfeb.pdf

Fischhoff DA, Bowdish KS, Perlak FJ, Marrone PG, McCormick SM, Niedermeyer JG, Dean DA, Kusano-Kretzmer K, Mayer EJ, Rochester DE, Rogers SG, Fraley RT (1987) Insect-tolerant transgenic tomato plants. Bio-Technology 5:807-813

Flint HM, Henneberry TJ, Wilson FD, Holguin E, Parks N, Buehler RE (1995) The effects of transgenic cotton, *Gossypium hirsutum* L., containing *Bacillus thuringiensis* toxin genes for the control of pink bollworm, *Pectinophora gossypiella* (Saunders) and other arthropods. Southwest Entomol 20:281-308

Gould F, Tabashnik B (1998) Bt-Cotton resistance management. In: Mellon M, Rissler J (eds) Now or never: serious new plants to save a natural pest control. Union of Concerned Scientists, Washington, DC, pp 67-106

Gould F, Kennedy GG, Johnson MT (1991) Effects of natural enemies on the rate of herbivore adaptation to resistant host plants. Entomol Exp App 58:1-14

Gould F, Anderson A, Jones A, Summerford D, Heckel GG, Lopez J, Micinski S, Leanard R, Laster M (1997) Initial frequency of alleles for resistance to *Bacillus thuringiensis* toxins in field populations of *Heliothis virescens*. Proc Natl Acad Sci USA 94:3519-3523

Hilbeck A, Baumgartner M, Fried PM, Bigler F (1998a) Effects of transgenic *Bacillus thuringiensis*-corn-fed prey on mortality and development time of immature *Chrysoperla carnea* (Neuroptera: Chrysopidae). Environ Entomol 27:480-487

Hilbeck A, Moar WJ, Pusztai-Carey M, Filippini A, Bigler F (1998b) Toxicity of *Bacillus thuringiensis* Cry1Ab toxin to the predator *Chrysoperla carnea* (Neuroptera: Chrysopidae). Environ Entomol 27:1255-1263

Huang F, Buschman LL, Higgins RA, McGaughey WH (1999) Inheritance of Resistance to *Bacillus thuringiensis* toxin (Dipel ES) in the European corn borer. Science 284:965-967

Jenkins JN, McCarty JC Jr, Buehler RE, Kiser J, Williams C, Wofford T (1997) Resistance of cotton with delta-endotoxin genes from *Bacillus thuringiensis* var. *kurstaki* on selected lepidopteran insects. Agron J 89:768-780

Koziel MG, Beland GL, Bowman C, Carozzi NB, Crenshaw R, Crossland L, Dawson J, Desai N, Hill M, Kadwell S, Launis K, Lewis K, Maddox D, McPherson K, Meghji MR, Merlin E, Rhodes R, Warren GW, Wright M, Evola SV (1993) Field performance of elite transgenic maize plants expressing an insecticidal protein derived from *Bacillus thuringiensis*. Biotechnology 11:194-200

Losey JE, Rayor LS, Carter ME (1999) Transgenic pollen harms monarch larvae. Nature 399:214

McGaughey WH, Beeman RW (1988) Resistance to *Bacillus thuringiensis* in colonies of Indian meal moth and almond moth (Lepidoptera: Pyralidae). J Econ Entomol 81:28-33

Metz TD, Dixit R, Earle ED (1995a) *Agrobacterium tumefaciens*-mediated transformation of broccoli (*Brassica oleracea* var. *italica*) and cabbage (*B. oleracea* var. *capitata*). Plant Cell Rep 15:287-292

Metz TD, Roush RT, Tang JD, Shelton AM, Earle ED (1995b) Transgenic broccoli expressing a *Bacillus thuringiensis* insecticidal protein: Implications for pest management strategies. Mol Breed 1:309-317

NC-205 (1998) Supplement to Bt Corn and European Corn bores: long-term success through resistance management. NSDA North Central Regional Research Committee.

Perlak FJ, Stone TB, Muskopf YM, Petersen LJ, Parker GB, McPherson SA, Wyman J, Love S, Reed G, Biever D (1993) Genetically improved potatoes: protection from damage by Colorado potato beetles. Plant Mol Biol 22:313-321

Pimentel D, Andow D, Dyson-Hudson R, Gallahan D, Jacobson S, Irish S, Kroop M, Moss A, Scheiner I, Thompson M, Vinzant B (1980) Environmental and social costs of pesticides: a preliminary assessment. Oikos 34:126-140

Regev U, Gutierrez AP, Feder G (1976) Pests as a common property resource: a case study of alfalfa weevil control. Am J Agric Econ 58:186-197

Stewart CN Jr, Adang MJ, All JN, Raymer PL, Ramachandran S, Parrott WA (1996a) Insect control and dosage effects in transgenic canola containing a synthetic *Bacillus thuringiensis* cryIAc gene. Plant Physiol 112:115-120

Stewart CN Jr, Adang MJ, All JN, Boerma HR, Cardineau G, Tucker D, Parrott WA (1996b) Genetic transformation, recovery, and characterization of fertile soybean transgenic for a synthetic *Bacillus thuringiensis* cryIAc gene. Plant Physiol 112:121-129

Tabashnik BE, Schwartz JM, Finson N, Johnson MW (1992) Inheritance of resistance to *Bacillus thuringiensis* in diamondback moth (Lepidoptera: Plutellidae). J Econ Entomol 85:1046-1055

Tapp H, Stotzky G (1998) Persistence of the insecticidal toxin from *Bacillus thuringiensis* subsp. *kurstaki* in soil. Soil Biol Biochem 30:471-476

GMO Releases in the Environment

JE. BERINGER[1]

It has been obvious for at least the last 15 years that genetically modified organisms would be released into the environment when used commercially. Why, then, are we in Europe in the middle of a period of intense debate about the safety of such releases? In part, the answer must be that we have foolishly decided to stigmatise biotechnology by requiring regulatory control solely on the basis that something is a GMO. Also it is very clear that the European Commission in particular greatly underestimated the growth of the technology and has been totally unprepared to handle public disquiet and the need for efficient and effective regulation. Regrettably, none of these problems is close to resolution, and indeed with the redrafting of Directive 90/220 they appear to be getting much worse. We must live with the consequences of ineptitude, so there is a need to consider why in Europe there is so much concern about environmental risks.

Before discussing what the risks might be it is important to ask the question, what is the environment? In Europe, unlike much of N America and Australia, most of the countryside is the product of human intervention. Over the past few thousand years we have removed most of the natural vegetation and provided agricultural environments in which many of the species we want to preserve are present primarily because they are well adapted to coexisting with agriculture. Or at least they were until the middle of this century, when countries in Western Europe started to intensify food production to provide sufficient food. The success of the intensification of agriculture is beyond doubt in that we have moved from severe food shortages to embarrassing surpluses. All changes in agricultural practice will inevitably affect those species that depend on existing practices, so changes to increase, or decrease, intensification change biodiversity. For example, the movement from spring to autumn sowing of crops has had an obvious detrimental impact on those species that require autumn stubble, and a positive impact on species that benefit from crops sown in the autumn. Likewise, the decline in consumption of meat and dairy products is leading to degradation of pastures and declines in many species of plants, insects, and birds.

What relevance does this have for GMOs? I believe that an understanding of the relationship between existing changes in agriculture and concerns about the release

1. School of Biological Sciences, University of Bristol, Woodland Road, Bristol, BS8 1UG, UK

of GMOs is critical if we are to tackle the problem of how to regulate releases and obtain consumer acceptance. The release of herbicide-tolerant crops provides a good example of the type of difficulty we face. Herbicides are an integral part of good farming practice, and even organic farmers are allowed to use manual weed control; but weeds are part of our biodiversity, are food for animals, and some are considered attractive in their own right. Herbicide-tolerant crops offer the opportunity to control weeds more effectively than at present, and are used by pressure groups as an example of how biotechnology is working to increase the intensification of agriculture. It is very unfortunate that the debate to date on herbicide-tolerant crops has generated far more heat than light. My dispassionate interpretation is that their use should allow farmers to reduce the number of herbicide applications and allow weeds to be present in crops for longer than under present regimes. To date, the herbicides involved are broad host range, ensuring that, when used, virtually all weeds are killed. Do we have any idea of the possible impact of losing all plants on those organisms feeding on them? Might we start to cause significant declines in the populations of many weed species, which will not set seed when sprayed with broad-spectrum herbicides? Such questions may seem trivial for countries in which a significant proportion of the landscape is non-agricultural and thus loss of biodiversity on farmland is not of great consequence. Suggesting that weed control should be discouraged seems inhuman and unreasonable from the point of view of agriculture in countries suffering from food shortages. However, in Europe, excess food production, wealth, and privilege mean that people have both time and resources to enjoy their environment. It is hard to argue that if society desires greater biodiversity it should not have it.

If indeed herbicide-tolerant crops do lead to increased intensification, countries that wish to enhance biodiversity will have great difficulty in approving the unrestricted use of such crops. Stringent controls that could be introduced because of a lack of adequate data on long-term impacts on biodiversity may well solve immediate political problems, but could in the future prevent us from benefiting from the effective use of GM crops to enhance the effectiveness of integrated pest management.

Insect resistance is another development that is likely to cause as many problems as it solves. Clearly, if a plant is naturally resistant to insect pests, there will be no need for routine applications of pesticides, with corresponding environmental and health advantages. From the point of view of the birds and other predators that feed on insects, the widespread cultivation of such plants will lead to considerable problems and surely to major declines in populations. We cannot, and should not, prevent insect control, but how will we in future ensure that it is not too efficient? Unfortunately for biodiversity, birds and other predators are not restricted to eating only the insects we call beneficial, so all losses are important. In passing, is it realistic to compare the impact of a GM insect-resistant crop with the impact of very efficient pesticide treatments, or on the absolute scale of how many insects such crops will allow to develop? The latter is the trend at present, but is hardly enabling GM crops to compete on a level playing field.

Concerns about insect-resistant and herbicide-tolerant crops will pale into insignificance when stress-tolerant crops come before regulatory authorities, and when applications are made to release genetically modified fish. The former have enormous potential to increase food productivity in some of the most impoverished areas of the world, and, in the case of frost tolerance, would be of great advantage to temperate agriculture. However, the potential to convert the crops concerned into invasive weeds would appear to be very great; would it not? For fish, our poor understanding of ecology will make risk assessments very difficult, particularly when the introduced genes are for enhanced growth rate or tolerance to cold. I am concerned that the near-hysteria in Europe may lead to a position in which very desirable crops for the developing world will not be developed, or will be developed very slowly, because of our probably very greatly exaggerated fears of possible environmental harm.

Another important problem facing those regulating releases is that of gene transfer from these crops. We have done a very poor job in educating the many critics of GM crops, and two important misconceptions have developed. The first is that cloned genes will behave differently from other genes and will be transferred among different species in ways that are not contemplated for other genes within the host genomes. The second results from a complete lack of understanding of natural selection. When these misconceptions are combined, one has the absurd situation that many people are convinced that not only can any cloned gene enter any host, but that such genes will also spread rapidly and widely within recipient populations.

A particular problem in the UK at present, that arises from concerns about gene transfer, is the decision made by the Soil Association that genetic modification has no place in organic farming and thus no level of cross-pollination is acceptable if produce is to be sold as organic. We know that pollen can be transported at least 5 km by wind or bees, and thus there is virtually no chance that organic crops could be grown if corresponding GM crops are used widely in UK agriculture, or elsewhere if this "organic standard" becomes widespread. A recent UK Appeal Court decision that internationally determined distances for seed certification provide for an acceptably low level of contamination was not accepted by the organic lobby, which nevertheless accepts a degree of chemical spray contamination. A resolution to this problem is needed urgently, particularly as it reinforces the perception that unsafe products of GM technology are being forced upon the public without controls in place to ensure that consumers who choose not to purchase GM foods can do so.

Where do we go now? I believe that there is an absolute need to wean regulators, governments, and the public from the misconception that GM products are intrinsically harmful. It is essential that industry becomes much more aware of existing concerns about the environmental impact of existing agricultural practices and introduces GMOs with appropriate and sensible assessments of benefits and disadvantages. There is no point at all in saying that a GM crop is no more harmful to wildlife than existing crops and treatments if countries are attempting to enhance biodiversity. Likewise, why is it that we hear so little about the potential for GMOs to enhance our ability to tackle environmental problems? How long will it be until we see the commercial release of GMOs whose use will encourage organic growers to concede that GMOs can have a very positive role in sustainable agriculture? I regret that the likely answer is: far too long; and that, as a result, the opponents to GMOs will continue to have the almost unrestricted freedom they presently have to "warn the public that they carry the risk and see none of the benefits".

Part VI
Ecosystems

James SCHEPERS

D. WALL and J. LYNCH
E. MALTBY et al.
H. KAWASHIMA et al.

Soil Biodiversity and Ecosystem Functioning

DH. WALL[1], JM. LYNCH[2]

The wealth of soil biodiversity is poorly known on both local and global scales, and thus the importance of soil organisms (microbes to invertebrates) has been largely disregarded in assessments of biodiversity consequences for ecosystem functioning. The huge abundance and numbers of species in soils as well as the diversity of methods required for sampling, extraction and identification of the organisms, have made assessments of diversity of all taxa at even one location extremely difficult. Despite that, we suggest that a better understanding of soil biodiversity and its relationship to critical ecosystem services is vital for planning for sustainable soils for the long term. A multitude of microbes, protozoa, invertebrates and vertebrates interact to maintain soil fertility, decay, recycling of organic matter, the breakdown of hazardous wastes, cleansing of water, the biological control of agricultural and human pests and the composition of the atmosphere. However, we have yet to identify which species play key roles in these processes, even those that may most directly and immediately further the human enterprise through influencing carbon sequestration, erosion control, the quality of food supply and biocontrol of pests. Much of the functioning in soil is carried out by communities of organisms and we need to know more about their structures. Urgency is added to this task because humans, through land use change and pollution, are impacting the biodiversity and functioning of soil organisms as well as that of organisms in interconnected habitats such as groundwaters, riparian and wetland systems, and aboveground habitats. The biotic groups and the ecosystem services we need to understand cross international boundaries and expertise on them is internationally distributed. International efforts are required to (a) synthesize available information on global soil biodiversity and functioning across ecosystems and (b) develop integrated research approaches and new technologies necessary to examine soil biocomplexity, if we are to acquire the information necessary to sustainably manage the soil as a primary natural resource.

1. Natural Resource Ecology Laboratory, Colorado State University, Fort Collins, Colorado 80523, USA
2. School of Biological Sciences, University of Surrey, Guildford, Surrey, GU2 5XH, UK

Soil Biodiversity – an Undervalued Resource

In 1937, Franklin D. Roosevelt, President of the USA, said: "A nation that destroys its soil, destroys itself" (Roosevelt 1937). History has shown that all nations have experienced significant land degradation resulting in a loss of an irreplaceable and primary natural resource, soil. With the Earth's accelerating increase in the human population, a sustainable future is dependent on international cooperation in prioritizing and developing the programs of interdisciplinary scientific research and the practical knowledge necessary to sustain our soils globally. We must learn what factors are critical to maintaining soil quality, assess the vulnerability of our soils to local, regional and global disturbances, and continue to integrate this knowledge into long-term maintenance of soil quality and for soil bioremediaton and reclamation.

Biodiversity is a major area of international research, because the diversity of individual genes, species and ecosystems provide society with a wide variety of extractable (e.g., food, fiber) and non-extractable (pollination, pest control) goods and services (Daily 1997). However, biodiversity research programs in soils have frequently been neglected, even though, for example, breeding programs to enhance crop productivity are dependent not only on the species diversity of crop plants, but also on the assemblages and interactions of soil invertebrates and microorganisms that maintain soil fertility. This lack of attention may be because many consider soil to be "dead" and a buffer or "dump", as oceans were once considered a "dump" for garbage and pollution. As with oceans, research has shown that soils have a tremendous wealth and abundance of species, yet to be explored, but which provide numerous services for humans (Wall and Virginia1999).

The Diversity of Life in Soil

Every taxonomic phylum that exists aboveground is represented belowground (Fig. 1) but species diversity is poorly known compared to above-ground habitats (Wall and Virginia 1999). In the USA alone, there are 12 000 types (or series) of soils that, combined with variation in climate, geology, vegetation and soil chemistry, provide millions of habitats for soil species (USDA 1990). The organisms in soil (microbes, protozoa, invertebrates and vertebrates) have an enormous range in size from microns to meters in length, widely differing life cycles and food preferences, and occur in vastly different habitats at temporal and spatial scales. Some organisms, such as nematodes, are aquatic and confined to water films around soil particles, while others, such as ants, live in air spaces between soil particles and migrate through meters of soil. Some species of bacteria are found only in the rhizosphere, while others occur in bulk soil (Lynch 1990, Beare et al. 1995). Earthworm species, critical to the maintenance of soil structure and hydrology, differ in their habitat preference, some living in surface soils and moving horizontally, and others preferring deeper soils (Blair et al. 1995). Taxonomic expertise is necessary to identify each group of organisms, particularly at the level of species or the molecular level, but worldwide, that expertise has declined substantially. Identification of soil biota and knowledge of their ecology is also hindered by their sheer abundance. In 1g of soil there can be more organisms than humans on Earth: 100 million bacteria, a million actinomycetes, 100 000 fungi, 10 000 microalgae, 100 000 protozoa, 100 nematodes and 100 000 other invertebrates, all representing a multitude of species (Dindal 1990).

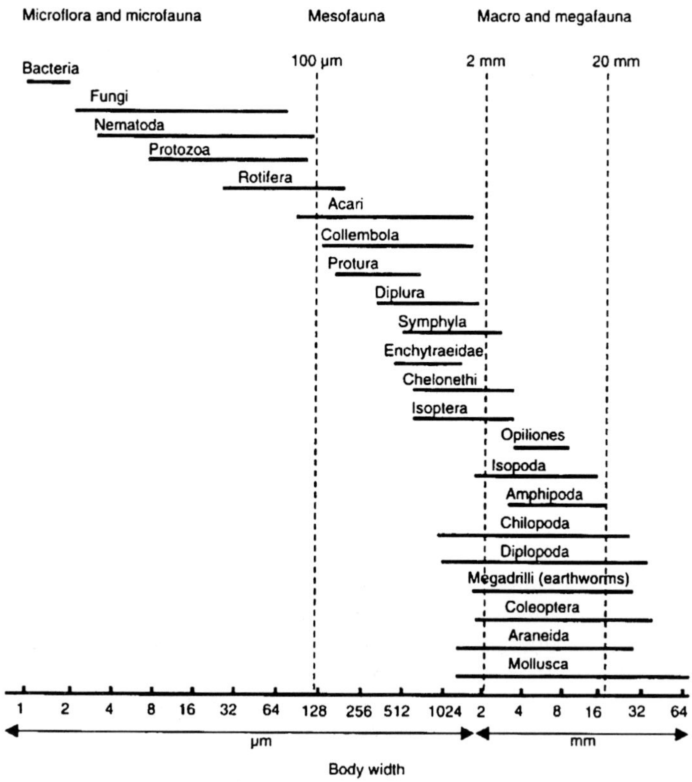

Fig. 1: Taxonomic phyla in decomposer foodweb in soils, ordered by body width. (Swift et al. 1979)

The diversity of species, their abundance, and the few taxonomic specialists and ecologists studying soil organisms and their interactions, has contributed to a taxonomic impediment, such that the identity of species necessary for soil ecosystem processes is largely unknown. However, the role of large numbers of species in the functioning of ecosystems has been inferred by identifying the role of a few species and assuming that species with similar morphology have the same role and are in the same functional group. For example, in a food web, the microbial nitrifier functional group can consist of thousands of genotypes of microbes. The microarthropod group of oribatid mites has a significant number of both unidentified and known species of similar morphology lumped within the functional group of fungal feeders. Research at the functional group level has shown that soil biota are critical for maintaining ecosystem processes that provide benefits or services to humans (e.g., maintenance of soil fertility, decomposition and recycling of organic matter, regeneration and renewal of soils, bioremedation, cleansing of water, biological control of agricultural and human pests, and the composition of the atmosphere.

Information on species in soil that are key to ecosystem processes has been very poor and has primarily been derived from studies in agriculture and other managed ecosystems (grasslands, forests). This is despite the fact that, for many years, individual species or interactions between species have been known to have important effects in agroecosystems. The economic loss of crops due to single pathogenic species has firmly established the importance of soil species in reducing net primary productivity (crop yields), a major ecosystem process. International quarantine efforts have documented numerous examples of 'invasive' soil pest species and their impact on productivity in managed and unmanaged lands. Commercial use of environmental biotechnology in agriculture has relied also on single species for genetically modified organisms (De Leij et al. 1999). The emphasis on single species for natural biological control or for commercial biotechnology should be reconsidered in light of evidence showing that ecosystem processes are governed by interactions of species within functional groups, not species in isolation. Research that explains how species within functional groups interact to affect ecosystem processes such as decomposition and nutrient cycling, and which species are key to those processes, can be potentially rewarding for managing our soils for the future.

How Does Soil Biodiversity Affect Ecosystem Processes?

Our knowledge of soil organisms tends to be focussed on detrimental species and not on the communities of beneficial invertebrates and microbes that might be threatened by land use change, pollution and global change. For example, one group of microscopic invertebrates, nematodes, known globally for the economic damage to agroecosystems caused by its plant parasitic species, has many more species involved in decomposition than in plant parasitism (Baldwin et al. 1999). Of the five functional groups (bacterial feeders, fungal feeders, omnivores, predators, and plant parasites) of nematodes, four are beneficial and involved in decomposition of organic matter (Yeates et al. 1993).

Table 1. Comparison of carbon storage and relative amounts of organic matter content in soils and sediments

		Sediments	
	Soils	Freshwater	Marine
Global coverage	1.2 × 108 km^2	2.5 × 106 km^2	3.5 × 108 km^2
Carbon storage	1500 Gt	0.06 Gt	3800 Gt
Organic content	High	Low	Low

On a global scale, decomposition of organic matter is one of the most important and intensively studied ecosystem processes that involve soil biota. Soils and sediments store the bulk of the world's carbon (Table 1), but the relative organic matter content is greater in soils. The organic matter enters soils through activity of decomposer organisms, and influences soil structure and the formation of aggregates, influences hydrology, contributes to the prevention of erosion, and affects the habitat for its multitudinous inhabitants. Soil organic matter also provides for retention of mineral nutrients necessary for plant growth. Without all the decomposition activities of the soil biota, the landscape might be several meters high in dead animals and plants and there would is limited recycling of nutrients for new plant growth and life. We may think that much of plant life supports aboveground animals through the herbivory food chain, but, the majority of terrestrial ecosystems have more than 90% of their Net Primary Productivity (NPP) directly returned to the soil for decomposition (Table 2), with the remainder entering the decomposition food chain as dead animals or feces. Even in agroecosystems, where humans harvest plant productivity for food, and in grasslands, where NPP is consumed by herbivores, 50% of NPP enters the soil directly. The high amount of organic matter in unmanaged systems suggests that the majority of the soil biota is involved in the breakdown of organic matter into nutrients for plant growth. Changes in management such as from a high intensity, high soil pesticide use in conventional monoculture agriculture, to no-tillage systems increases the amount of soil organic matter and the composition of the soil biota involved in decomposition (Bardgett and Cook 1998). The quality (chemical composition) of the plant organic matter (in conjunction with climate and soil chemical and physical properties) determines the species assemblages in decomposer foodwebs and the rate of decomposition. Research can determine whether, in a given agroecosystem, the species composition, community structure and their interactions can return nutrients for plant growth efficiently enough to sustainably support no-till agriculture.

Table 2. Net primary production (NPP) and the proportion that directly enters the soil detrital system for the world's terrestrial ecosystems

Ecosystem type	World NPP (dwt. 10^9 tons year^{-1})	% of NPP to detritus	World detritus (dwt. 10^9 tons year^{-1})
Tropical forest	40.0	95	38.0
Temperate forest	23.4	95	22.2
Boreal forest	9.6	97	9.3
Temperate grassland	4.5	50	2.2
Desert scrub	1.3	95	1.2
Extreme desert	0.07	97	0.1
Agricultural land	9.1	50	4.6

Decomposition occurs through the interactions of many organisms and does not occur quickly. Numerous species participate in the ripping, shredding, tearing, and digesting of the chemical components of a leaf or twig, and at each step complex compounds are broken down into inorganic labile or recalcitrant organic compounds. The many organisms and steps involved in decomposition in unmanaged systems is bypassed in conventional agriculture by the use of fertilizers, which when used inappropriately resulted in increased in NO_X fluxes to the atmosphere (Matson et al. 1998) and excess nitrates in groundwater with effects on neighboring and distant ecosystems. Learning how to exploit soil diversity involved in decomposition processes could be beneficial for reducing fertilizer application and their associated environmental and economic costs. For example, Brimecombe et al. (1999) have suggested that plants which support a larger bacterial feeding nematode population in the rhizosphere benefit from a higher availability of inorganic nitrogen, since bacterial feeding nematodes are key in accelerating, by predation of bacteria, the turnover of microbially mobilized nitrogen and organic matter (Bardgett and Griffiths

1997). At the ecosystem scale, Hunt et al. (1987) showed that bacterial feeding nematodes and protozoa are significant contributors to the 30% total net nitrogen mineralisation of an arid grassland.

Another extremely promising benefit from a greater understanding of soil biota is development of indices to assess disturbance. For instance, scientists are able to use the abundance, composition and life histories of the five nematode functional groups to indicate disturbance to soils (Bongers 1990, Freckman and Ettema1993, Bloemers et al. 1997).

The importance of entire suites of organisms for decomposition processes and the study of cohorts of organisms to assess disturbance, as described above, elegantly indicate that we cannot simply consider those organisms directly involved in a process (e.g., the nitrogen-mineralizing bacteria for production of nitrogen) as the only species of importance. Instead, research should be directed towards increasing our knowledge of how the interactions of species in soil communities and the functional groups within soil food webs contribute to ecosystem functioning.

Research and Policy Implications

Soils and soil biodiversity have been of increasing interest to policy makers and the international scientific community. The Intergovernmental Panel on Climate Change (Cole et al. 1996; Watson et al. 1996) and scientists (Heal 1997; Ingram and Wall-Freckman 1998; Wall et al. 1999) have examined factors contributing to changes in atmospheric composition and noted the urgency of determining the role of the soil biota, particularly as plants under elevated atmospheric CO_2 scenarios appear to allocate carbon to roots with trickle down effects to soil food webs. Many international scientific groups such as the Global Change in Terrestrial Ecosystems Program, the Scientific Committee on Problems of the Environment and the European Union Terrestrial Ecosystems Initiative, have made research and analyses of data on soils and soil functional groups under global change, a priority.

Examining soil biodiversity as a means of stabilizing soils and C and N flux and soil fertility in agriculture is also of importance to the Convention on Biodiversity (articles 7 and 8) (http://www.biodiv.org/chm/conv/default.htm/). The biodiversity sciences are being coordinated under the umbrella of DIVERSITAS an international program integrating biodiversity research. Soil and Sediment Biodiversity is one of six Special Target Areas of Research: areas that have been previously neglected and are cross-disciplinary issues concerning the conservation and sustainable use of biodiversity. Scientists are examining gaps in our knowledge of global patterns of biodiversity, linkages between species and ecosystem processes in subsurface habitats, consequences to ecosystem functioning when key soil species are lost, and identifying the impact of invasive soil species. These syntheses are revealing a need for research that integrates across many sciences, such as using GIS maps of soil chemical and physical factors and diversity of vegetation to predict areas of high soil biodiversity and incidence of soil pests. Research combining traditional disciplines with increasingly spectacular technologies such as the internet, molecular biology, GIS, and use of stable isotopes for species level determinations of C and N flux, will allow predictions from individual farm to regional levels.

In this chapter, we have attempted to convey the importance of examining soil biodiversity for managing the world's soils for the future. We cannot continue to disregard evidence that soil biodiversity, the many species in different functional groups, influences ecosystem processes, and that disturbances, such as land degradation, impact not only individual species but, as species interact, impact the ecosystem services they provide. An international research emphasis on identifying soil species assemblages and community interactions that influence ecosystem functioning across fields, regional and global scales, offers a new and perhaps more sustainable means for managing agricultural ecosystems for the future.

References

Baldwin JG, Nadler SA, Wall DH (1999) Nematodes – pervading the earth and linking all life. In: Raven PR, Williams T (eds) Nature and human society: the quest for a sustainable world. National Academy Press, Washington, DC (in press)

Bardgett RD, Cook R (1998) Functional aspects of soil animal diversity in agricultural grasslands. Appl Soil Ecol 10:263-276

Bardgett RD, Griffiths BS (1997) Ecology and biology of soil protozoa, nematodes and microarthropods. In: Van Elsas JD, Trevors JT, Wellington EHM (eds) Modern soil microbiology. Marcel Dekker, New York, pp 129-163

Beare MH, Coleman DC, DAC Jr, Hendrix PF, Odum EP (1995) A hierarchical approach to evaluating the significance of soil biodiversity to biogeochemical cycling. In: Collins HP, Robertson GP, Klug MJ (eds) The significance and regulation of soil biodiversity. Kluwer, Dordrecht, pp 5-22

Blair JM, Parmelee RW, Lavelle P (1995) Influences of earthworm on biogeochemistry. In: Hendrix PF (ed) Earthworm ecology and biogeography in North America. Lewis, Boca Raton, Florida, pp 1-31

Bloemers GF, Hodda M, Lambshead PJD, Lawton JH, Wanless FR (1997) The effects of forest disturbance on diversity of tropical soil nematodes. Oecologia (Berlin) 111:575-582

Bongers T (1990) The maturity index: an ecological measure of environmental disturbance based on nematode species composition. Oecologia 83: 14-19

Brimecombe MJ, De Leij FAAM, Lynch JM (1999) Effect of introduced *Pseudomas fluorescens* strains on soil nematode and protozoan populations in the rhizosphere of wheat and pea. Microb Ecol (submitted)

Cole CV, Cerri C, Minami K, Mosier A, Rosenberg N, Sauerbeck D (1996) Agricultural options for mitigation of greenhouse gas emissions. In: Watson RY, Zinyowera MC, Moss RH (eds) Climate change 1995: impacts, adaptations and mitigation of climate change: scientific-technical analyses contribution of working group II to the Second Assessment Report of the Intergovernmental Panel on Climate Change. Cambridge University Press, Cambridge

Daily GC (1997) Natures services. Societal dependence on natural ecosystems. Island Press, Washington, DC

De Leij FAAM, Hay DB, Lynch JM (1999) Natural investment in biodiversity. In: Raven PR, Williams T (eds) Nature and human society: the quest for a sustainable world. National Academy Press, Washington, DC

Dindal DL (ed) (1990) Soil biology guide. John Wiley, New York

Freckman DW, Ettema CE (1993) Assessing nematode communities in agroecosystems of varying human intervention. Agricult Ecosyst Environ 45:239-261

Heal OW (1997) Effects of global change on diversity-function relationships in soil. In: Wolters V (ed) Functional implications of biodiversity in soil. Proc workshop organized by the Department of Animal Ecology (University of Geissen) within the framework of the TERI science Plan. European Commission, Brussels http://www.biodiv.org/chm/conv/default.htm/. Convention on Biological Diversity

Hunt HW, Coleman DC, Ingham ER, Elliott ET, Moore JC, Rose SL, Reid CPP, Morley CR (1987) The detrital food web in a shortgrass prairie. Biol Fertil Soils 3:57-68

Ingram J, Wall-Freckman D (1998) Preface to: Soil biota and global change special issue. Global Change Biol 4:699-702

Lynch JM (1990) The rhizosphere. Wiley, Chichester

Matson PA, Naylor R, Ortiz-Monasterio I (1998) Integration of environmental, Agronomic and economic aspects of fertilizer management Science 280:112

Roosevelt FD (1937) Letter to the Governors of all States, requesting implementation of Standard State Soil Conservation District Laws

Swift MJ, Heal OW, Anderson JM (1979) Decomposition in terrestrial ecosystems. Blackwell, Oxford

USDA. 1990 Soil series of the United States, including Puerto Rico and the U.S. Virgin Islands. Their taxonomic classification. United States Department of Agriculture, Washington, DC

Wall DH, Adams GA, ParsonsAN (1999) Soil biodiversity under global change scenarios. In: Sala O, Chapin FSI, Huber-Sannwald E (eds) Scenarios of future biodiversity. Springer, Berlin Heidelberg New York (in press)

Wall DH, Virginia RA (1999) The world beneath our feet: soil biodiversity and ecosystem functioning. in P. R. Raven and T. Williams (eds) Nature and human society: the quest for a sustainable world. National Academy Press, Washington, DC (in press)

Watson RY, Zinyowera MC, Moss RH (eds) (1996) Climate change 1995: impacts, adaptations and mitigation of climate change: scientific-technical analyses contribution of working group II to the 2nd Assessment Report of the Intergovernmental Panel on Climate Change. Cambridge University Press, Cambridge

Yeates GW, Bongers T, De Goede RGM, Freckman DW, Georgieva SS (1993) Feeding habits in soil nematode families and genera: an outline for soil ecologists. J Nematol 25:315-331

Linking Wetland Science to Policy: Meeting the challenge with special reference to water quality issues

E. Maltby, MSA. Blackwell, CJ. Baker[1]

Wetland ecosystems perform functions which give rise to goods and services for direct and indirect human use, environmental quality and biodiversity. Yet policies for wetland protection and management, especially in Europe and the developing world, have been based primarily on wildlife criteria and traditional nature conservation approaches. This chapter examines the wider functional significance of wetlands with particular reference to water quality benefits and the case for new policies to capture greater societal advantages from their protection. Examples are given at various scales of the effort to reduce the gap between scientific knowledge of how wetlands work and the new methods emerging to improve decision making and ensure stronger linkage to policies.

Introduction

Wetlands represent some of the most abused ecosystems on the planet and are posing some of the most contentious questions to both scientists and policy makers. This chapter presents some of the features and the case for the better protection and management of wetlands with particular reference to the maintenance and/or enhancement of water quality. This emphasis is simply illustrative of a wider functional significance of wetland ecosystems. Despite the growing scientific evidence for their importance to the health and welfare of human communities and wildlife as well as key aspects of environmental quality and stability, the special significance of wetlands is only gradually permeating societal decision-making. This is due, at least in part, to a lack of linkage between the science and policy instruments.

Wetlands comprise complex and varied ecosystems which occupy the transitional areas between terrestrial and aquatic systems. They occupy about 6% of the world's land surface (Maltby and Turner 1983) and occur on all continents. They include both freshwater-and saltwater-influenced systems and may comprise herbaceous vegeta-

---

1.Wetland Ecosystems Research Group, Royal Holloway Institute for Environmental Research and Department of Geography, Royal Holloway University, UK.

tion (e.g. marshes and fens), woody vegetation (e.g. cypress and mangrove swamps) or may lack higher plant cover (e.g. mudflats). More than half of the global resource is characterised by the accumulation of peat (partially decomposed plant remains which build up progressively under waterlogged conditions to depths which may exceed tens of metres). Although peat-based wetlands are most characteristically presented as the bogs, fens and muskeg of the temperate oceanic and high-latitude northern regions, they are now known to occupy extensive areas of the humid tropics where peat swamp forest accounts for perhaps 15% of the global peatland carbon store (Immirzi et al. 1992; Maltby and Immirizi 1993). The size and extreme densities (e.g. 600-1500 tC ha^{-1} within the upper 1 m, Armentano and Menges, 1986) of the carbon store make peatlands an important part of the global carbon cycle. Significant and rapid depletion and release to the atmosphere can occur as a result of drainage and alteration of the wetland conditions necessary for stabilisation of the store. Yet there is generally no explicit incorporation of a wetland or peatland component in climate change models or related policy initiatives such as the Kyoto protocol.

Wetlands have played key roles in the development of human society (Maltby 1986). Notwithstanding this, human effort, historically, has focused on the alteration of wetlands. Wetland loss and degradation continues worldwide. With the exception of the boreal zone, probably more than half of the historic area has been drained and converted to other uses, which preclude their functioning as wetland ecosystems. The United States has lost an estimated 54% (87 Mha) of its original wetland area. From the 1950s to the 1970s, losses averaged 185 000 ha $year^{-1}$, more than half from the lower Mississippi Alluvial Plain and Louisiana. Ninety nine per cent of Iowa's natural marshes have been lost since colonial times (Tiner 1984). Estimates of wetland loss in Europe are equally large as exemplified by The Netherlands 55% (1950-1985); France 67% (1900-1995); Germany 57% (1950-1985); Spain 60% (1945-1995); Italy 66% (1938-1984) and Greece 63% (1920-1991). In Britain, between the middle of the 19th century and 1978, 84% of lowland raised bog was lost through afforestation, agricultural reclamation and commercial peat cutting. Much of the remainder has been severely damaged by burning and draining, leaving only 6% of the original area intact (Nature Conservancy Council 1986).

Cases of wetland destruction continue to be instigated which are of global significance. One of the most notable recent examples is the drainage of the marshlands of Southern Iraq. Between 1984 and 1985 and 1991 and 1992 the area of permanent lakes and marshes was reduced from just under 1.5 M ha to 0.84 Mha (57% of the 1984/5 figure). The total wetland area had declined by about a quarter from just under 2 M ha to about 1.44 Mha (Maltby 1994). By the end of 1993 the additional construction of a "moat" feature prevented flooding of the central marshes from the Tigris distributaries and diversion of the Euphrates has led to significant drainage of the southern marshes. Recent satellite imagery (1997) shows the conversion of the central marshes to salt pans and major reduction in the area of the southern (Hammar) marshes. The total extent of surviving wetland is now probably less than 0.5 Mha, just 25% of the 1984 extent and more or less confined to the Haweizeh marshes east of the Tigris. The future of the residual area is at best uncertain. The decision to drain the marshlands, overtly stated to achieve intensified agricultural production, has taken no account of the significance of the wetlands in water supply and quality, supporting fisheries (including those of the northern Gulf), wildlife (including migratory and endemic species) and the unique culture of the marsh Arab or Madan (Maltby 1994).

Irrigated agriculture has yet to be developed in the drained area and experts believe that this will never be realised because of the scale involved and the inevitability of salinisation problems associated with the location. Yet the example serves as salutary illustration of the force majeur which can supercede consideration of scientific opinion, international conventions, and recognition of both on-site as well as off-site impacts. The political, ethnic and military rationale for the loss of one of the globe's most important wetland resources can only be speculated here.

The Policy Environment

Several salient features have characterised the wetland policy environment:

1. Wetlands have been considered generally as components of nature conservation, sustainable development, natural resource or other environmental policy. Hence wetland-specific issues or considerations are often diffused or overpowered by other interests.
2. The historic policy framework for wetlands, especially in Europe, has been set by the nature conservation lobby rather than related to the water policy agenda. This is changing together with the obvious recognition of the importance of water for the genesis, development and functioning of wetlands, e.g. the 1992 legislation in France on water and wetlands (Loi sur l'eau No. 923, J.O. 04.01.1992).
3. The rapid increase in attention to wetland policies and/or strategies in both developing as well as developed countries in tandem with appreciation of their functional benefits. Thus whilst in 1987 only five signatory nations to the Ramsar Convention indicated they were involved in any sort of national wetland policy, strategy or plan initiative, by March 1996, 55 nations had national wetland strategies or plans and 27 of the 92 Ramsar Contracting Parties indicated they were engaged in development or implementation of wetland policies (Rubec et al. 1998).
4. Despite the strength of this trend there has been very limited attention to matching the scientific/socio-cultural/economic case for wetland protection and "wise use" (sensu Ramsar Convention) with appropriate policy instruments.

Throughout much of the developed world, the primary instruments for wetland protection have been based on wildlife and a traditional nature conservation approach utilising criteria such as rarity, uniqueness, population abundance or particularly good examples of ecosystem or habitat type. Until recently, in Europe neither national nor supranational policies have separated out wetlands for special consideration from the wider nature conservation agenda. Typical of the approach at the European level is the Habitats Directive (92/43/EEC), which has established a network of Special Areas of Conservation (SACs) to promote biodiversity. Member states are required to protect these sites and manage them, having regard for impacts from anthropogenic activities that will affect their conservation status. In the UK this has been achieved through networks of protected areas such as National Nature Reserves (NNRs), Sites of Special Scientific Interest (SSSIs) and through agri-environmental schemes such as Environmentally Sensitive Areas (ESAs). There is often lack of linkage between the wider functional benefits of wetlands and the environmental policy framework such that protected area status or management actions may not realise the anticipated benefits for conservation of the environment. In Europe the current lack of linkage between the science of how wetlands function in terms of water quality enhancement and/or maintenance and the precise location of buffer zones in the landscape is a clear indicator of the poorly coordinated relationship between science and policy (see below).

The forthcoming European Union Water Framework Directive (Council Directive 6404/99) provides an opportunity for more effective translation of the emerging wetland science base into European policy. This can be achieved through measures to support key articles such as a methodology for determination of the ecological and chemical status of surface water fluxes into and out of wetlands (Art 10), development and implementation of tools and methodologies for linking information on wetland functioning directly to River Basin Management Plans (Art 16) and development and implementation of methodologies for involvement of all interested parties in the wise use of wetlands as required by such plans (Art 17).

Policy in the United States has been driven increasingly by the acknowledged significance of wetland functioning including water quality conditions. A complex regulatory framework has developed around the US Federal Clean Water Act (1977). To meet the requirements of such legislation, environmental managers need tools and methodologies to assist in decision-making. In the United States a plethora of techniques now exist which are variously described as functional analysis or evaluation systems. However, these techniques are often flawed because of deficiencies in the science base and the inadequate knowledge of the people using them (Brinson 1991). Recognition of the utilitarian values of wetlands through ecosystem functioning is an important mechanism for the engagement of public support for their more effective conservation in the developed world. However, it is an aspect even more relevant in the developing world, where the lives of human communities are often directly dependent on wetlands for food, building materials, water supply and other services. This functional context is now embodied in the emerging new wetland policies in developing countries such as that of Uganda and Kenya.

The Ramsar Convention (1971) on Wetlands of International Importance Especially as Waterfowl Habitat brought these ecosystems firmly into the global policy arena. However, as the full title of the Convention indicates, the original driving force was the wildlife and especially the bird conservation lobby. Part of the appeal to the conservation community of the first international agreement to deal with a single ecosystem group was its global outreach and realisation of the importance of maintaining the links in the migratory chain extending across national, regional and continental boundaries. Nevertheless, the uptake was initially concentrated in the developed world. For most developing countries the need to protect wetlands for the conservation of migratory birds was considered an ill-affordable luxury in the face of more immediate problems such as hunger, poverty and international debt. A turning point occurred at the 1987 Conference of Parties in Regina, when emphasis and criteria for recognition of internationally important wetlands was revised to take more account of their role in supporting human populations. This led to an increase in signatories from the developing world and a progressive increase in the emphasis on the wider functional importance of wetland ecosystems beyond the traditional nature conservation viewpoint.

Despite the increased global profile of wetlands, at least three key issues must still be addressed by the scientific and policy-making communities:

1. Continued loss and degradation of wetlands without fully informed public debate and approval.
2. Lack of clearly focussed and effective tools to establish a sound operational basis for innovative wetland policies of wetland protection and management.

3. Lack of mechanisms to bridge the gaps between natural science, socio-cultural, economic and diverse sectoral interests to improve the decision-making processes affecting wetlands.

Wetland Processes and Functioning

Some of the specific reasons for the importance of wetlands which have raised scientific, conservation, political, socio-cultural and economic interests are listed in Table 1.

Table 1. Some underlying reasons for the importance of wetlands (various sources)

Groundwater recharge/discharge
Flood flow alteration
Sediment stabilisation/shoreline anchoring
Sediment/toxicant retention
Nutrient removal/transformation/water quality
Biogeochemical cycling
Fisheries support
Production of biomass/food chain support
Wildlife habitat
Heritage
Science
Recreation

The general functional relationships occurring in wetland ecosystems are outlined in Fig. 1. Hydrological, biological, chemical and physical processes occurring naturally in wetlands (e.g. infiltration, primary production, decomposition, reduction/oxidation, sedimentation) result in functions such as groundwater recharge/discharge, nutrient transformation/sink/source/store and ecosystem maintenance.

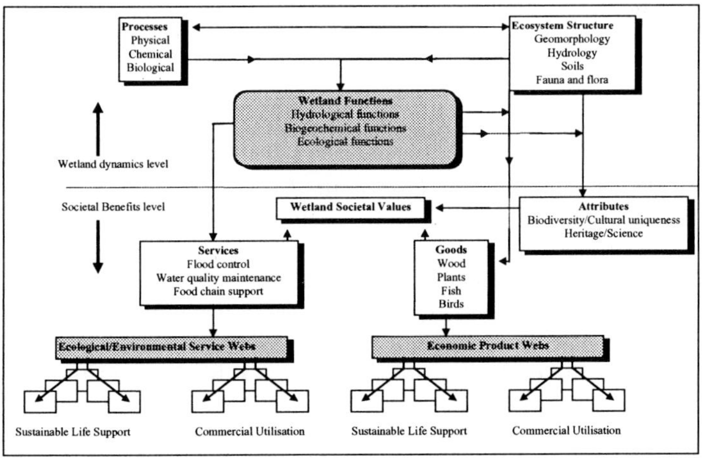

Fig. 1: Relationship between wetland functions, processes, structure, attributes, goods, services and values, and the resultant economic, ecological and environmental webs that functions support. (Maltby et al. 1996)

Particular combinations and levels of functioning produce and are linked in feedback loops to structural characteristics of the ecosystem. For example, where primary production exceeds decomposition, organic matter builds up and may accumulate progressively as peat. The extent and characteristics of the peat depend on the topographic and hydrological conditions under which formation occurs but the development of the peat body itself may give rise to new hydrological, water quality and ecological conditions (Ingram 1982; Wheeler 1984).

Structure and function are together responsible for the generation of products (e.g. forest goods, fisheries, wildlife and grazing animals) and services (e.g. flood protection, water quality enhancement/maintenance, and carbon storage). At the ecosystem level wetlands possess important attributes such as biodiversity, cultural uniqueness (e.g. landscape and associated traditional life styles such as that of the Madan of Southern Iraq) and heritage or scientific significance (notably through preservation of archaeological remains and paleo-environmental records, e.g. pollen, diatoms, sediment and other indictors of environmental change).

This simplified view of wetland functioning belies an enormous complexity of processes together with their spatial and temporal dynamics on which scientific research has focussed on understanding the salient relationships and driving forces. Although the science base is still inadequate in explaining precisely how different wetland ecosystems actually work, increasingly we understand how different factors (controlling variables) determine functioning (Table 2). The recognition of such controlling variables is vitally important if science is to assist in the better decision-making regarding the protection and management of wetlands. Not all wetlands perform all possible functions or give rise to the same combination of goods and services that may be valued by civil society. Also wetlands generating those similar functions may not do so to the same extent. An urgent priority for the scientific community is to assist decision-makers with procedures which realise the optimal functional benefits of wetlands under the various scenarios prescribed by society. The starting point for this must be a robust and verified empirical science base but it is clear that other social and economic considerations will also determine final decisions.

One of the key functional benefits of wetlands arises from their roles in biogeochemical cycling and in particular their ability to interact with sediment. Wetlands are often able to filter and hold more than 90% of the suspended solids and sediments from agricultural runoff prior to discharge into streams and lakes (Maltby 1986). Periodic wetting and drying cycles and the associated shift between aerobic and anaerobic conditions enables biochemical transformations to take place in wetland soils and sediments, which together with physical processes are highly significant in the determination of both chemical and biological water quality (Table 3).

Recognition of the possible controls by natural wetlands in the landscape on water quality has motivated many investigations of the potential of using both natural and constructed wetlands for the removal of pollutants, especially nutrients, from both diffuse sources and in wastewater treatment (e.g. Godfrey et al. 1985; Hammer 1989; Richardson 1985; Kadlec and Knight 1996) (Table 4).

Table 2. Examples of specific functions, processes and controlling variables relevant to water quality control

Function	Processes maintaining functions	Controlling variables
Nutrient retention	Plant uptake of nutrients (N and P)	Nutrient input; vegetation type; degree of disturbance
	Storage of nutrients (N and P) in soil organic matter	Nutrient input; presence of peat; soil water regime; vegetation type; landform
	Adsorption of N as ammonium	Nutrient input; potential for interaction with ammonium; soil composition; soil water regime; soil pH
	Adsorption and precipitation of P in the soil	Nutrient input; soil type; soil water regime; landform
	Retention of particulate nutrients	Nutrient input; flooding regime; particulate nutrient load; evidence of deposition; evidence of erosion
Nutrient export	Gaseous export of N (and)	
	Denitrification	Nutrient input; potential for interaction with nitrate; soil carbon; soil oxygen status; soil pH; soil temperature
	Ammonia volatilisation	Nutrient input; soil environment; exposure
	Nutrient (N and P) export through land use management	Nutrient input; vegetation type; retention of water; degree of disturbance; land use and management
	Export of nutrients (N and P) through physical processes	Nutrient input; hydrological regime; landform; evidence of erosion

Table 3. Overall functional roles and examples relevant to water quality, nutrients and contaminants with significant potential impact on water quality

Functional Role	Definitions	Examples relevant to water quality
Sink	Wetlands acts as sinks when there is an actual and irretrievable loss of nutrients or materials	Conversion of nitrate to nitrous oxide and nitrogen emitted to atmosphere
Store	Wetlands act as stores when there is an accumulation of nutrients or materials which may be permanent or temporary	Fixation of nutrients and/or contaminants in biomass or accumulating organic matter especially peat
Source	Wetlands act as sources when there is a release of nutrients or materials from the ecosystem	Release of phosphate or toxins to drainage waters under reducing conditions
Filter	Wetlands act as filters when nutrients or materials are physically retained from through-flowing water	Removal of sediment and attached nutrients and/or contaminants from through-flow waters
Transformer	Wetlands act as transformers when nutrients or materials are altered as a direct result of the prevailing ecosystem conditions	Conversion of nutrients from inorganic to organic forms or vice versa

Table 4. Examples of the efficiency of natural and created wetlands for N and P removal

Function	Removal efficiency (%)	Wetland Type	Reference
N-Removal			
	89	Natural wetland, USA	Peterjohn and Correll (1984)
	56-100	Natural wetland, New Zealand	Cooper (1990)
	80-90	Created wetland, USA	Kadlec and Knight (1996)
P-Removal			
	65	Natural wetland, Switzerland and France	Pommel and Dorioz (1995)
	27-88	Natural wetland, Estonia	Mander et al. (1991)
	98	Created wetland, Norway	Jenssen et al. (1993)
	20-42	Created wetland, Norway	Braskerud (1994)

The Example of Nitrate Dynamics

Nitrate is a potential major problem in the aquatic environment because it is highly soluble and it represents a large proportion of land-based fertiliser applications which have increased significantly in recent years. Above certain concentrations it may be hazardous to human health and in excess contributes to ecological harm, especially through eutrophication. Wetlands can play a key role in reducing the nitrate flux to groundwater and especially surface waters. As in other ecosystems, this might be achieved through plant uptake, but the particularly luxuriant growth that may occur in wetlands, often without the limitations imposed in other ecosystems by dry season water stress, means that nitrate-nitrogen storage can be particularly effective. If harvesting of the vegetation subsequently occurs, then removal from the system is complete and the plant uptake process has resulted in a sink function. However, in addition, the waterlogged soil or sediment environment can provide an ideal medium for denitrification. This is the microbially mediated reduction of nitrate by facultatively anaerobic bacteria abundant in wetlands. The result is conversion of nitrate to gaseous nitrogen compounds which are lost from the wetland system to the atmosphere. Notwithstanding the possible contributions of any nitrous oxide component to the greenhouse effect, there is major potential benefit in protection of ground and surface waters from nitrate pollutions. This possibility has been part of the rationale for the maintenance or establishment of wetlands as buffer zones especially in river marginal areas.

Buffer Zones

A buffer zone is an area of natural or seminatural vegetation situated between agricultural land and a surface water body, and acting to protect the water body from harmful impacts such as high nutrient, pesticide or sediment loadings that might otherwise result from land-use practices. It offers protection to a water body through a combination of physical, chemical and biological processes, and some of the most beneficial processes for water quality improvement occur optimally in wetlands. The degree to which this protection is provided depends on a number of factors including the size, location, hydrology, vegetation and soil type of the buffer zone (Dosskey et al. 1997; Leeds-Harrison et al. 1996), as well as the nature of the impacts by which the water body is threatened. A buffer zone established to protect the quality of a surface water body may provide additional benefits such as bank stabilisation, floodwater detention and wildlife habitat, depending on its location. Appropriate location of a buffer zone is essential to achieve optimal performance of a desired function.

In the UK, the current protocol for the implementation of buffer zones is described by both the Environment Agency (1996) and the Ministry of Agriculture, Fisheries and Food (1997). Generally it is recommended that buffer zones should be situated in riparian areas adjacent to main water courses and extend to between 5 and 30 m in width. The Water Fringe Habitat Scheme Option (MAFF, 1997) provides one of the few sources in the UK from which funding may be acquired for the implementation of riparian buffer zones.

Models for buffer zone designs such as REMM (Riparian Ecosystem Management Model; Lowrance et al. 1998) and RiMS (Riparian Management Systems; Isenhart et al. 1995) also focus on riparian buffers, although these models have been developed specifically for application in the USA. While such an approach in Europe is likely to provide substantial environmental benefits, such a rigid approach to the establishment of wetland buffer zones in riparian regions, particularly for the protection of surface water quality, may be impractical, inefficient and scientifically flawed.

While riparian buffer zones are often highly effective at removing nitrate and other potential pollutants from diffuse sources such as shallow groundwater or surface runoff (Ambus and Christensen 1993; Cooper 1990; Haycock and Burt 1993; Weller et al. 1994), they are often bypassed where natural hydrological flows are intercepted by ditches or drains which are common features in many landscapes. In such cases, a riparian buffer zone can be rendered ineffective for nitrate removal from water draining agricultural land. It may prove more effective and cost-efficient to establish buffer zones in wetland areas associated with ditches or areas of discharge that may be acting as zones of enhanced denitrification (Blackwell and Maltby 1998).

Kismeldon Meadows, a river marginal wetland in the River Torridge catchment, southwest England, provides an example of such a situation. A ditch draining improved agricultural land and passing through the wetland has become disrupted at several locations along its course, forming areas of overland flow (see Fig. 2). Monitoring of nitrate concentrations in the ditches has indicated that more than 90% of the nitrate in the ditch water is removed regularly in these zones, with the lowest observed removal efficiency being 60% (Blackwell 1997). These areas, which occupy only a small percentage of the total wetland area, are therefore acting as effective buffer zones for the removal of nitrate. The establishment of a conventional riparian buffer zone along the banks of the River Torridge in accordance with current policy recommendations would have no impact on the quality of the water in the ditch, as it bypasses any potential conventional buffer zone area, discharging directly into the river. The maintenance of these alternative buffer zone locations along the ditch system is essential if nitrate pollution is to be minimised. Targeting buffer zones away from the riparian margins and not restricting them to riparian strips therefore may be most effective for water quality maintenance and improvement.

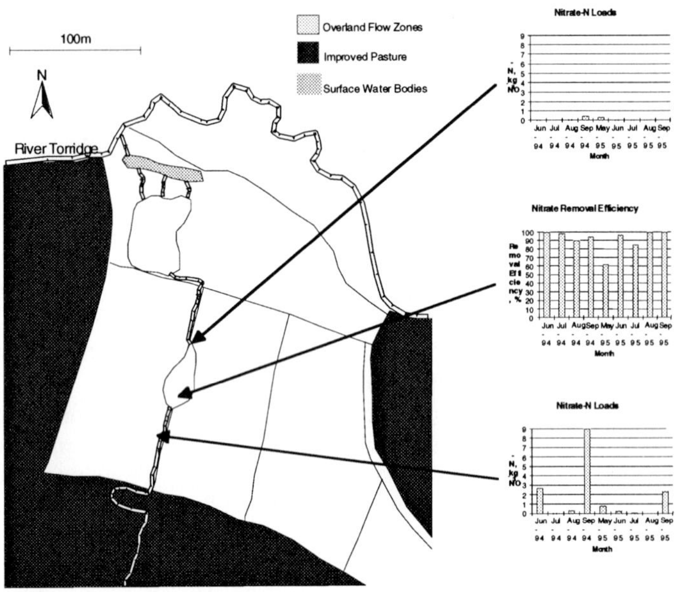

Fig. 2: Kismeldon Meadows: an example of alternative wetland buffer zone locations for the removal of nitrate from agricultural runoff

Similar conclusions have been drawn from work by Haycock and Muscutt (1995), who reported that while 85% of some subcatchments of the River Avon in Hampshire in the UK were served by effective riparian buffer strips, 60% of polluting material in the river was delivered by roads and drains that effectively bypassed them.

A more flexible approach to buffer zone location is required if their use is to prove efficient, and resources are to be directed towards buffer zone establishment in the most effective areas. A useful tool for assisting in the identification of wetland buffer zones for specific functions is the FAEWE (Functional Analysis of European Wetland Ecosystems) procedures (Maltby et al. 1996a). These enable the semiquantitative assessment of different landscape units, referred to as hydrogeomorphic units (HGMUs), for specific functions, including N storage and removal. These procedures are being used to functionally assess landscape units outside riparian regions in the NICOLAS (Nitrogen Control by Landscape Structures in Agricultural Environments) project (NICOLAS 1997), which is currently developing a European version of the REMM model, designed to incorporate the use of alternative wetland buffer zone locations.

The example of wetland buffer zones to protect surface waters from nitrate pollution illustrates the challenges which remain:

1. To link the precise nature and patterning of wetland functioning to appropriate policy instruments.

2. To recognise the need for spatial congruency between key wetland processes, the routing of pollutants and techniques for water quality protection.
3. To determine the capacity of natural systems to handle diffuse pollution loads without significant or irreversible ecological harm such as loss of habitat or biodiversity. The use of wetlands for treatment of concentrated pollution or wastewater should be restricted to artificially created wetlands – see Kadlec and Knight (1996).

Constraints

The use of wetlands in the management and protection of water quality is limited in at least three main ways:
1. Incompatibility with maximising the performance of other ecosystem functions e.g. biodiversity, production or of different components of water quality, e.g. what favours nitrate reduction may increase mobility of phosphates.
2. Inequitable distribution and difficulties in evaluation of the costs and benefits associated with any implementation strategy.
3. The science base is still ineffectively translated into operational tools for optimal efficiency of application of policy.

Incompatibilities

The effect of potential incompatibility between water quality control and maintenance of other functions is very well illustrated in the current problems facing the Florida Everglades. Release of high concentrations and loads of phosphorus from the Everglades Agricultural Area (EAA) with drainage waters flowing south into previously "intact" wetlands has resulted in significant alteration of the ecosystem. The increase in nutrients has favoured the establishment and rapid spread of cattail (*Typha domingensis*), developing dense continuous stands which have replaced communities such as sawgrass (*Cladium jamaicensis*) and more open water-dominated habitats characterised by a variety of submerged and floating aquatics.

The wetlands receiving the contaminated waters are acting as a significant check on the downstream movement of phosphates through their incorporation into plant biomass and the surface peat. Unfortunately, the consequence is significant ecological alteration of the area impacted resulting in loss of habitat, biodiversity and accelerated buildup of peat, probably exceeding by a factor of 2 the rates at sites unimpacted by nutrient runoff. The new cattail habitat offers poorer conditions for fish and other wildlife and the increase in peat accumulation may favour the anaerobic (reducing) conditions under which mercury (from original marine sediments, agricultural or atmospheric sources) may become methylated. This form of mercury is highly soluble and represents an increased hazard to animals (and humans) higher up the food chain. Potentially toxic levels of mercury have been identified in both fish and the highly endangered Florida panther (Lord 1993). Increasing effort is being devoted currently to the resolution of the water quality problems in Everglades which involve inter alia the reversion of agricultural land to wetland nutrient removal systems to reduce the P levels entering the Everglades (South Florida Water Management District 1997).

A further example of the practical incompatibility in utilising wetlands for the reduction of nutrient flux to surface waters is that the fluctuating reducing/oxidising (redox) condition may be ideal for nitrate removal by denitrification but additionally may increase the release of any phosphate or other contaminants held as part of metal complexes sensitive to changes in redox potential (e.g. Fe^{3+} transformed to Fe^{2+}, resulting in release of previously bound phosphate). We are still relatively ignorant of these detailed interactions and of the flux of the wide range of potential pollutants (including heavy metals, pesticide residues and microbiota) in relation to wetland environment and process dynamics.

Distribution of Costs and Benefits

An inherent problem arises from the fact that water quality benefits are incurred largely downstream and are not generally directly realised by those landowners whose actions are required to maintain or establish wetlands for that purpose. Policy instruments are still largely inadequate in addressing this fundamental problem. However, recent experience in the UK with a pathfinder project called Tamar 2000 SUPPORT (Sustainable Practices Project on the River Tamar) Project (funded under EU Objective 5B, EAGGF Measure 5.1) has demonstrated that a non-prescriptive, voluntary approach, working with farmers and other landowners, can achieve significant improvements largely by showing how wetlands can be part of an overall best land and water management strategy that is able to bring about savings of expenditure and reduced risk of pollution (and prosecution costs).

Translation of the Science Base

Examples such as managing and optimising the use of wetlands as buffer zones for water quality demonstrates the clear need for operational tools to enable non-experts to apply the results of fundamental scientific research to practical problems and within appropriate policy frameworks. The cost and complexity of field-based environmental research and involvement of the multidisciplinary teams now required to tackle the range of relevant scientific, social and economic dimensions prohibits empirical and site-specific research-based solutions to support all management or policy decisions. This is particularly true if policy is to be applied at significantly larger scales such as river basins.

Our inability to satisfactorily translate the available science and especially to link it to relevant social and economic contexts has certainly impeded progress in guiding the decision-making process to enable the more effective protection and wise use of wetlands.

Towards Solutions

Decision Support Systems

Better planning tools can help evaluate the incompatibilities between different wetland functions and assist decisions which enable the better distribution of costs and benefits. For a number of years now, functional analysis tools have been used in the United States to assist in the implementation of wetland relevant legislation such as Section 404 of the Clean Water Act. In Europe a system has been under development (Maltby et al. 1994, 1996a) that parallels more recent conceptual developments in the United States (Brinson 1993). This system uses information on hydrology and geomorphology (hydrogeomorphology) to identify units (hydrogeomorphic units or HGMUs) of uniform functioning in wetlands. Desk information and field observation are used to delineate these HGMUs and then characterise them, using a range of different environmental data (e.g. soils, land use and management, geomorphology, hydrology, ecology). This information is interrogated by a series of decision trees for hydrological, ecological and biogeochemical functions. A level of functioning can then be ascribed to each of these units providing a spatial representation of functioning and degree of functioning within a wetland. The system has the additional benefit that functional outputs can be easily translated into economic values using a number of standard ecological-economic techniques. Incorporation of monetary values into planning and decision making systems can assist in the appreciation of the benefits and costs of wetland functioning and provide information in a form that can assist in their reallocation.

Integrated/Strategic Catchment Management

There is increasing emphasis on the importance of a catchment approach in order to achieve satisfactory maintenance and/or improvement of water quality. This is manifest in the forthcoming European Commission Water Framework Directive, which will introduce catchment management throughout Europe. The Directive emphasises biological as well as chemical water quality goals and introduces the importance of ecological functioning as well as structure. Unfortunately, however, the Directive "gives little advice on policy integration for catchment management... and fails to provide for public involvement in the process of ecosystem management" (Pollard and Huxham 1998). Without integration of the various policies that influence land use, it is difficult to imagine the directive alone being immediately effective in delivering major improvements in water quality.

A holistic approach is required to catchment management which inter alia incorporates considerations of land use, water resources, ecosystem functioning (especially wetlands), socio-cultural practices and economics based on a shared vision of the desired environmental quality. This is implicit in the ecosystem approach advocated by the Convention on Biological Diversity (UNEP/CBD/COP/4/Inf.9).

The views and support of various stakeholders such as farming communities, fishery owners, conservation groups, others with recreational interests and diverse commercial and public-interest groups are essential to take into account if the ecosystem approach to management is to be implemented successfully. In the Tamar 2000 SUPPORT Project, staff (referred to as advisers) have worked on a one-to-one basis with landowners to develop integrated farm management plans contributing not only to cost savings for the farmer but also significant environmental benefits (e.g. fencing and new drinking troughs to reduce stock-induced bank erosion and in-stream sedimentation; restoration of wetlands as nutrient buffers). It will be some years before the degree of success of the project can be properly evaluated in terms of improvement of physical, chemical and biological water quality and socially important indicators of success such as numbers of migratory salmon. Nevertheless, Table 5 gives an idea of the level of success attained in terms of on the ground action.

Table 5. Key outputs and achievements of the Tamar 2000 SUPPORT Project, Phase I, 1996-1998

Training and advice	8 Farm advisors provided 3 River advisors provided 300 integrated management plans produced for individual farms
Environmental improvements	312 ha of river corridor restored and replanted 23 km of fencing erected for river bank protection 5 wetland treatment areas developed 15 km of ditches managed for pollution control 2 demonstration sites established 79 salmonid spawning sites desilted 20 buffer zones established
New enterprises introduced	Liaison between environmentally friendly schemes and landowners incorporated into management plans Management plans provided for development of 19 coarse fishing ponds Workshops provided on tourism marketing and training
Technical improvements	Production of a practical, integrated land-use manual for use by farmers
Wetland technical improvements	GIS based functional inventory of wetlands in catchment produced Handbook for the identification, creation and management of buffer zones
Employment	150 jobs sustained 10 full-time and 38 part-time jobs created

Benefits of Stakeholder Engagement

Whilst an important element in realising the worth of wetlands is achieving and maintaining high water quality in a developed world context, the potential benefits to stakeholders is a policy imperative in the developing world. It is well illustrated in the Mekong Delta in Vietnam, where drainage and conversion of former wetland forest dominated by *Melaleuca* spp. to rice paddy has resulted in the oxidation of potential acid sulphate soil materials, resulting in the generation of sulphates, aluminium and other toxic elements and compounds into drainage waters. The resulting acidity is frequently less than pH 2 impacting drinking water quality, fisheries and rice yields (Maltby et al. 1996b; Safford et al. 1997). Through close cooperation with local farmers it has been possible to reestablish small plots of *Melaleuca* wetland to improve the quality of water which can be recirculated to the rice paddy. The raised water table in the wetland maintains the soil conditions which prevent oxidation and further increase in acidity. The pH of the water is ameliorated further through interactions with the decomposing *Melaleuca* leaf litter by mechanisms currently under more detailed investigation (Ni DV, pers. comm.). Wetland restoration thus has the potential to prevent further acidification and also ameliorate already degraded waters with beneficial effects on adjacent rice production. Such a solution, however, can be implemented only by a close liaison between the researchers and the farmers, creating mutual trust and making available to the rural poor sufficient credit to enable the system to become established and operational.

Conclusion

Scientific research increasingly is demonstrating the functional importance of wetlands. This is significant for the welfare of human communities as well as wildlife and environmental quality. The results of this research are gradually permeating policy frameworks. In particular, wetland policies are evolving from an emphasis on nature conservation to wider functional aspects. Nevertheless, much still needs to be done if society is to best manage the already much depleted and degraded wetland resource. There is still a wide gulf between defining the "wise use" of wetlands and actually implementing actions to optimise their benefits.

Policy-makers can assist in bridging the gulf by recognising the wide range of external factors such as tax structures, agricultural subsidies, energy shortages, technology export, development aid and interagency conflicts which often play determining roles on the future of specific wetlands worldwide. The scientific community can help by providing the knowledge base and better tools to assist decision-making. In the case of water quality issues, this means better identification of the spatial dimensions of wetland functioning at the landscape and catchment scale together with well-targeted restoration and management systems appropriate to maximising desired water quality benefits. It means working with stakeholders and using an ecosystem approach (*sensu* Convention on Biological Diversity) to link science, sociocultural and economic considerations. Only when robust and easy to understand tools of functional analysis are available, which bridge traditional conflicts between ecology, economics and immediate social priorities, will it be possible to adequately deliver the innovative policy instruments necessary for improved wetland management and protection.

Acknowledgements. The authors would like to acknowledge the assistance of David V. Hogan in the production of this chapter. Much of the work described was supported by EU funding under the FAEWE (ECDG XII STEP-CT90-0084) and PROTOWET (EC DG XII ENV4-CT95-0060) projects. Thanks also to the Amar Appeal, ODA and WWF for support that enabled the S. Iraq study.

References

Armentano TV, Menges ES (1986) Patterns in the change of carbon balance of organic soil wetlands of the temperate zone. J Ecology 74(3):755-774

Ambus P, Christensen S (1993) Denitrification variability and control in a riparian fen irrigated with agricultural drainage water. Soil Biol Biochem 25(7):915-923

Blackwell MSA (1997) Zones of enhanced denitrification in river marginal wetlands. Unpublished Ph.D. Thesis, Geography Department, Royal Holloway University of London, UK

Blackwell MSA, Maltby E (1998) Zones of enhanced denitrification. In: McComb AJ, Davis JA (eds.) Wetlands for the future. Proceedings of INTECOL's V International Wetlands Conference. Gleneagles Press, Adelaide, Australia

Braskerud B (1994) Sedimentation of phosphorus and soil particles in constructed wetlands. In: Persson R (ed) Agrohydrology and Nutrient Balances. Proc NJFseminar 247, Div Agric Hydrotech, Swedish University of Agricultural Sciences, Uppsala, Sweden. Communications 94, 5:66-69

Brinson MM (1991) Landscape properties of pocosims and associated wetlands. Wetland 11:441-465

Brinson MM (1993) Changes in the functioning of wetlands along environmental gradients. Wetlands 13:65-74

Cooper AB (1990) Nitrate depletion in the riparian zone and stream channel of a small headwater catchment. Hydrobiologia 202:13-26

Dosskey MG, Schultz RC, Isenhart TM (1997) Agroforestry Notes 1-5. Internet web page http://waterhome. tamu. edu/texasyst/agroforestynotes/afnote3.htm

Environment Agency (1996) Understanding buffer strips: An information booklet. EA, Bristol

Godfrey PJ, Kaynor ER, Pelczarski S, Benforado J (eds) (1985) Ecological considerations in wetlands treatment of municipal wastewaters. Van Nostrand Reinhold, New York

Hammer DA (1989) Constructed wetlands for wastewater treatment: municipal, industrial and agricultural. Lewis Chelsea, Mich

Haycock NE, Burt TP (1993) Role of floodplain sediments in reducing the nitrate concentration of subsurface run-off: a case study in the Cotswolds, UK. Hydrolog Proc 7:287-295

Haycock NE, Muscutt AD (1995) Landscape management strategies for the control of diffuse pollution. Landscape and Urban Planning, 31:313-321

Howard GW (1996) Experiences in wetland policy development. In: Ntiba M, Gichuki N and Chumo N (eds). Sectoral policies relating to wetlands in Kenya. Proc Workshop Karen, Nairobi 11-12 April 1995, NES and UNO/RAF/006/ GEF. 77pp

Immirzi P, Maltby E, Clymo RS (1992) The global status of peatlands and their role in the carbon cycle. Friends of the Earth. London, UK

Ingram HAP (1982) Size and shape in raised mire ecosystems: a geophysical model. Nature 297:300-339

Isenhart TM, Schultz RC, Colletti JP, Rodrigues CA (1995) Design, function, and management of integrated riparian management systems. In: Proc Nat Sympo Using Ecological Restoration to Meet Clean Water Act Goals. USEPA. Chicago, IL. March, 1995, pp 93-102

Jenssen PD, MΩhlum T, Krogstad T (1993) Potential use of constructed wetlands for wastewater treatment in northern environments. Water Sci Technol 28(10):149-157

Kadlec RH, Knight RL (1996) Treatment wetlands. CRC Press, Florida, US

Leeds-Harrison PB, Quinton JN, Walker MJ, Harrison KS, Tyrrel SF, Morris JM, Harrod T (1996) Buffer zones in headwater catchments. Report on MAFF/ English Nature Buffer Zone Project CSA 2285. Cranfield University, Silsoe, UK

Lord LA (1993) Guide to Florida environmental issues and information. Florida Conservation Foundation, Winter Park, Florida

Lowrance R, Altier LS, Williams RG, Inamdar SP, Bosch DD, Sheridian JM, Thomas DL (1998) The riparian ecosystem management model: simulator for ecological processes in riparian zones. Proc First Fed Interagency Hydrologic Modeling Conference, Las Vegas, NV, April 1998

Maltby E (1986) Waterlogged wealth. Earthscan, London, 200 pp

Maltby E (1994) An environmental and ecological study of the marshlands of Mesopotamia, Amar Appeal Trust

Maltby E, Turner RE (1983) Wetlands are not wastelands. Geographical Magazine, Vol. LV:92-97

Maltby E, Hogan DV, McInnes RJ (1996a) Functional analysis of european wetland ecosystems – phase I (FAEWE). Ecosystems Research Report 18. Office for Official Publications of the European Communities, Luxembourg, pp 448

Maltby E, Burbridge P, Fraser A (1996b) Peat and acid sulphate soils: a case study from Vietnam. In: Maltby E, Immirzi CP, Safford RJ (eds) Tropical lowland peatlands of Southeast Asia. IUCN, Gland, Switzerland. 187-198

Maltby E, Immirzi P (1993) Carbon dynamics in peatlands and other wetland soils: regional and global perspectives. Chemosphere 27(6):999-1023

Mander Ü, Matt O, Nugin U (1991) Perspectives on vegetated shoals, ponds and ditches as extensive outdoor systems of wastewater treatment in Estonia. In: Etnier C, Guterstam B (eds) Ecological Engineering for wastewater treatment. Proc Intern Conf Stensund Folk College, Sweden pp 271-282

Ministry of Agriculture Food and Fisheries (MAFF) (1997) The habitat scheme, water fringe areas, information pack. MAFF Publications, London

Nature Conservancy Council (1986) Nature conservation and afforestation in britain. Peterborough. Nature Conservancy Council

NICOLAS (1997) Nitrogen control by landscape structures in agricultural environments. European Commission DGXII, Research Project: 1997-2000: Project Number ENV4-CT97-0395

Peterjohn WT, Correll DL (1984) Nutrient dynamics in an agricultural watershed: observations on the role of a riparian forest. Ecology 65:1466-1475

Pollard P, Huxham M (1998) The European water framework directive: a new era in the management of aquatic ecosystem health? Aquatic Conservation – Marine and Freshwater Ecosystems 8(6):773-792

Pommel B, Dorioz JM (1995) Movement of phosphorus from agricultural soil to water. International workshop Phosphorus Loss to Water from Agriculture, (Sept 27th - 29th), Teagasc, Johnstown Castle, Wexford, Ireland. pp 14

Richardson CJ (1985) Mechanisms controlling phosphorus retention capacity in freshwater wetlands. Science, 228:1424-1427

Rubec C, Mafabi P, Mahy M, Nathai-Gyan N, Phillips B, Pritchard D (1998) A framework for developing and implementing national wetland policies. Ramsar Convention Bureau, Gland, Switzerland

Safford RJ, Ni DV, Maltby E, Xuan VT (eds) (1997) Towards sustainable management of Tram Chim National Reserve, Vietnam. Proc Workshop balancing economic development with environmental conservation. Royal Holloway Institute for Environmental Research, 144p

South Florida Water Management District (1997) Everglades annual report. SFWMD, West Palm Beach, Florida

Tiner RW (1984) Wetlands of the United States: current status and trends. U. S Fish and Wildlife Service

UNEP/CBD/COP/4/Inf.9 (1998) Report of the workshop on the ecosystem approach. Lilongwe, Malawi. 26th-28th January 1998

Weller DE, Correll DL, Jordan TE (1994) Denitrification in riparian forests receiving agricultural discharges. In: Mitsch (ed) (1994) Global wetlands: old world and new. Elsevier Science

Wheeler BD (1984) British fens. In: Moore PD (ed) European mires. Academic Press, London, pp 237-281

Global Distribution of Arable Land, Cereal Yield and Nitrogenous Fertilizer Use

H. KAWASHIMA[1], K. OKAMOTO[2], K. OHGA[1], JM. LYNCH[3]

Visible information on world agriculture can help to understand the current world agricultural status. The global maps for cereal yield and nitrogenous fertilizer application are composed of data sets provided by USGS (United States Geological Survey) and the FAO (Food and Agriculture Organization of the United Nations). Arable land is not distributed uniformly over the earth. Much arable land is located in Europe and Asia. The cereal yield is relatively high in the temperate zone, while that in the Tropics is relatively low. The use of nitrogenous fertilizer is high in Western Europe and East Asia, but quite low in Africa, except in the Nile basin. Agricultural activity in the Northern Hemisphere is higher than that in the Southern Hemisphere. Maps of arable areas and agricultural activity indicate that there is enough room to increase cereal production in parts of Africa, Australia and South America.

Introduction

The distribution of arable land area is not uniform over the earth. The cereal yield and fertilizer application are also not uniform. The proportion of arable land area per person is relatively small in Asia and Europe, while it is large in North America, South America and Oceania. The arable land area per capita is 2.75, 0.65 and 0.40 ha in Australia, Brazil and the USA, respectively. In the UK, China and Japan, it is 0.10, 0.11 and 0.03 ha, respectively (FAO 1998). Although these figures are readily available, there are few reliable maps showing the global arable land distribution. Matthews has presented an arable land distribution map (Matthews 1983). The resolution of the map was 1.0 degree of latitude and 1.0 degree of longitude. The agricultural land use intensity was classified to five levels. Kawashima et al. presented a world distribution map of fertilizer discharge from fields based on the map by Matthews (Kawashima et al. 1998).

1. The University of Tokyo, Graduate School of Agriculture and Life Sciences, Japan
2. National Institute of Agro-Environmental Sciences, Tsukuba, Japan
3. School of Biological Sciences, University of Surrey, UK

Recent progress in satellite remote sensing techniques gives more precise information on the world land use. USGS provides the land use map whose resolution is 1 km. Adopting the FAO data with the USGS map, we generated two maps which show the global distribution of crop yield and nitrogenous fertilizer use. These maps can help us to understand the current agricultural status and investigate the world food supply in the 21st century.

Method

The geographical data for the distribution of world land is provided by USGS (USGS 1997). The Interrupted Goode Homolosine Projection is used to show the globe by USGS. Resolution is 1×1 km at the equator. The land use is divided in 24 categories, one of which is water bodies. We extracted five types of land which are closely related to agriculture: (1) dryland cropland and pasture, (2) irrigated cropland and pasture, (3) mixed dryland/irrigated cropland and pasture, (4) cropland/grassland mosaic and (5) cropland/woodland mosaic.

We transformed the Interrupted Goode Homolosine Projection to the Equirectangular Projection, since the geographical information data set for international borders is provided on the Equirectangular Projection (UNEP/GRID-Tsukuba 1994). We used the data sets of cereal yield and nitrogen fertilizer consumption for 205 countries and regions (FAO 1998). Cereals include wheat, rice, maize and other grains. The total cereal crop divided by the total cereal-planted area gives the average cereal yield in a country. The amount of nitrogenous fertilizer consumption per year divided by the arable land area gives the nitrogen fertilizer input per unit of arable land area. The area included permanent crop land areas.

In order to elucidate the intensity of agricultural activity, we assumed that the whole area is used for agriculture in (1) dryland cropland and pasture, (2) irrigated cropland and pasture and (3) mixed dryland/irrigated cropland and pasture regions. Fifty percent of the area is assumed to be used for agriculture in (4) cropland/grassland mosaic regions and (5) cropland/woodland mosaic regions. Although the actual yield is 4.0 tonnes ha^{-1} in the cropland/grassland mosaic region and the cropland/woodland mosaic region, the yield is displayed as 2.0 tonnes ha^{-1} on the map.

Results

Arable land area, cereal yield and nitrogen fertilizer input in major countries are shown in Table 1. We show the top 30 countries plus the UK and Japan for arable land area. The USA has the largest arable land area in the world. Within the 30 countries, cereal yield is the highest in France. Nitrogen input per unit of arable land area is the highest in China. Currently, China consumes the greatest amount of nitrogenous fertilizer.

Figure 1 shows the global distribution of arable land area and nitrogen fertilizer use, while Fig. 2 shows the distribution of cereal yield in the world. In Figs. 1 and 2, it is clear that arable land is not distributed uniformly over the earth. Most of the arable land is in Europe, East Asia, the southern part of the Russian Federation and the central USA. There is little arable land area in Africa and Australia. Most of the irrigated land is in Asia.

Nitrogenous fertilizer input is high in Western Europe and East Asia. Although nitrogen input is high in East Asia, the cereal yield is relatively low. The use of fertilizer is low in all of Africa, except the Nile basin in Egypt. Fertilizer input in South America is also currently low. Fertilizer input is closely related to shallow groundwater pollution by nitrate and to eutrophication in lakes and bays. Figure 1 shows the regions where aquatic environments are polluted by agricultural activity.

Table 1. Cereal yield and nitrogen fertilizer consumption in the top 30 arable countries, plus the UK and Japan

Country	Arable (1996)	Yield (1988)	N. Fertilizers (1996)	N. Input (1996)
	× 1000 ha	tonne ha^{-1}	1000 tonne $year^{-1}$	kg ha^{-1}
USA	177 000	5.62	11 184	63.2
India	169 700	2.21	10 301	60.7
China	135 072	4.81	25 779	190.9
Russia	132 980	0.93	1 100	8.3
Brazil	65 500	2.60	1 197	18.3
Australia	50 221	1.95	718	14.3
Canada	45 500	2.71	1 670	36.7
Ukraine	34 211	2.33	610	17.9
Nigeria	32 909	1.21	95	2.9
Kazakhstan	32 030	0.64	63	2.0
Indonesia	30 987	3.81	1 925	62.1
Mexico	27 300	2.30	950	34.8
Argentina	27 200	3.81	500	18.4
Turkey	26 946	2.08	1 147	42.6
Pakistan	21 600	2.18	1 985	91.9
Thailand	20 445	2.33	811	39.7
Spain	20 129	3.34	1 153	57.3
France	19 461	7.36	2 525	129.7
Iran	19 400	1.92	685	35.3
S. Africa	15 825	1.85	405	25.6
Poland	14 452	3.03	950	65.7
Sudan	13 000	0.56	70	5.4
Germany	12 064	6.31	1 758	145.7
Ethiopia	v11 950	1.11	57	4.8
Italy	v10 768	5.00	894	83.0
Myanmar	10 138	2.80	130	12.9
Romania	9 882	3.03	270	27.3
Morocco	9 661	1.11	180	18.6
Philippines	9 520	2.24	427	44.9
Bangladesh	8 820	2.73	997	113.0
UK	6 133	7.11	1 346	219
Japan	4336	6.06	512	118.1

Figure 2 shows that cereal yield is high in Western Europe. The yield in the central USA, China, Indonesia and Argentina are also high. In contrast, the yields in India and South America are relatively low. The yield in all of Africa, except the Nile basin, is very low. Most of the high-yield areas are in the temperate zone and the subfrigid zones. The yield in the tropical zone is relatively low. Generally speaking, the net primary production (NPP) in the tropical zone is higher than that in the temperate zone, and there is a good relationship between NPP and the yield. The yield in the tropical zone should, therefore, be higher than that in the temperate zone. However, the current yield distribution contradicts this. Most of the developing countries are in the tropical zone, while most of the developed countries are in the temperate and subfrigid zone. The state of social development as well as climate determines the current cereal yield.

Conclusion

The global maps for agricultural activity show that the distribution of world agriculture is not uniform and there seems to be a possibility of improving agricultural production in the Southern Hemisphere. This visible information can help us to understand world food supply in the 21st century.

Reference

FAO (1998) Statistical database. http://apps.fao.org/cgi-bin/nph-db.pl. The Food and Agriculture Organization of the United Nations, Rome

Kawashima H, Okamoto K, Bazin MJ, Lynch JM (1998) Nitrogen fertilizer and ecotoxicology: global distribution of environmental pollution caused by food production. In: Lynch JM, Wiseman A (eds) Environmental bio-monitoring – the biotechnology – ecotoxicology interface. Cambridge University Press, Cambridge, pp 208-219

Matthews E (1983) Global vegetation and land use: new high resolution data bases for climate studies. J Climate Appl Meteorol 22:474-487

UNEP/GRID-Tsukuba (1994) GRID global data sets: documentation summaries. Center for Global Environmental Research, National Institute for Environmental Studies, Tsukuba

USGS (1997) Global land cover characterization. http://edcwww.cr.usgs.gov/landdaac/glcc/glcc.html. United States Geological Survey, Reston

1996 Nitrogen Fertilizer Input

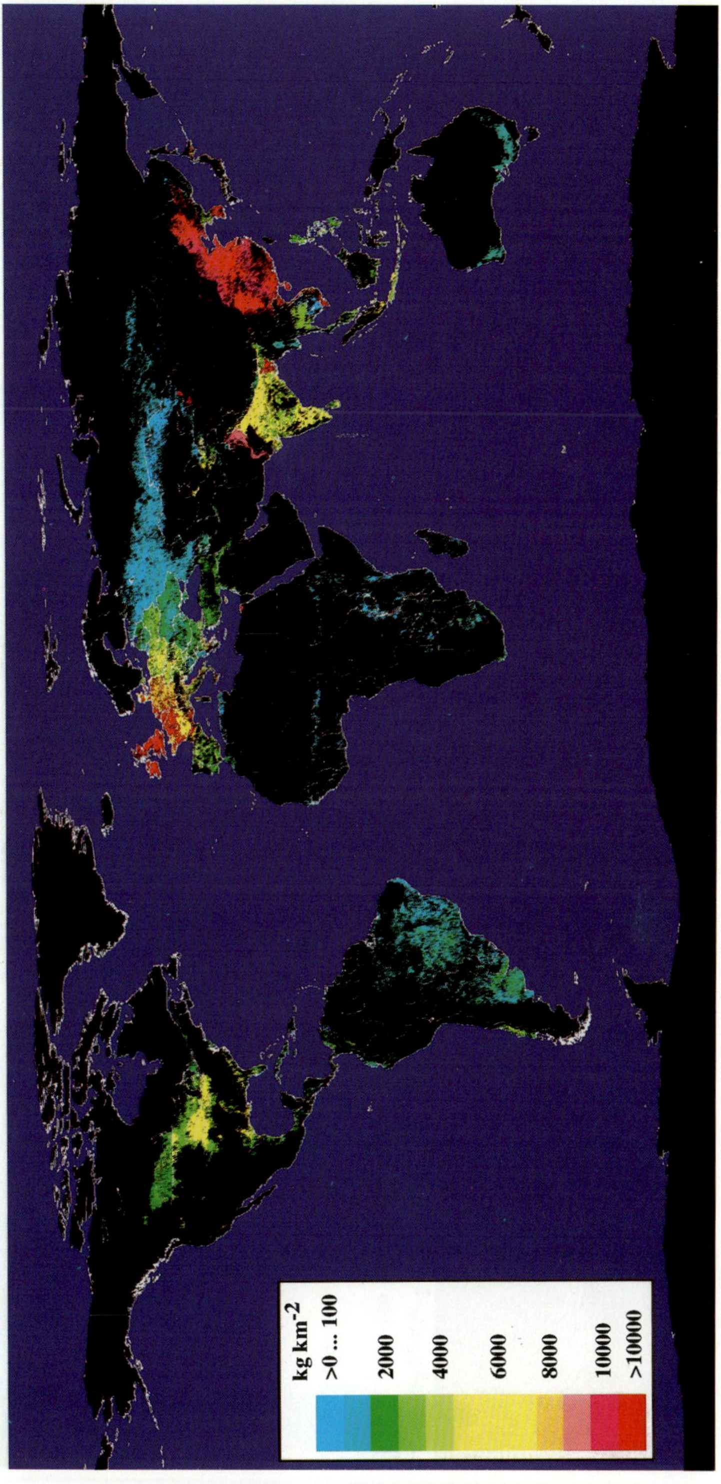

Fig. 1: Distribution of nitrogenous fertilizer application over the earth

1998 Yield

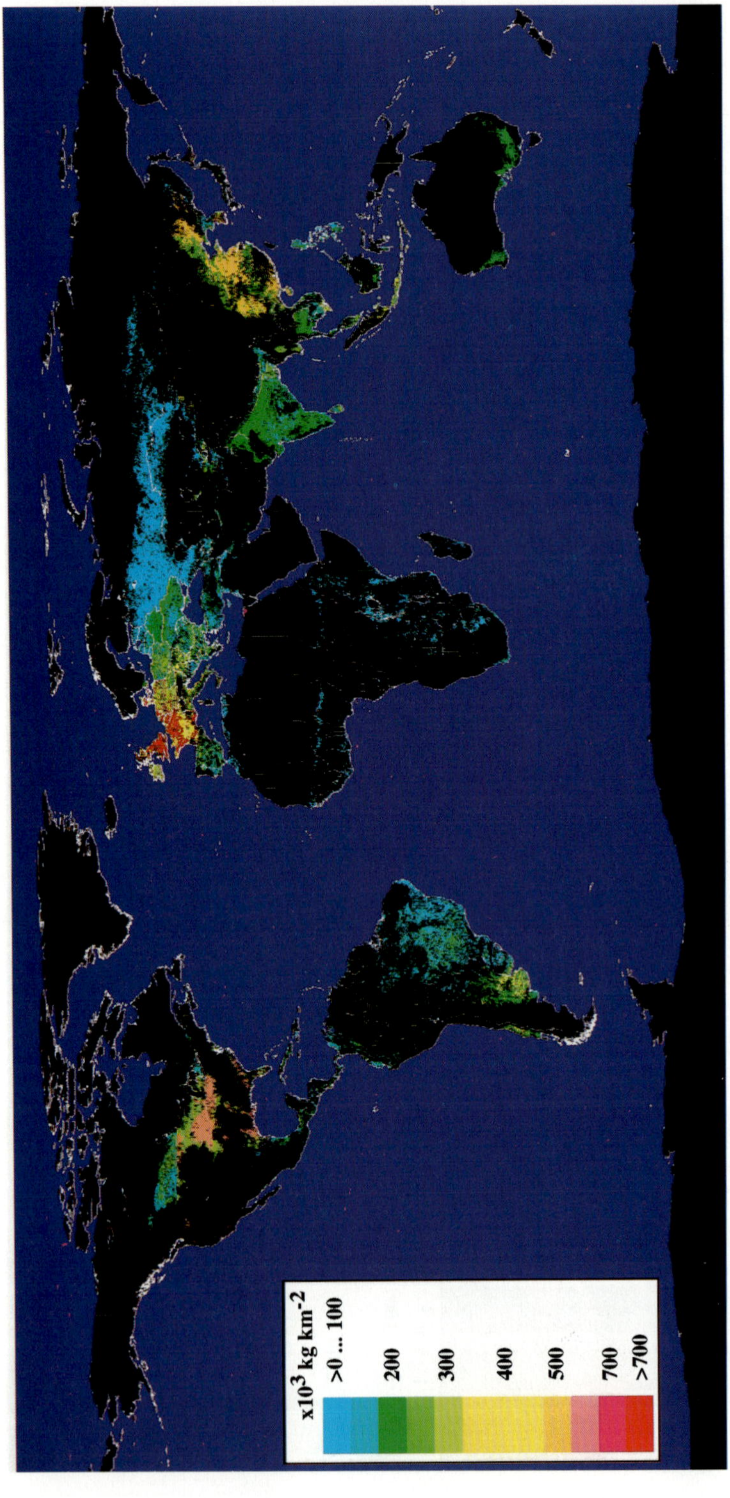

Fig. 2: Cereal yield distribution over the earth

Managing Organisms and Ecosystems

JC. MUNCH, D. WERNER (Germany)

Professor Werner (Germany) as chairman of the panel welcomes the key experts Professor J.J. Lipa from Poland, Professor R.A. Olson from Norway and Dr. A. Tomlin from Canada. The title of this panel already indicates one basic question of all four themes of our project Biological resource management for sustainable agricultural systems: to what extent are we allowed to "manage" organisms and ecosystem and how much are we forced to do so, to keep 6 000 000 000 people on this planet alive. Even more important for agricultural and ecological changes is the fact that this figure will increase by 1 billion people within the next 12 to 15 years, the equivalent of the whole population on earth in 1820. This means also that the pressure on the production systems and agricultural research and policy will increase, not decrease.

Professor Lipa works on biological control at a governmental institute in Poznan (Poland) and is chairman of the Polish plant protection committee. He emphasizes, to the topic *Biological Control*, that speakers of the conference indicated that large companies are not interested in development and marketing of products for biological control and of biopesticides. However, medium-sized companies commercialize products against nematodes and insects, mostly for use in greenhouses. On a meeting in Vienna, it was obvious that there is progress in control of pests by technological improvement of biological control. There is a demand for the products in eastern Europe, at least for *Agrobacterium* products against insect pests in greenhouses.

To the topic *Transgenic Crops*: the time for these products is coming and we have to be patient. Opposition is evident. Risk assessments and risk limitations have to be given by public organizations, not only by the companies involved. The development for transgenic crops is promising for plant protection, especially in view of some negative experiences with chemical plant protection.

To the topic *Management of Ecosystems*: it is evident that the role of trees in agricultural systems is not sufficiently acknowledged. In Poland, valuable experience has accumulated so the OECD program should be extended because of the necessity of new plant protection strategies in sustainable agriculture. Also information on landscape development is of interest. This is a speciality of the Polish Academy of Sciences, and Poland works in excellent co-operation with Germany, France and other countries. In a functioning agricultural landscape, the role of the forest is of interest.

Forest and trees are refuge habitats for birds and other animals, that can survive and offer a special richness to the agricultural system (including biological control).

A symposium in Poland on the use of insecticides deals with insect communities in uniform and mosaic landscapes: insects are more abundant in mosaic landscapes. The shelter belts are positive for diversity in agricultural landscapes and also for birds. These and other subjects are of topical importance and under discussion in Poland.

Professor Olsen from Norway works on microbial ecology, also on arctic microbiology, with a main topic in management of soil microbial communities and biodiversity in agroecosystems.

To the topic *Microbial Communities in Soils*: we have to ask if manipulation of communities is really possible. The diversity in soil samples is not well known, and is different for various soil samples. Microorganisms are present in large numbers and the diversity in soils is great. There is also the question of the physics and chemistry of soil in correlation to soil microorganisms; for example the role of the soil atmosphere in the niches, microhabitat soil, and the activities of organisms in these niches are not wellknown.

Transfer of methods from medical microbiology to ecosystem microbiology has now established adequate methods. The role of diversity (diversity is primarily a political slogan) is obviously not clear: is diversity necessary for stability? Critical loads are not defined for organisms in complexes matrices.

The soil ecosystems should be conserved, but research is operating on damages and impacts. Is a damage negligible as far as it is reversible? Organism communities in soils are a gold mine. Their role for soil quality and sustainable agriculture is not well understood.

There is no public knowledge on soils microorganisms. Also, we have at present not enough experience to estimate if soil systems may be destroyed by the introduction of genetically manipulated microorganisms. More risk assessment studies must be carried out.

Dr. Tomlin from Canada has a toxicology background and draws attention to the effects of tillage and pesticides on nontargeted organisms in soils. Can the effects of management be measured? We need more knowledge on effects of ecosystems processes in soil matrices (transports, concentrations of elements, functions of microorganisms).

Many questions are still open, for example toxic effects of *Bt* on nontargeted organisms or herbizide effects (noneffects) on soil structure and on conservation of populations.

Agriculture has to use the best technology to be sustainable. There is a growing need for food, but regulations are necessary to keep the production in the future sustainable.

In the discussion, eight participants were involved for three topics. It was argued that the lectures did not cover in detail the relationship between soil biodiversity and soil functioning. It is obviously difficult to study microbial communities and microbial reactivity in the soil at the same time. Potential activities are expressed.

To the nontarget effects the lack of knowledge was discussed. A scientific approach with realistic worst cases should be encouraged. It was noted that we have more questions to this topic now than at the beginning of the conference.

To the topic *Risk Management with Use of GMM*: in almost all OECD countries studies are underway. Argentina, Australia and Germany were mentioned. Of course, risk management is difficult for developing countries. The development of GMM has not only pest management as a goal, but also the creation of crops with new qualities. Regulations for GMM are certainly disproportional to the real risks. The greatest impact may be on public trust. Regulation should be limited to companies. Risk management is the regulation of a biotechnology. What are the probability levels for risks? The history of regulations may serve as a lesson. The connection of science and policies in the OECD is very helpful in this respect.

Final Panel on Science, Policy-Making and Society: Where do they meet?

In preparation for the discussion, the panel considered the following questions:

1. Are the most urgent concerns of society being taken into account in agricultural research programmes, and what would you personally consider as the most pressing concerns?
2. Is agricultural or agro-food research sufficiently connected with other fields of research (health, ecology, etc.) to address these issues?
3. Are the results of research expressed in ways that can be easily understood and "translated" by the media? By the competent authorities? By professional organisations? How do scientists consider that it is possible for them to improve the expression of their results in a way which meets the "clients'" requirements, especially when there are conflicting results from scientific research?
4. How are the concerns of society expressed to the research community (by governments, parliaments, other bodies, the media)? Is the present institutional structure adequate for a dialogue between all interested parties and for facilitating the decision-making process at the national level? At the international level? Have you any practical suggestions for improving this "architecture"? In this regard, what are the respective roles of civil society on the one hand, and of public authorities on the other?
5. How can policy-makers reconcile often conflicting scientific advice and information with pressures from consumers and NGOs, etc., in designing "good policies"?

Keynote Address

Mr. Moe, Deputy Secretary-General, OECD, in an overview of the previous day's discussions, expressed a concern that we have not always been as effective as we could be in turning the good science we have done into useful material which can be used by decision-makers. There is, indeed, a need to explore the connection between science, policy-making and society.

He felt that we as began this debate, we should have in mind the discussions of the previous panels; and he wished, in particular, to call attention to three themes: biotechnology, sustainable development and the linkage with agricultural policy.

As he knew well from his work as Chairman of the Internal Coordination Group on Biotechnology, the OECD has been working on biotechnology since 1982, on projects covering new technologies in agriculture and food production as well as a sound scientific approach to the environmental and food safety assessment of these new products. Despite this painstaking work, however, the issue of the safety assessment of transgenic foods has become an increasingly public concern. Is part of the reason for this because we have failed in some way to transmit the hard scientific work that has been done to the policy-makers and public?

Work being done on the management of soil microorganisms, the biocontrol of plant disease, as well as the management of pesticidal crops, should, at least in theory, lead to the decreased use of pesticides and other chemicals in agriculture systems, thus having a beneficial impact on human health and the environment.

The above issues are becoming increasingly important within the context of agro-food policies. It was obvious at the meeting of the Committee for Agriculture at Ministerial Level, which took place in March 1998, that Ministers for Agriculture are paying more attention than ever to technological developments, consumer concerns, and to the need for sustainable agriculture.

Although this is not the appropriate time and place for a discussion of the overall problematic of agricultural policy reform, which is under continuing analysis and discussion within the OECD, the following three remarks seem relevant to the work of this panel: (1) issues regarding food safety or sustainable agriculture have both domestic and transboundary dimensions; (2) such issues are at the crossroads of various policies and therefore require an integrated approach; (3) these issues require a new type of dialogue between governments, parliaments, industry, consumers and other non-governmental organisations in a dialogue in which the media are also active participants.

The Chairman, Mr. Viatte, Director of Food, Agriculture and Fisheries, OECD, opened the discussion by stating that the issue of how the OECD would deal with the various aspects of biotechnology, including genetically modified organisms (GMOs) was critical, and thus the timing of this Conference was ideal.

Professor Werry, who chaired the panel on the first day, evoked three observations and one conclusion. The first observation was that in pursuit of sustainable agricultural production, the many subdisciplines in agricultural research were operating in an uncoordinated way. There was no common operational understanding of sustainability, and, consequently, no balanced and strategic employment of research efforts. The second observation was that there was no interaction between the agricultural sciences and the scientific community as a whole. Third, it was not clear for whom the agricultural science community was working e.g. the farmers, the processing industry, or consumers: agricultural sciences were not connected to the stakeholders they were supposed to be working for. Hence his conclusion that as soon as the agricultural sciences connected among themselves; did away with sectorial fences; left their ivory tower; and committed themselves to the benefits and interests of the stakeholders, then they would become relevant for the politicians.

Ms. Reed, on behalf of Senator Kerrey, presented the outcome of the discussion in the panel on production systems. The Senator had indicated that in the United States, the public did not have a great trust in scientists, but they had even less trust in politicians. Different degrees of acceptance of GMOs in the United States and Europe were discussed. One point made was that if scientists cannot explain the benefits or value of new technologies to society, then the technology will not be accepted. Unless those benefits and challenges can be adequately communicated, they will fail in gaining public acceptance and, ultimately, policies beneficial to the agricultural sector will not be adopted. Genomic research has significant potential to improve agricultural production both in terms of enhancing qualitative and quantitative traits in crops and livestock and in terms of alleviating pests and pathogens in foods. General agreement existed that public education is necessary to overcome rational and irrational fears, in order to best utilise new scientific and technological developments.

Professor Werner then presented the results of the last panel. Two aspects for possible inclusion in the programme for the years 2000-2004 were identified: landscape management, including agroforestry, and the still widely acknowledged gap in the understanding of soil resources and soil management. The OECD was in a good position to coordinate global research, also with many non-members and other scientific institutions. Finally, a third point emerged from the experience of the panel he chaired, and that is the interface between science and policy. The knowledge that has led to huge gains in production in OECD countries over the past 50 years has come from science, so scientists should not constantly have to defend their work. Science has produced the largest returns from investment in most OECD countries.

Mr. Doornbos, President of the International Federation of Agricultural Producers – the World Farmers' Organization – noted that farmers still see great potential in science for improving agricultural production. Many of the problems found in connecting science and policy were linked to the commercialisation of science. Governments have withdrawn or cut back significantly in their involvement in research and development and the best science is increasingly becoming privatised and patented. Scientists in multinational companies need to get results out fast into commercial application to keep up share prices, so more and more scientific search was money-driven. Perceptions are that technology is being applied before the implications are fully understood, or at least faster than the general public or the regulators can cope with it.

He also drew attention to recent initiatives of IFAP: first, the setting up of an international agri-food network that brings together on a regular basis the professional organizations of the different sectors in the food chain, at the international level, including input suppliers, farmers, cooperatives, commodity sectors and the food industry; the second is trying to have direct contacts with the biotechnology companies themselves. A third is the Global Forum on Agricultural Research, a forum that brings together national and international research institutions with farmers' organizations, governments, industry and NGO representatives.

On the side of the public authorities, there was a need to be able to reassure consumers that the food they ate was safe for human health and produced by methods that were safe for the environment and not cruel to animals. This requires governments to give a greater commitment to public funding of basic research to better understand biological processes. It also requires government to be clearer about policy goals, and if society wants more landscape and wildlife habitat and less traditional farm production, then farmers should be rewarded for these services. Finally, industry needs to show greater sensitivity to the demands of society, and particularly the need to eliminate poverty in developing countries.

Dr. Toet of Nestlé noted that most people in Europe no longer know how their food is being produced. Consumers perceive agricultural research as aiming only at higher production, which they do not like. What they want are quality products and they are not particularly interested in the technology behind it, although origin seems to be claiming more and more attention. Since food shortages are not a problem in Europe, consumers are not interested in biotechnology aimed at increasing production.

Dr. Dixon spoke about public perceptions, the media and education. The food producers and marketers had a large role to play in this; but so did the scientists, and the team that produced Dolly did a great deal of work to prepare the ground in advance, talking to the media; undertaking public education efforts and going to science fairs. Scientists had to work hard to get their "stories" onto the media agenda in the right way at the right time, and this was largely not the case. The OECD has compiled excellent reports on biotechnology, but they did not reach the right people, the "gatekeepers", the media editors and ultimately the politicians, who are much influenced by the media. In his view, the OECD needed to be supported by greater resources to get its scientific policy reports across to the media and ultimately to the politicians.

Professor von Wartburg, from Novartis, noted that any new technology, including GMOs, biotechnology, etc., required a certain period of gestation before gaining public acceptance. For decades, the pharmaceutical industry had to deal with the regulatory hurdle, but once that hurdle was cleared, it was plain sailing and one had a saleable product. Those days were over, because there was now the hurdle of public acceptance. There was now a sequence of science, then policy (rules and regulations) then politics (public acceptance). As he said, "I have faced many a science, or even an industry minister, who has said to me, I can agree with you in terms of what you are telling me on a rational basis: if I have to look at my situation as a politician, I can simply not maintain your position, because I won't be there the next day."

Professor Poutiainen cited the Framework Programme V of the EU Programme. Two things were important and should be underlined: the transparency of the resource work and results to the public, to those making regulations and to the policy-makers, and then the translation of their results to these different people who are finally making the decisions or using the results. This needs continuous dialogue and he thought the OECD was a very good forum for this kind of discussion. He emphasised the importance of quality control covering the whole production chain in order to guarantee the confidence of the consumers for safe and ethically acceptable products.

Mr. Vialle, Managing Director of the French National Institute for Agricultural Research made a number of recommendations regarding the research community: first, research should reach and learn to communicate with society, and more specifically, to convey goals; second, the public should trust government research. The debate should therefore be open, transparent, and not based on a single doctrine even if this causes public concern, because it only takes one person to challenge 200 – everybody knows that research is not democratic – and for doubts to emerge. His third recommendation is that those in agricultural research need to go beyond technology and hard science and focus heavily on economics, social science and the humanities. His fourth recommendation was that researchers should view issues through the eyes of the public at large. Every research programme should include and earmark funding for an impact study. What kind of findings will be reached? How can they be communicated? What are the implications of what researchers are doing? This should be a prime concern. His fifth point, more specifically in the of field agricultural science, was about enabling consumers and researchers/producers to communicate, i.e. via product standardisation. He noted that we all lived in a democratic society where elections are not confined to yes/no referenda but offer a choice of more than one candidate. By the same token, consumers should not be confined to a simple choice between products that do or do not meet standards. There could conceivably be a wider range of options; for instance environmental standards X, Y or Z, each with its own characteristics, as well as internal characteristics for differentiated products, and then, of course, there are personal preferences – some people do not eat pork, others are vegetarians and still others, who are also entitled to their opinions, refuse GMOs –. His sixth and last recommendation was that much needs to be done to develop product standardisation, an area in which the OECD might play a major role, and to foster broader, deeper, dialogue. So his sixth point was the need for training and information, without which there will be no mutual understanding.

Mr. Vialle wished to conclude with a few words. First, he believes that the problems we are encountering today, with GMOs, for instance, are understandable. Man has always had fears about food. In 18th century France, the land of philosophers, France banned the growing of potatoes because they were thought, quite mistakenly, to carry leprosy. So food scares have always existed. Second, everything happens so much faster nowadays. When wheat first arrived in France, about 3000 years B.C., it took about ten generations for it to travel 100 km. In the 18th century maize arrived, this time from the west. It took only one and a half generations to cover the same distance, but people still took time to adapt. Now with GMOs, the distance is covered in a tenth of a generation. Things are moving fast and peoples' fears are understandable.

Consequently, in the entire area at issue here, our concern is with science alone; the key word is comprehension and so, again, there must be a special focus on economics, the humanities and the social sciences where science is concerned and on information where the media are concerned.

As for the OECD, it should perhaps realise that there are all kinds of approaches, that anything to do with food is rooted in culture, and in some cases, religion. Consequently, a yes/no, all-or-nothing approach to this problem is probably too simplistic and a culturally hegemonic approach is the right answer. People are all free to act in a number of areas and need to given the means to make their own choices. Here, a great deal of reflection is needed on standardisation, information and diversified solutions.

Dr. Schofield from Unilever noted that there have been a lot of food scares and each successive revelation of a scientific or technological threat has caused a pervasive corrosion of confidence and disillusionment on the part of the public with science and scientists. Both governments and industry are increasingly distrusted and the statements they make appear to be damage- -limitation exercises rather than a proactive approach to innovative science and technologies. The stakes are very high. On one side, people think new technologies are being introduced for purely economic reasons. On the other, failure to actually see the creative potential of science and technology can lead to economic stagnation; and given the challenges faced in the developing world in terms of food security and expanding populations, one could be "throwing out several babies with the bath water".

Dr. Shannon, Chief Scientist, Ministry of Agriculture, Fisheries and Food, United Kingdom, observed that science and policy move forward at very different speeds, sometimes leapfrogging each other. As he noted, one of the problems is that today's policy is always right until tomorrow; and when tomorrow comes, one does not have the research to support the new policy one wants to introduce. He suggested that there was a strong case for ensuring that research is not tied too closely to a particular policy being pursued at a particular time. In other words, science should be policy-relevant, not policy-dominated; and should explore a range of policy options. This, however, may be difficult to sell to a government or policymakers in that the current policy is the one they are going to pursue, and to say one needs to invest in a range of other policy options is not something they can sign up to readily.

Mr. Silverglade of the International Association of Consumer Food Organizations stressed that people do not necessarily react logically or rationally to food issues and that there can be an emotional and cultural dimension, too. In the United States, one third of all cases of heart disease are related to diet. In the EU, the top concern for consumers and consumer organizations is genetic engineering, and in the developing countries, the top concern is that food research address the basic problem of producing enough food. He noted that at Codex Alimentarius meetings, the United States and France continue to fight about the need to pasteurise cheese and other dairy products, and this debate is based very much on food culture. Even the scientists disagree vehemently with one other. These emotions have to be confronted, and that is what the scientific community has to do at large. That means speaking about these issues in terms not just of hard science, but of the social sciences, and the underlying consumer fears.

The Chairman then opened the floor to a general debate among panellists and participants.

The first intervention drew the distinction between animal products and new plant or pharmaceutical products. The speaker noted that transgenic pigs – producing more pig meat while consuming 20% less feed – had been on the market for years in the United States and Australia, and similarly cattle and dairy products. He wanted to make the point that the existing control systems were very adequate for modified animal products. Control systems for transgenic plants did not yet exist.

Mr. Doornbos spoke again of the need, in a globalising world, for an international body. The OECD could help such an organization get off the ground, but in the end it should not be linked to the OECD, where it could be dominated by trade negotiators, but rather perhaps to the World Health Organization. Scientists, like the trade negotiators, farmers, consumers and politicians, need a body in which they can find common ground on what is safe and what is not.

Mr. Viatte remarked that he would like to keep in reserve the very important point made by Mr. Doornbos about the need for coordination or a possible body, or at least increased coordination among existing bodies to ensure what is really safe.

Professor Lynch from the University of Surrey (UK) pointed to the existence of the International Institution of Biotechnology, which could already serve as an international coordination body. He went on to point out that the Church of Scotland (and the Catholic Church in Rome) were both publishing books on the ethics of biotechnology.

Another speaker was concerned about another international organization, which already exists and tries to set international standards in trade, the CODEX Commission. She came from Canada and was concerned that in June the CODEX

Commission was set to vote on bovine somatotropine; but the reality is that there is still considerable scientific debate about this product. For example, in January, the Canadian government voted to ban this product on veterinary grounds. Several weeks ago, a EU scientific committee issued a report raising further concerns about the impact of these drugs on human health. Of course, on the other side of the equation, the US government approved the product and it has been on the market there for several years. The US argue that it is safe, and so does the company that produces it. The point was simply that an international organisation, the CODEX, is about to approve the safety of the drug, while there remains a strong divergence of scientific opinion, so concerns about so-called convergence of regulatory standards remain.

There followed a lengthy discussion of the significance of "Dolly", the cloned sheep, and the ethical issues raised in cloning animals, particularly if the cloning was being done in an agricultural context aimed at human consumption. For a participant from the Czech Republic, the border was not clear between genetically modified animals used for human heart transplants or genetically produced insulin or genetically modified animals destined for consumption as food. He thought there was a role for the OECD in dealing with these questions since it included members from both Western and Eastern Europe.

Mr. Vialle intervened again to note that one speaker had raised the idea of finding neutral ground for what might be called scientific truth. In his view, the very concept was a problem. First, science is not smooth sailing; it is a troubled world, stirred by passion, controversy and rivalry. Science is never calm. When Pasteur discovered microbiology, any history of science will tell how bitter and protracted the discussions were. Nor is it democratic, since one person can triumph over everyone else, e.g. Galileo. Unanimity among scientists verges on the suspicious: recall what one powerful institution said about the nuclear industry: there are no problems, no difficulties. Where there is only one doctrine or line of thought, after a number of years there is no longer a question of whether it is right or wrong. An institutional mistrust develops.

Another speaker returned to Professor Werry's comments. The Netherlands has led the world in a systems approach to agriculture and the importance of agricultural scientists working with the social sciences. He did not believe it was the case that all scientists were working in very isolated units.

Mr. Cantley, formerly of the OECD and now with the EU, intervened to point to the risk to the food security of the world if the regulatory, or perhaps the excess of regulatory structures adopted in food-surplus countries, were to be exported worldwide. This would put at risk the food security of millions, which could itself be ethically very questionable.

In response to questions, Dr. Dixon returned to the issue of relations between scientists and the media, particularly news editors. He cited the example of genetically modified food, noting that OECD reports 7 or 8 years earlier had ventilated this issue, but they were not effectively targeted to reach the news editors. So, today, the information is coming from Greenpeace or Monsanto, who have "set the agenda". Mr. Ryder from the CSIRO in Australia added an example of science and media relations there, where proactive science writing did largely defuses debates.

Another speaker noted that science produces new knowledge, and this new knowledge has to be assessed, and usually it is assessed in the end by regulatory authorities based on legislation which is inevitably politicised. He thought the

OECD was somewhere in the middle between science and politics; it was the least politicised body to assess new wisdom and new knowledge, and it should continue to do so. It really brought together people from various disciplines in a way free from politics, and was really geared to bringing science into a policy process.

Dr. Toet noted that the idea of a consulting group had been mentioned several times. In order to be successful, such a group should accommodate regions in the world that are at different stages of development and have different needs; it should incorporate different stakeholders.

Dr. Dixon returned to the issue of the dissemination of OECD reports, and the failure, in his view, to get them to the right people at the right time, citing the Cleaner Industrial Processes report. It was not just a matter of putting them in the post, but of being much more proactive, spotting opportunities, contacting the key gatekeepers in the media, persuading them, enticing them, getting them interested in themes and ideas in a very active way. However, he did not underestimate the amount of work that this would involve.

Mr. Doornbos repeated his call to the OECD and to other international bodies to examine whether an independent scientific body could be established. He also reiterated that it was particularly in developing countries that a lot of technology was needed. The demand for food there could be doubled in the next 25 years.

Dr. Schofield reminded participants that the CODEX Commission was not, in fact, a meeting of independent scientists, but a meeting of governments, advised by scientists. She cited the European Medicines Evaluation Agency as a good example of incorporating scientific advice in the regulatory arena and of getting products on to the market quickly. She posed the question of where people wanted to be in agriculture in the year 2020. The place of biotechnology in overall sustainable agricultural policies, in both the developed and the developing world, had to be understood and she encouraged the OECD to continue to contribute to broad science-based recommendations in the longer term.

Dr. Shannon noted that the public felt that new technologies were being rushed on it, so anything the OECD could do to actually give some time for people to reflect on new technology and become more familiar with it would help. He did not favour a large independent scientific body, because there were a lot of those already. On the media, it seemed to him that the media were very good at ventilating issues; if one had good answers to the questions the media were asking, then the issue died very quickly.

Mr. Silverglade did not consider the Codex Alimentarius to represent good science since so many votes were very close. He said the consumer had to be given choice and that was where labelling came in. Labelling of genetically engineered foods was the best means of giving the consumer choice, and he regretted that the US was opposed to labelling. Last, he thought the aspect of consumer emotions had to be addressed, the process opened up, and the OECD's invitation to participate in this meeting was a step in that direction.

Professor Werry urgently invited the OECD to continue its initiative to confront the scientists with the policy-makers and the other stakeholders of science.

Professor Poutiainen thought the OECD should concentrate on the economic and socio-economic impact of new technologies in agriculture, for instance the cost and benefit ratio of different new processes, and that these aspects should be underlined in the new Programme for 2000-2004.

Mr. Viatte made four "telegraphic" points in summing up the discussion. First, there was a need for continuing improved international coordination: what was discussed may not necessarily respond to the needs of tomorrow, so that discussions must take place in a dynamic context. Second, a very, very strong point has been made about the need for increased coordination among sciences, not only between agricultural research sectors, but also between hard science and soft science, between agricultural research itself and economic, social and ethical aspects. The third conclusion was the need to develop the dialogue with all the stakeholders involved, so as to have a response to society as a whole and not just a series of different responses from one sector or another – and here the media had a very important role to play. His fourth and last point was that, although this meeting took place in the OECD framework, the non-OECD countries must be involved in the debate, and the capacity did exist at the OECD to develop this dialogue with non-OECD members.

Mr. Moe concluded by expressing his view that economics was not a soft science, but as hard a science as any others. He was very satisfied with the advice the OECD had received respecting the media and the need for a more proactive approach regarding OECD publications. Second, the OECD can convene people from all sectors, governments, science and also expand the dialogue with non-member countries. However he did not wish to hide the fact that resources and contributions from member countries were being consistently reduced, so the OECD's possibilities were limited.

Printing: Druckhaus Beltz, Hemsbach
Binding: Buchbinderei Schäffer, Grünstadt